S0-BEC-500

Manufacturing, Technology, and Economic Growth

Carlos Sabillon

M.E. Sharpe
Armonk, New York
London, England

Library of Congress Cataloging-in-Publication Data

Sabillon, Carlos, 1967– .
Manufacturing, technology, and economic growth / Carlos Sabillon
p. cm.
Includes bibliographical references and index.
ISBN 0-7656-0513-9 (c : alk. paper).
1. Economic development. 2. Technology—Economic aspects.
3. Industrialization. I. Title
HD82.S2 1999
338.9—dc21 99-27549
CIP

Printed in the United States of America

The paper used in this publication meets the minimum requirements of
American National Standard for Information Sciences
Permanence of Paper for Printed Library Materials,
ANSI Z 39.48-1984.

BM (c) 10 9 8 7 6 5 4 3 2 1

This book is dedicated to the millions of people who every day endure poverty, material and moral hardship, unemployment, underemployment, war, violence, and disease. This book is dedicated as well to those who throughout history have advocated the cause of free markets, free trade, fiscal rectitude, sound monetary policies, respect for the environment, proper governance, and democracy.

Contents

Preface

This book is intended to cater to the needs of policy makers. Therefore, it utilizes both a language that is accessible to them and an abundance of relevant historical examples. As policy makers are mainly interested in concrete results, abstract notions and theoretical positions are reduced to the minimum. In order to reinforce its empirical approach and practical goals, this book makes a decisive effort to present clear policy recommendations for the attainment of fast and sustained economic growth in an environmentally responsible way.

This volume is also intended for economists, social scientists, and businesspeople who are interested in exploring new approaches and ideas toward finding alternative solutions to the problems of poverty, unemployment, underemployment, and ecological sustainability. The text's jargon-free and user-friendly language also allows for a reading that the general public can understand.

With the goal of understanding the essence of economic growth, the book offers an overall synthesis of the world economy during the second half of the twentieth century. Only the most relevant passages of this period are included in order not to divert the reader from the central thesis. At the end of the book, several charts are provided containing the main macroeconomic indicators of the countries analyzed.

Acknowledgments

The research for this book was undertaken from 1991 to 1998. During this time, the author concentrated exclusively on this activity.

This volume was made possible thanks to financial assistance from numerous sources. The first stage of research was financed by the author although the Swiss National Fund for Scientific Research provided a one-year grant in mid-1992. The Graduate Institute of International Studies in Geneva, Switzerland, supplied a few additional funds, and during the rest of the period the research continued with financial assistance from friends and relatives.

The author is particularly thankful to Professor Norman Scott of the Graduate Institute of International Studies for dedicating so much of his valuable time to revising and commenting on the numerous manuscripts that led to this book. His useful suggestions contributed noticeably to improving the quality of the text.

In Munich, Germany, where a smaller part of the research was conducted, the author wishes to thank Professor Karlhans Sauernheimer and Professor Franz Gehrels of the Ludwig Maximilians University for the time they took to revise and comment on the initial manuscripts. Professor Peter Tschopp of the University of Geneva also made valuable comments that led to significant improvements. Professor George Viksnins of Georgetown University in Washington D.C. made interesting comments, and several other economists from international organizations also made useful comments on the ideas presented. The remarks of Carlos R. Montoya were particularly helpful.

In the preparation of this volume, the author has benefited greatly from the comments of those who revised the manuscript and from the numerous books and other publications that were consulted. The author is particularly thankful to mainstream economists for their lucid and extremely detailed understanding of the market economy. However, the central thesis presented in this book did not originate with any of the publications that were consulted or any of the comments that were made about the manuscript.

The author wishes also to thank Dafne Perhat, Pablo Aviles, and Omar Alkadhi for the long hours they allocated for typing, processing, and other formal preparations that made this book possible. The financial assistance that this research effort received from Herman Vogelsang, Leslie Wittwer, Reynaldo Sabillon, Krimil Davila, Carlos R. Montoya, and Yolanda Calzada, was significant, and the author is very grateful for their collaboration.

Introduction

This volume will analyze the period from 1950 to the present, and will attempt to extract from the empirical evidence an alternative explanation of the causes of economic growth. It will concentrate on analyzing the nations in East Asia, Europe, and North America that attained the fastest rates of growth during this period. Even though the ideas underlying this book are unorthodox, it will not contest the basic tenets of mainstream economics. The alternative approach that will be introduced in this volume is only operational within the framework of a free market economy. The maintenance and reinforcement of capitalist institutions are therefore of utmost importance for the application of this alternative.

It will be demonstrated that orthodox economic thinking is correct in the vast majority of its contortions, but the reason its policy applications rarely fulfill the expectations of most nations of the world is that it has not identified all the factors responsible for economic growth. Mainstream economics has done a praiseworthy job of identifying most of the variables that intervene in the growth process, of developing the mechanisms for harnessing inflation, and of developing the means for measuring economic variables with great precision. However, it has evidently not succeeded in identifying the factor that is more strongly linked to the phenomenon of economic growth.

It will be argued that the factor that is fundamentally responsible for the generation of economic growth has been largely overlooked by economists and social scientists. It will be held that this factor is intimately linked to the manufacturing sector. Manufacturing shall be understood as all economic activity that does not fall into the category of primary sector activities, construction, and services. At present, manufacturing occurs almost exclusively in factories.

The traditional division of economic sectors is not compatible with the purposes of this volume. Traditionally, the economy has been divided in three sectors, with the secondary sector identified with industry. Orthodox definitions of industry include several components. The most important one is usually manufacturing, but it is also accompanied by construction, mining, and at times even transportation and telecommunications.

Under the orthodox division of sectors, manufacturing does not have a place of its own, and the fundamental reason is that existing economic theories do not assign manufacturing a predominant role in the generation of economic growth.

For the purposes of this book, manufacturing will be classified separately, mining will be included in the primary sector, construction will be placed in a separate category, and services will continue to remain independent.

The term "manufacturing" will therefore not be used as synonymous with industry. The word "industry" will be avoided in order to diminish misunderstandings, and "factory" will be used as a synonym for manufacturing.

Manufacturing and Growth

The empirical data from the second half of the twentieth century supplies abundant evidence that the rate of manufacturing output during this period correlated consistently with fluctuations in the economy of the most important nations in the West, East Asia, and Russia.

Throughout history, numerous economists and schools of thought have pointed to the effects that industry has on growth, but these efforts have failed to present a consistent argument on the subject. In this volume, it will be held that manufacturing, and not industry, is the determinant sector for the generation of growth.

It must be emphasized that what is relevant about this sector is not its size as a share of gross domestic product (GDP), as so many economists have contended, but the average annual rate by which it expands. Sometimes fast economic growth occurred while the manufacturing sector accounted for a small share of GDP, and on other occasions stagnation prevailed with an equally small factory sector. At times, fast growth took place while a nation already possessed a large manufacturing sector, and at other times stagnation occurred with a large sector. History makes it clear that it is not possible to establish a correlation between any given size of the manufacturing sector and fast economic growth. However, in every case of a fast expansion of the economy, it was always accompanied by a fast rate of factory output.

Of even greater importance to the goals of this book is what made the growth of manufacturing possible. History supplies considerable evidence that manufacturing production almost never managed to expand unless there was government support for it. It is extremely rare to find, during the second half of the twentieth century or even further back in time, a spontaneous growth of the sector.

It is important to note that the vast majority of this support has gone largely unnoticed, for it was supplied in an indirect way by means of fiscal, financial, and nonfinancial incentives. It did not go unnoticed when governments supported the sector in a very direct way by constructing production facilities, financing the whole operation, and taking care of daily activities. However, direct efforts in the form of state manufacturing enterprises were few, compared with indirect ones.

There is ample empirical data showing that differing levels of state support for the sector cause proportionate rates of manufacturing output. Throughout the recent history of the West, East Asia, and Russia, a weak factory promotion effort from the state coincided with a weak rate of output of this sector and a slow rate of GDP. Strong support from policy makers coalesced uniformly with a fast rate of factory output and a fast rate of economic growth. It is important that this subject is not interpreted in a bipolar way. The issue is not whether there is government support for this sector, but the level at which it is offered. Even during the Middle Ages, it is possible to find evidence of support for manufacturing.

Investment and Technology Creation

The evidence suggests that manufacturing is by far the most investment-intensive of all sectors of the economy, and it is because of this intrinsic characteristic that manufacturing has so rarely expanded without government support and has so systematically stagnated in the absence of it. Manufacturing requires very large amounts of investment while the other sectors require considerably less. In addition, primary-sector activities, services, and construction require much shorter periods of time to recover an investment than manufacturing.

With these characteristics, it becomes inevitable for would-be investors to constantly shy away from the sector. Unless the government changes this natural state of affairs by giving incentives for manufacturing production, thus guaranteeing a profitable venture, investment in the sector remains almost nonexistent. The government therefore plays a determinant role in the generation of economic growth because the state alone has the capacity to provide incentives. Only with incentives can costs of factory production be reduced, and only then can the private sector be induced to channel its resources in this sector. Therefore to invert an investment-hindering natural structure of things, policy makers must supply the sector constantly with support.

This book intends to demonstrate that manufacturing is so investment-intensive because of the exceptional capacity of this sector to create technology. Abundant empirical data seems to indicate that manufacturing is perhaps the sole sector with the ability to generate technology. The evidence allows for the establishment of a strong correlation between the output of manufacturing and the creation of technology. From 1950 to 1997, whenever governments increased their support to the sector and output rose faster, technology made its appearance in greater amounts. Over and over again, the two variables fluctuated in unison, and there seems to be little doubt about the causality linkage. Because government support for the sector is so clearly within the discretion of the people dictating policy, it becomes evident that manufacturing was the cause and technology the effect.

Very indicative of the above was the fact that a large share of the investment in the sector during this period was made for the sake of fabricating armaments. For centuries most intellectuals and policy makers have seen investment in weapons as a bad allocation and a waste of resources. However, just about every time that war and similar events forced much larger investment into the production of armaments and delivered a faster overall rate of manufacturing output, technology advances were more prolific. If investing in military manufacturing was such a wasteful activity, technology should have been less prominent than in the periods when weapons production was lower. However, during the periods when armament production was higher and overall manufacturing expanded faster, not only did technology tend to grow faster, but overall wealth was also created at a faster pace.

The peace dividend that the era of détente in the 1970s was supposed to bring for the members of NATO and the Warsaw Pact never materialized. GDP rates for both groups of countries decelerated noticeably compared with the preceding two decades. And the even bigger peace dividend that the end of the Cold War was supposed to bring also failed to materialize for both camps. GDP figures in the 1990s were slower than before in Western nations and catastrophic for ex-communist states. The resources used for the making of weapons could have easily been transferred to civilian manufacturing. However, because the bulk of economists and policy makers do not believe manufacturing plays a determinant role in growth, they concluded that if those resources were reallocated to services, construction, or primary activities, the economy would function just as well or even better. The result was a decelerated rate of factory output for the West and a contraction of output for the ex-Council of Mutual Economic Assistance (CMEA) countries.

Another phenomenon reinforcing the thesis that this sector is causally responsible for technology is that of productivity. Ever since the first records on this matter, manufacturing productivity systematically attained the fastest rates, exceeding all other sectors by a large margin. In East Asia, Western Europe, the United States, or Russia, the phenomenon has been repeatedly observed. The service sector, which has been seen increasingly by economists as determinant for the generation of growth in developed economies, has systematically shown extremely low productivity rates. During the second half of the twentieth century, in all of these countries, factory productivity has been almost exponentially faster than that of services.

This book will also show how overall productivity always grew the fastest when the factory-promotion efforts of the state were strong. Total factor productivity in the numerous examined nations consistently grew at a slow pace when support for the sector was weak.

Analyzed over the long term, manufacturing has proven to be, by far, the most productivity-intensive sector. Since technology is the fundamental variable determining productivity, it becomes inevitable that the sector that is most intimately bonded to technology is also the one with the most productivity-enhancing characteristics. Primary activities, services, and construction seem only to have the capacity to passively absorb the technology created by manufacturing. Because these sectors are mere technology recipients, their productivity performance inevitably ends up being inferior to that of manufacturing.

It is also worth noting that from 1950 to 1997, as well as before that period, new technology systematically came to life in the form of manufactured goods, such as printing presses, steam engines, spinning gins, telephones, light bulbs, automobiles, airplanes, pharmaceuticals, medical equipment, refrigerators, microwave ovens, TV sets, electronics, composite materials, microchips, computers, spacecrafts, satellites, biotechnological medicines, and video conference machines. Since the appearance of the first patent system in England in the sixteenth century, patents in England and practically everywhere else were almost always directly tied to a manufactured good.

History suggests that faster growth of technology is the inevitable result of faster growth of this sector. This ultimately finds itself expressed in a faster pace of innovation, a faster rate of technology imports, or a faster pace of both. A government that finds itself in possession of the policy tools that can accelerate the pace of economic growth finds itself automatically in possession of the tools to accelerate the pace of technological development.

A strong factory-promotion policy for a nation that is significantly lagging technologically behind many others has over time translated into the capacity to import the existing stock of world technology at a much faster pace. The endorsement of such a policy in a highly developed nation has mostly expressed itself in the capacity to accelerate the pace of innovation. When factory output has grown rapidly in a developed nation, there has also been faster growth of technology imports in the form of patents, machinery, equipment, and the like.

Government Policy

The macroeconomic policy implication of such historical facts is that state support for manufacturing becomes the fundamental effort that any government must undertake in order to attain fast and sustained economic growth. Such a policy at first glance would seem incompatible with orthodox economic thinking. However, a logical interpretation of the empirical evidence demonstrates that policies in support of free trade, privatization, liberalization, deregulation, low inflation, balanced budgets, Central Bank independence, and small government have no problem coexisting with manufacturism. Policies that put the emphasis on education, job training, infrastructure, and not on relying on foreign aid are also fully compatible with a strong factory promotion effort. History suggests that orthodox policies combined with a strong government support of manufacturing deliver the best possible results.

The empirical data leads also to the conclusion that decisive promotion of the sector is not a policy limited to nations that are lagging behind the most advanced manufacturers. Contrary to what the infant-industry school of thought asserts, practically all nations, regardless of their level of development, constantly need to supply their manufacturing sectors with a strong dose of support. The infant-industry school asserts that nations that are lagging developmentally should subsidize their manufacturing sectors for a certain period of time until they catch up with the most advanced. The moment the support stops, manufacturing production tends to stagnate and, along with it, the rest of the economy. Logically analyzed, the empirical evidence demonstrates that producing manufactured goods in vast amounts and at a very fast pace is not either exclusive to nations endowed with a certain number of characteristics.

This book intends to demonstrate that practically all nations, independent of their culture, their endowment with natural resources, their level of development, and their record of production, can attain fast and sustained rates of manufacturing output.

Least-developed nations, middle-income countries, transition economies, and most-developed nations can attain fast and sustained rates of factory output, for it is within the exclusive jurisdiction of each government to decide the level of support it offers to its manufacturing sector. From the perspective of manufacturism, economic growth is fundamentally an endogenous phenomenon.

History also shows that producing factory goods in a cost-effective way and with high levels of quality is something that can be achieved by all countries. On this matter, what is relevant is the level of competitive pressure that falls over the manufacturing sector. In order to attain the highest levels of efficiency, competition must be increased to the uppermost levels, and on that matter mainstream economics has very clearly defined the policy mechanisms that deliver the best results.

The evidence clearly indicates that a manufacturist policy is fully compatible with the bulk of what mainstream economics proposes to governments. A misunderstanding of this thesis, however, is likely to take place in the minds of many people who have invested much time and energy into orthodox ideas. The debate in the academic community and in policy-making circles has revolved for so long around levels of government intervention in the economy that, at first glance, a policy calling for strong support for manufacturing seems to demand a very interventionist approach.

Communism was, from the start, a system that aimed to distort market forces in an encompassing way, and precisely because of that, political tensions with capitalist countries became very high. In particular after World War II, tensions between the East and the West became so intense that they drowned the economic discussion with a heavy ideological burden. The Cold War drove these tensions to extreme levels and along with it came an even more bipolar vision of economics in which the fundamental focus of attention resided on how much state intervention was applied. For the left, more state intervention was a positive undertaking, and for the right, it was a negative one.

If the analysis had been conducted on a more objective basis, it would have become very evident that it was impossible to present a clear correlation between either of those two extreme positions and the empirical data. Many tried to explain reality with a middle-ground position, but that did not succeed either in being consistent with the facts. The position of the left was obviously the one that had more problems being consistent, but the other two were nonetheless very far away from succeeding. It was evident that the fundamental causal variable of growth resided somewhere else.

Had the analysis been conducted on purely logical terms it, would have become evident that a manufacturist interpretation of reality had no problem bonding in a coherent way with the facts. It would have also become evident that a very strong promotion of this sector is not dependent on a high level of state intervention in the economy.

The empirical data of numerous nations suggests that a strong factory-promotion policy is fully compatible with a small overall government intervention in the economy. A small budget can easily coexist with strong and even very strong support of this sector. The case of Hong Kong and Singapore during the second half of the twentieth century demonstrate the above. In these cases, a small budget as a share of GDP was mostly utilized to promote private manufacturing, and that coincided with impressive rates of factory output and impressive rates of sustained economic growth.

Ideology and Public Policy

This volume will demonstrate that practically all of the support that governments have supplied to the manufacturing sector was out of ideological motivations. Never throughout history has a government decreed policies in support of the sector for scientific reasons; that means that never has a government believed manufacturing was the key to economic growth and technological innovation. Quite on the contrary, most of the time when governments gave support to manufacturing, it was because of national security concerns. During all that time, they were convinced that expenditures in armaments and related goods were hampering economic progress.

On other occasions, driven by jealousy, governments attempted to catch up in economic development with a competitor or a neighbor by producing the manufactured goods the other possessed. Balance-of-payments concerns also drove many governments to promote manufacturing because the goods from this sector had the capacity to generate a fast increase in exports and to substitute imports. In their efforts to promote high-technology goods and more employment, policy makers also have supplied the factory sector with a certain degree of support. Policies of economic autarchy and policies attempting to promote the least-developed regions of a country have driven the state to offer support to this sector. Another motivation has been postwar reconstruction efforts.

In each of these cases, the state was not zeroing in the sector nor perceiving it as the bottom line of economic growth, but in a indirect, unintended, and unconscious way it ended up promoting it. Precisely because manufacturing output was not seen as the bottom line, governments did not concentrate their efforts wholeheartedly in promoting the sector.

The degree of support from one government to another as well as from country to country has varied significantly. Even though historically the overall levels of support were very low, there have been some episodes in which the state promoted factory output in a very decisive way. The best examples are to be found in East Asia during the second half of the twentieth century, coinciding with the fastest rates of economic growth ever. Since the reasons that drove East Asian policy makers to support manufacturing were as ideological as any previous efforts in other countries, it is highly unlikely that the uppermost limits of support for the sector were reached. Therefore, it is logical to assume that if factory promotion efforts are pushed to the limit, much faster rates of manufacturing output are possible and for even more sustained periods of time.

That means it is perfectly possible for countries to attain sustained rates of economic growth faster than the 10 percent achieved by several nations in East Asia. Many economists and policy makers believe that developed countries do not need to attain very fast rates of growth, for they have already overcome poverty. It is hard to see why the citizens of Organization for Economic Cooperation and Development (OECD) countries would not want to experience a yearly double-digit rise in real incomes. However, even if that would be the case, it is evident that OECD countries are in need of much faster GDP rates than the present ones. With the slow rates of the 1990s, unemployment and underemployment will continue to rise.

As for developing countries, it is evident that in order for them to overcome poverty as fast as possible, a sustained rate of 10 percent is not fast enough. Practically all nations in sub-Saharan Africa are in desperate need of rates similar to the ones the Special Economic Zones (SEZs) in China attained during the 1980 to 1997 period. These zones which have a population larger than fifteen nations in the world, grew on average by 26 percent per year in that period. Much evidence suggests that such a feat can be reproduced not just in sub-Saharan countries but also in the north of Africa, in South Asia, in Latin America, and in all other least-developed and middle-income nations of the world. Manufacturing output in the SEZs of China in that period grew by about 35 percent annually, and that coincided with a massive level of government support for this sector.

Transition economies such as those resulting from the former Soviet Union are desperately in need of double-digit growth in order to improve living conditions rapidly. These countries would very much like to recuperate the lost ground due to the massive depression of the 1990s, and they would also like to reach standards of living similar to those of OECD countries as soon as possible.

It is worth noting that the ideological motivations that drove governments to support manufacturing during the second half of the twentieth century were major errors of policy that were not worth pursuing. Of all of these motivations, war and national security concerns proved to be the worst reason for supporting the sector. Historically, most people saw expenditures in armaments as resources that could have been much better utilized in other activities. They were absolutely right from the perspective that if such large amounts of funds had been invested in civilian manufacturing, the benefits for society would have been much larger.

History, however, rarely witnessed such a transfer of resources. Whenever governments felt that they did not need to invest in the production of weapons, resources were almost never re-channeled into civilian manufacturing, and the end result was an overall lower level of manufacturing production. A slower pace of this sector coincided almost always with a slower pace of technological development and a slower pace of wealth creation.

Had governments understood that manufacturing is the bottom line of economic growth, they surely would have concentrated exclusively on promoting the sector in a direct way and not as a by-product of inconsistent policies. This book attempts to present an adequate understanding of the strong causality relationship between manufacturing and growth, so that governments will largely renounce their policies in support of weapons production. Once such a phenomenon is fully understood, governments should transfer most of their defense resources into civilian manufacturing, and allocate still more resources into civilian manufacturing than the ones that at present flow into this field.

The best defense against military aggression is fast and sustained economic growth. In an environment of prosperity and high living conditions, the possibilities of war become almost nonexistent. Poverty, material hardship, or a rapidly deteriorating economy have historically been fundamentally responsible for leading nations into war. So long as fast and sustained economic growth is attained, nations have little to fear, and the more nations invests in civilian factories, the faster the economic growth.

The other ideological motivations that have driven nations to support the sector are also not worth pursuing. Trade protection because of balance-of-payments concerns is perhaps the most noteworthy of the nonarmament motivations. The desire to constantly experience a trade surplus drove many nations to give fiscal, financial, and nonfinancial incentives to the manufacturers of export goods. However, the desire for surpluses also made governments raise trade barriers, which systematically had a negative effect on manufacturing and on the economy as a whole. Trade barriers ultimately proved to be detrimental to the cause of growth for they reduced the level of competitive pressure on manufacturing and therefore reduced the sector's capacity to generate technology and raise productivity. The fiscal, financial, and nonfinancial incentives were positive, but the trade barriers were not.

Attempts to catch up in economic development with a competitor or a neighbor were not reflective of a rational motivation. The most important thing for a government is not reaching development parity with another nation, but improving the living conditions of the population it governs as much as possible. When governments have succeeded in catching up with their competitors, they have tended to diminish considerably the level of support to manufacturing. Under those circumstances, the economy has decelerated and problems have mounted. Nations need to supply abundant support to manufacturing constantly and not just for a certain period of time, for the moment the support vanishes, the economy performs proportionately and unemployment starts to rise.

When governments have attempted to promote high technology goods and create employment, they were actually doing the right thing, for it is manufacturing that is fundamentally responsible for the generation of technology and of employment. However, technology promotion efforts have mostly concentrated on the few manufacturing fields that at a certain moment in time possessed the latest technology. Since manufacturing (as a sector and not just a few of its fields) is naturally endowed with the capacity to generate technology, it is evident that the best policy for accelerating the pace of technological innovation is one that offers across-the-board strong support to the whole sector.

The same goes for the policies that most governments have utilized for their job-creation efforts. Most of these efforts had little to do with manufacturing because they supplied subsidies across the board to all sectors, therefore being of a limited effect. Because manufacturing is fundamentally responsible for economic growth, and growth is the main factor determining the creation of employment, it is obvious that the best job-promotion policy a government could adopt concentrates exclusively on promoting manufacturing.

Because there has never been a clear understanding of the causal relationship between this sector and the creation of wealth, governments have frequently undertaken policies that ended up being noticeably harmful. That is particularly the case when countries decree trade protection, although this is not the only way harm is done to the sector and to the economy as a whole. In an effort to increase rapidly manufacturing production, governments frequently have resorted to creating state companies. Private-sector enterprises, however, proved again and again after World War II to be superior over public-sector ones in making a more rational use of resources. They were more cost-efficient and the quality of their goods was above that of state firms. At times, the state went even so far as to nationalize existing private manufacturing companies. This book will show that this was a grave error.

A rapid increase in manufacturing output, and even in the particular fields of this sector that governments wanted more to promote, could just as well have been attained by means of fiscal, financial, and nonfinancial incentives. The evidence suggests that it all depended on how much and how many of these fiscal, financial, and nonfinancial incentives were supplied to private manufacturers. The higher the overall level of these incentives, the more any particular field of manufacturing would have seen its output rise. There was never a need to create state enterprises.

Governments have also attempted to promote manufacturing by trying to secure profitable returns for producers through the creation of monopolies and cartels. Trying to secure profits for manufacturers is a positive undertaking, but reducing internal competition is not. The empirical evidence demonstrates that the creation of monopolies and cartels diminishes significantly the level of competitive pressure on the sector and on the rest of the economy, and ends up hampering productivity considerably.

There also have been numerous forms of regulation on prices, labor, and capital that not only failed to stimulate the sector but even did harm to it and the rest of the economy as they decreased the level of competition.

Market-Driven Incentives for Manufacturing

This book will demonstrate that the only efforts that ended up proving to be useful were those of a fiscal, financial, and nonfinancial nature. Under the category of fiscal incentives, governments offered tax exemptions and tax reductions to private manufacturers. Under the category of financial incentives, they supplied grants and induced private banks to lend more to the sector. The grants were basically destined to cover part of the costs of constructing the factory, of acquiring the machinery and equipment, of training the workforce, and to absorb part of the costs of research and development (R&D).

In the category of nonfinancial incentives, the state gave manufacturing entrepreneurs free or subsidized land for the factory. It also absorbed the costs of constructing the adjacent infrastructure totally or partially. The state also promoted the type of infrastructure that was manufacturing-intensive. Instead of building highways, for example, it developed railroads and made sure that practically all the trains were fabricated domestically. Utility inputs such as electricity and water were supplied at subsidized prices. The state also concentrated on promoting the fields of education that were tightly linked to this sector, such as engineering, natural sciences, and mathematics. On top of that, governments offered similar incentives to foreign manufacturing capitalists.

Democracy and the Environment

This book also will demonstrate that a decisive factory-promotion policy is the one most capable of developing democratic institutions and promoting the rule of law. The history of the West and East Asia presents a very strong correlation between manufacturing and political institutions. The rapid growth of the sector's output has been systematically accompanied by a rapid increase in political liberties and a rapid rise of the rule of law. A high overall development of the sector has also regularly coincided with the existence of relatively developed democratic institutions and a developed judicial system.

Much evidence suggests that there is a causal link between manufacturing and these variables. A rapid creation of wealth is the bottom line for rapidly improving living conditions. When living conditions improve, a population becomes less restless as its basic necessities become satisfied. As a population becomes less violent and unpredictable, governments are inevitably moved to become less authoritarian and more democratic because it is no longer possible to justify repression.

Overall levels of education have always also increased rapidly when wealth is created at a rapid pace. A more educated population is simultaneously driven to become more politically conscious of its rights. The following pages will therefore attempt to show that the most rational effort for the promotion of democracy and the rule of law is a decisive policy in support of manufacturing.

The evidence also suggests that enthusiastic support for this sector is highly compatible with a decisive policy of environmental protection. The empirical data indicates that countries that attain fast and sustained rates of manufacturing output are more capable of protecting the environment because the increased wealth resulting from the sector provides society with the means to carry out such a protection. Without first attaining a certain level of wealth, nations find it very difficult to take care of the environment because their priorities center exclusively on overcoming poverty. It is mostly when nations overcome poverty that they develop the interest and the means to worry about the living conditions of plants and animals.

Nevertheless, a manufacturist policy has the potential to reduce environmental degradation in undeveloped countries. This book intends to show that if these countries would concentrate on producing factory goods, they could immediately stop exploiting their natural resources irrationally. At present, the governments of these countries believe that their comparative advantage lies with primary-sector activities and, as a result, they clear vast forest areas to exploit the wood and to make way for agricultural land. If they would come to understand that the best mechanism for overcoming poverty lies in promoting factories as much as possible, abundant and better employment possibilities would immediately appear. Since these jobs would offer a better life to the millions of people who currently live off the land, governments would have no problem declaring vast areas of their territory as national parks for purely conservation purposes.

Manufacturing, Technology, and Economic Growth

1
The Second Half of the Twentieth Century in East Asia
The Case of Japan

On the Causes of East Asian Growth

East Asia astonished the world during the second half of the twentieth century as it grew at a pace that nobody had thought possible during the preceding history. No region of the world had ever grown as fast for such a sustained period of time, and nobody could have imagined that a region that performed so badly in all of the preceding history (with the exception of Japan) would outperform everybody else by so much.

Not all nations of the region, however, experienced fast economic growth. Some actually continued to move at a snail's pace. When attention is concentrated on just the fast growers, their performance becomes the more impressive. The unprecedented GDP figures of these nations occurred amid a diverse confluence of events and very varied circumstances, which massively blurred the situation. Policy-making priorities among the governments of these economies were so varied that they went from one extreme to the other. This wide kaleidoscope of policies and the economic theories that backed them up created a gigantic morass of data that seemed highly contradictory and unable to satisfactorily explain the causes of such a phenomenon.

As time moved along and the economic prowess of the region increased, there were more and more efforts to explain the fast growth and extract the policy recommendations that could supply other nations with similar results. Of these efforts, the most thorough and encompassing was conducted by the

World Bank in the early 1990s. Like all the other efforts, this one also could not uncover a set of policies common to all of these nations that was consistent with the macroeconomic statistics.[1]

The study concentrated on analyzing the degree to which state intervention was behind the rapid growth of these economies, and partly because it analyzed the phenomenon from this perspective, it did not see a number of other aspects that were more important. Communism, from the start, was a system that aimed to widely distort market forces, and precisely because of that, political tensions with capitalist countries became very high. Particularly after World War II, tensions between the East and the West became so intense that they drowned the economic discussion with a heavy ideological burden. The Cold War drove these tensions to extreme levels and brought an even more bipolar vision of economics, in which the fundamental focus of attention rested on how much state intervention was applied. For the left, more state intervention was a positive policy undertaking, and for the right, it was a negative one.[2]

After analyzing eight of the best performers in East Asia, the World Bank concluded that state intervention was not the cause of the spectacular growth of these economies. The World Bank was right, because this factor was not common to all. Although seven of the nations had applied a significant level of state intervention, one of them had not, excluding the possibility of a causality linkage. Hong Kong's economy inhibited the establishment of a correlation, because during the second half of the twentieth century, the British colony was one of the most market-driven economies in the world. Some studies even classified it as having the least intervention of all the world's economies.[3]

However, had the study not been so fixated with this matter, it would have probably discovered that in spite of the numerous and strong differences among the eight economies, there was nonetheless a factor that was common to all of them. The study would have also noticed that this common denominator was as well consistent with the macroeconomic statistics of other economies in the region that the study left out. The study analyzed only Japan, Taiwan, South Korea, Hong Kong, Singapore, Thailand, Malaysia, and Indonesia. The effort left out China, which during the 1980s attained a very fast rate of growth, as well as Vietnam and Laos, which by then had already attained a few years of fast growth.

The common denominator to all of these economies was manufacturing. In all of these nations—including China, Vietnam, and Laos—the years of fast growth coincided with fast rates of factory output. Even more important was that, in all of the economies where factory output grew rapidly, there were also government policies that supplied a very strong level of support to this sector. Hong Kong, which had inhibited the state intervention-growth correlation, added up consistently with the manufacturing-growth thesis. During this period, the colony was indeed one of the most market-driven economies in the world, but the colonial government was also among those that supported manufacturing in the most enthusiastic way. In an indirect way, by means of abundant financial, nonfinancial, and fiscal incentives, the colonial authorities channeled massive domestic and foreign resources into this sector.

In other countries such as Japan, the three other NICs, Thailand, Malaysia, and Indonesia, there was also abundant indirect state support for manufacturing, although a large share was supplied directly. In countries such as China, Vietnam, and Laos, the bulk of the support for the sector was very direct. State factories, whether owned by the central, provincial, or local authorities, accounted for the vast majority of manufacturing output. In these countries, which by the mid-1990s were still officially communist, there was also government support in the form of fiscal, financial, and nonfinancial incentives to private manufacturers.

Japan's Economic Recovery

Japan was not the country in the region that attained the fastest rates of growth during the second half of the twentieth century. Every one of the NICs had faster GDP figures. However, Japan was the largest economy in the region and also the OECD country with by far the best performance. On top of that, Japan managed to improve its performance relative to the preceding fifty years by a large margin. Economic growth had averaged about 4.5 percent annually during the 1900 to 1949 period and, although that was the fastest rate in the world, the figures of the 1950 to 1997 period were even faster. In these years, the economy averaged about 6.3 percent annually. (See tables at the end of the book.)

These events were probably the most outstanding and conclusive evidence for debunking the argument that once a nation had attained high levels of wealth it could no longer grow as fast as during its initial stages of development. By 1949, Japan was by far the most developed nation in all of Asia, and on top of that it had a level of development similar to most nations in Western Europe. If there were at least some scientific value in such an argument, Japan should have attained slower rates of economic growth during the 1950 to 1997 period. That was not the case, and the fact is that GDP figures exceeded not only those of the past, but also the expectations of policy makers and economists all over the world.

The World Bank study was right in pointing out that state intervention is not the cause of economic growth in East Asia or anywhere else. A comparison of the two halves of the twentieth century in Japan reveals very clearly the absence of a correlation. The first half was characterized by a higher level of state intervention than the second, but it was the second that attained the fastest rates of growth. Capital, labor, and prices were more distorted during the first half of the century, state enterprises as a share of total output were more numerous, and cartels were more abundant. During the years 1900 to 1949, the private sector was largely under the heel of the government and the military, while during the 1950 to 1997 period it was given a much higher degree of liberty.[4]

Although the argument about the inability to grow fast once development is attained broke down when confronted with the events of the second half of the twentieth century, and the one about state intervention also failed to add up consistently, the one about manufacturing did bond coherently with the facts. The faster GDP figures correlated with faster rates of factory output. While factory output averaged about 6.2 percent annually during the 1900 to 1949 period, in the following half century it averaged about 7.5 percent. That, in turn, coincided with more enthusiastic factory promotion efforts from Tokyo during the second period.

In overall terms, state intervention decreased during the 1950 to 1997 period, but support for manufacturing rose. The vast majority of the support was supplied indirectly by means of fiscal, financial, and nonfinancial incentives, and practically no state factories were created. Market forces were much more operational and that was actually a very positive undertaking, but the fact is that the sector also was promoted more decisively than formerly.

The stronger level of support was the result of a confluence of politically driven motivations and ideological visions of the political class. These motivations were very strong in the first decades, then became considerably less pronounced in the 1970s, and progressively decreased thereafter. This correlated with very fast rates of factory output in the first decades, with a significant deceleration (although still fast) in the 1970s, and a further deceleration in the following decades. This trend in its turn was paralleled by impressive GDP figures in the first two decades, by a significant deceleration in the 1970s, and by a further slowdown of growth in the following decades.

The fears that drove the United States in early 1948 to reverse its original postwar policies with respect to Japan were proven well founded. In October 1949, the Communists took over all of mainland China and the next year the North Korean Communists invaded the capitalist south of the peninsula. With these events, Washington became even more convinced that the economic recovery of Japan was indispensable in its fight against communism in the region. The policy of allowing the archipelago to produce whatever it wanted as long as it was not weapons was maintained and reinforced. Up to 1947 the Allies had suppressed all of heavy manufacturing in order to assure that weapons would not get produced. The Soviet invasion of Czechoslovakia in 1948 changed all that.[5]

In addition, Washington pressured the newly created international financial organizations to lend abundantly to Japan. American commercial banks were also persuaded with incentives to do likewise. The Americans decided to allow Japan to maintain restrictions on U.S. imports while giving Japanese products unrestricted access to the U.S. market. Another way in which Washington tried to assist Japan to recover was by buying as much as possible from the archipelago for the supply of Allied troops during the Korean War in the early 1950s.[6]

However, the fundamental measures that made fast growth possible were the work of Tokyo. Japanese policy makers wanted, above all, fast economic growth in order to repair the massive damages of the war, and were consequently prepared to do anything to achieve that goal. Since by then the manufacturing sector was the largest and the country was relatively poor in natural resources, promoting factory production seemed the only alternative. The large scale bombardments of the war had largely targeted factories, and getting these production establishments back into production was also seen as a priority.

Japan's policy of promoting exports had appeared since the late nineteenth century, mostly because of balance-of-payments concerns, and since then a number of circumstances had driven policy makers to support mostly manufacturing because of its vast potential to generate exports. After World War II, Tokyo was determined to promote exports even more enthusiastically than ever before because by then foreign exchange reserves were almost nonexistent. The end of Japan's imperialist ventures also put an end to the idea that the country could attain wealth by stealing it from other nations, and therefore drove Tokyo to the conclusion that the only way to attain wealth from abroad was by exporting as much as possible. The idea to export substantial amounts was reinforced by the American open door policy to Japanese goods.

As a very proud nation, policy makers decided as well that if Japan would not be a military superpower, it would at least be second to none on civilian matters. The desire to catch up on civilian domains with the United States, which was by the end of the war the most developed country in the world, became another strong motivation for supporting manufacturing because the bulk of the American superiority found its expression in factory goods.

The outcome of the war therefore delivered a very strong determination to support manufacturing, and in consequence the government took a number of measures so as to channel a larger share than formerly of the overall resources of the nation into that sector. The share of GDP allocated to the sector was the largest in the history of the country, and was achieved by decreeing an unprecedented amount of fiscal, financial, and nonfinancial incentives to the producers of factory goods.

The government borrowed abundantly from abroad and transferred the money to state financial institutions such as the Japan Development Bank, the Long Term Credit Bank of Japan, the Japan Import-Export Bank, and the Ministry of Finance's Treasury Investments and Loans Authority. The foreign borrowing was so large that by 1955 Japan was the second largest borrower from the World Bank and one of the largest foreign borrowers in the world. There was also abundant domestic borrowing (notwithstanding the low level of savings), which was also transferred to these institutions.[7]

These institutions channeled the bulk of their funds to manufacturing by means of grants, loans with interests rates below those set by the market, and loans with periods of repayment that were longer than those determined by the market. Some of the money was also used for direct investments in the sector through public corporations. By means of incentives and regulation, the government also drove private banks to lend the majority of their assets to factory producers. Commercial banks were allowed only to carry a narrow range of functions, of which the majority had to do with the sector. Lending to nonmanufacturing sectors was largely limited and interest rates were fixed at artificially low levels for investment in the sector.[8]

Subsidized interest rates from commercial banks were not limited to manufacturing, but this was the sector that benefited the most from this policy. The more the government wanted to promote a factory field, the greater the financial subsidization from public and private banks. Even though the government's financial institutions supplied a very large share of the funds that manufacturing consumed during the 1950s, it was the commercial banks that financed the majority of manufacturing investment. It is very clear, however, that if Tokyo had not applied to them a stick-and-carrot policy, their lending decisions would have been radically different. By the late twentieth century, Tokyo gave them the liberty to do almost as they pleased, and their lending decisions became fundamentally geared toward the nonmanufacturing sectors.[9]

Taxes were also decreed at very low levels for this sector, and the factory fields that the government wished to favor the most received even more tax benefits. During the 1950s, the tax burden on manufacturing in Japan was the lowest among OECD countries. Aside from the very favorable fiscal incentives, there were also a number of nonfinancial incentives such as subsidized rates of electricity, water, and other utilities for manufacturers. There was also the supply of fully equipped factory installations at subsidized rents, free land for the factory, and the supply of technology created at government laboratories at subsidized prices.[10]

There were other measures that the government took to promote the sector, although it appears that these policies were not effective. Tariff and nontariff barriers, for example, were raised to very high levels during the 1950s and Tokyo also turned a blind eye toward the formation of cartels. Up to the end of the war, cartels had existed in the form of *zaibatsus,* and although compulsorily dissolved in 1946 by the Americans, by the 1950s they were again in existence in the form of *keiretsus.* The existence of cartels was not limited to manufacturing, nor were the trade barriers only for factory goods, but it was this sector that policy makers were basically targeting. They believed these policies increased the profit margins of firms in this sector and therefore promoted larger investments. They were wrong. Cross-country comparisons demonstrate that even massive investments in factories are not at all dependent on trade barriers or cartels. Hong Kong, for example, during the 1950s attained a much faster rate of factory output than Japan, even though not the slightest trade barrier was applied and in spite of a very competitive market that was largely free of cartels.[11]

Other policy errors that did not deliver the desired results involved the financial measures that were taken. Because of nationalist reasons, for example, the government did not allow foreign banks to operate in Japan. There were no logical grounds for such a measure. Had they been allowed and regulated so that they would channel the bulk of their assets to manufacturing, it is highly likely that many more resources would have been mobilized into factories. American banks in particular were by then much more efficient, and direct competition with them would have surely prodded Japanese banks to improve productivity. As with trade barriers and cartels, such a measure seems to have only harmed the economy. If the government wanted to accelerate the pace of factory output, then it should have increased the level of fiscal, financial and nonfinancial incentives even more. That is what Hong Kong did during the 1950s, and it achieved faster manufacturing growth and faster GDP figures. Hong Kong also did not put any limits for the establishment of foreign banks in the territory.

Notwithstanding the numerous policy errors during the 1950s, Tokyo supported manufacturing much more enthusiastically than ever before and that correlated with unprecedented rates of factory output and GDP. Factory output during this decade averaged about 11.1 percent per year and GDP expanded by about 9.1 percent. Never before had the economy grown so rapidly and never before had the state given as much support to the sector.[12]

Many argued that this impressive growth was mostly the result of the reactivation of the existing production capacity that had been largely paralyzed because of the war. However, by 1952, Japan had fully recovered and GDP was already at parity with the previous peak level of the early 1940s. The growth of the rest of the decade was therefore strictly the result of net increases over the highest production levels of the past. It is obvious that something else was the cause of growth. The 1960s, with even faster rates of growth, further invalidated the reactivation thesis.

Others asserted that the fast growth was the result of the social, economic, and political reforms that the Americans undertook. In November of 1945, the American occupying force initiated a thoroughgoing land reform, the emperor was dispossessed of most of his assets and his political power, and the zaibatsus were broken up. A three-power division of government was established, female suffrage was legalized, labor was permitted to organize freely, the educational system was restructured, and the economy was liberalized.[13]

It is worth noting that the economy had already started to grow before these reforms were decreed and that after their approval, between 1946 to 1947, growth did not improve by much. Only since 1948 did GDP expand rapidly. Had these reforms been fundamentally responsible for the strong performance of the economy, then stagnation should have prevailed in the months preceding the approval of the reforms and rapid growth should have been attained immediately thereafter. While it is not possible to establish a correlation between the reforms and the performance of the economy, a very tight parallelism is nonetheless observed between manufacturing and growth. There was modest support for the sector from the end of the war until 1947, and from then on, support was very strong. Factory output and GDP shadowed these differing levels of support.[14]

What the Americans did was to transplant their social, economic and political institutions through these reforms, and therefore it becomes very hard to explain why these institutions attained much better results in Japan than in the United States. The Americans occupied Japan from 1945 to 1952, but their influence was still very strongly felt during the rest of the 1950s. While GDP growth averaged 9 percent for Japan during this decade, in the United States it only averaged about 4 percent. The Japanese were not particularly thrilled about these reforms, as the rapid restoration of the zaibatsus in the form of keiretsus demonstrated. If their lack of experience in dealing with these institutions is added, then it becomes even harder to understand how these institutions could deliver for Japan a rate of growth more than twice as fast as that of the United States. Japan's growth during the 1950s was also faster than that attained by the United States in any other decade of its history. If these institutions were determinant for growth, then the country that most practiced them at least should have at some moment during its history attained the most impressive rates of growth in the world. That was not so.[15]

It was also stated that Japan's impressive postwar growth depended substantially upon a large backlog of suitable human skills and talents. If the high educational levels and work experience that were acquired up to 1945 were responsible for the impressive postwar performance, then it becomes hard to understand why before World War II those rates were never attained.

In the 1950s and the following decades, Japan put much emphasis on education and many argued that the backlog of skills plus the new efforts on education were the root of the fast growth. If there had been a causality relationship then, the economy should have at least maintained its fast rates of growth over time. That was not the case. By the 1970s, for example, the educational and skill levels of the population were much higher than during the 1950s. The country, however, experienced a very pronounced deceleration of its GDP figures. In the following decades, the educational and skill levels improved still more, but the economy continued to move in the opposite direction. GDP figures decelerated more and more. The correlation was once again missing.[16]

There is, however, one aspect about the strong educational efforts that Tokyo decreed that seems to have assisted the process of growth. Since the government was particularly interested in promoting manufacturing, emphasis was made on promoting the fields of education that were more directly linked with this sector: mathematics, engineering, and natural sciences. Social sciences, which are mostly linked to services, were largely neglected. A ready supply of engineers, technicians, and blue-collar workers trained in these fields obviously played a positive role for manufacturing, but as the later example of Japan itself and other nations demonstrated, this variable played only a marginal role. By the 1990s, Japan had an abundance of people trained in these fields and in spite of that it was only capable of attaining a modest rate of growth. It is also worth noting that Hong Kong during the 1950s largely neglected education and job training in all fields, but it nevertheless managed to attain faster GDP figures than Japan.[17]

The 1960s went a step further to substantiate the thesis of manufacturing. Since 1948, Japan had been importing machinery and equipment at a furious pace in order to feed the rapid growth of factories. Exports since this date had diversified considerably relative to the prewar period, but by 1957 Japan was still incapable of producing a large amount of manufactures in an internationally competitive way. It was not just machinery and equipment, but also numerous other goods that were inputs to other manufacturers, as well as end products.[18]

Since Japan had been importing at a breathtaking pace and much faster than what it had exported, a very large trade deficit appeared. The massive amounts of capital that had been imported were fundamentally utilized for buying foreign capital goods, and by 1957 foreign exchange reserves were close to zero. So in that year, the Ministry of International Trade and Industry (MITI) came out with a plan to produce domestically the goods which were fundamentally responsible for the large trade deficit. They included machinery and equipment, motor vehicles, refined oil, petrochemicals, artificial fibers, electrical goods, electronics, airplanes, and several others.[19]

Since the strong motivations of the afterwar years driving the state to support manufacturing had not decreased, Tokyo decided also to largely maintain the level of support to the factory fields that until then had been strongly promoted. In order to increase the number of grants and subsidized loans to promote the new fields, Tokyo borrowed more from abroad and domestically. It also gave incentives and pressured commercial banks more so that they would lend more and at more favorable terms to these fields. On top of that policy makers supplied more fiscal and nonfinancial incentives. Taxation, for example, rose on consumption while factory producers experienced a small reduction in the taxes they paid. The increased support for manufacturing was sustained throughout the 1960s and this situation coincided with an accelerated pace of factory output and a faster rate of GDP.[20]

Factory production averaged about 13.4 percent during this decade and GDP expanded by about 11.3 percent. During the 1960s, trade barriers were almost as high as during the preceding decade and cartels were as numerous as in the 1950s. However, economic growth became even more impressive. Abundant historical evidence strongly suggests that these competition-hindering practices were harmful for the economy, and what this situation actually ended up doing was to give even more credibility to the thesis of manufacturing. If it is assumed that a free trade policy and abundant domestic competition are essential for growth, then the situation of the 1950s and the 1960s becomes impossible to explain. If, on the other hand, it is assumed that the essential variable for growth is government support for manufacturing, then the phenomenon becomes fully understandable.[21]

Many argued that Japan's impressive growth was largely the result of the very strong work habits of the population. Average weekly working hours were indeed higher than in the other OECD countries, but the difference was not large enough to justify the large growth gap between Japan and the other developed nations. During 1950 to 1969, for example, average GDP in Japan grew by about 10 percent while in the United States it averaged only 4 percent per year. In Japan, the average workweek was about fifty hours while in the United States it was about forty-three. That was too small a difference to be able to explain the large growth gap. However, when it comes to manufacturing, the gap between both countries was more than large enough to justify the growth difference. While average factory output in Japan was about 12 percent per year, in the United States it was only about 5 percent.[22]

It was also asserted that exports were the engine of growth. Exports indeed grew at an impressive pace since the late 1940s and faster than GDP. However, the recent history of Japan had demonstrated that exports did not always move in tandem with the economy. In the 1930s, for example, as well as in the early 1940s, there had been fast economic growth while exports had been largely stagnant. The historical evidence of Japan and numerous other nations seems to suggest that exports are one of the many effects of growth and not its cause, and that is why sometimes the correlation breaks down.

There is one way, however, in which exports seem to have improved the performance of the economy. In the postwar period, the destination of Japanese exports took a different route than formerly. Before the war the large majority of exports went to Asia. Just China, India, and Indonesia during the 1930s, for example, absorbed more than 40 percent of Japanese exports. With the open-door policy that Washington decreed for Japanese goods, the disappearance of the Chinese market due to its communist autarchic policies, and the adoption of protective policies by most of the newly independent nations in Asia, exports were forced to take a new direction. Since the 1950s, the majority of exports went to OECD countries and in the following decades the share increased.[23]

As Japan was forced to sell in the markets that demanded the highest levels of technology, quality, and cost-effectiveness, it was forced to redouble its efforts so as to satisfy these demands. The end result was a large improvement in productivity. These new markets in North America above all, but also in Western Europe, were very far away and transport costs were much higher. This situation therefore inflicted a higher level of pressure to cut costs as much as possible.[24]

Although the economy benefited from this higher level of competitive pressure, it is evident that the benefits were only marginal. By the 1990s, Japan was exporting to OECD countries almost as large a share as during the 1950s and 1960s, and in spite of that, economic growth had massively decelerated to very low levels. After having averaged about 10 percent between 1950 to1969, GDP rates managed to average only about 2 percent between 1990 to1997. It is evident that some other variable was the determinant factor for the fast growth of GDP and productivity during the postwar decades. Productivity had also massively decelerated by the end of the century.[25]

Many economists and policy makers have long believed that low inflation is a precondition for economic growth. In the last years of the war and in the immediate postwar years, Japan endured hyperinflation. However, by the early 1950s, prices were brought to very low levels and were maintained low in the following decades. Many thought that such a price-control success was largely responsible for the impressive GDP figures for the years 1950 to 1969. Inflation averaged about 4 percent annually during this period.[26]

There is an abundance of empirical data that very convincingly demonstrates that working to attain very low levels of inflation is a very positive policy undertaking. However, a consistent correlation between inflation and growth is impossible to find. For example, during 1946 to 1949 inflation in Japan was extremely high, averaging about 200 percent annually. If there were some form of causality relationship between prices and growth, then the country should have experienced a terrible contraction during those years or at best stagnation. That, however, was not the case. The economy grew and it did it at an impressive pace. Average GDP growth was about 11 percent per year.[27]

Japan was not the only country that experienced this phenomenon. Several other nations such as Taiwan during the 1950s, South Korea during the 1960s and 1970s, and Brazil during the 1970s also experienced high inflation while simultaneously attaining fast growth. However, there is much evidence that if prices had been kept at very low levels, economic growth perhaps would have been faster in all of those countries. While the inflation-growth thesis does not add up consistently with the facts, the manufacturing thesis succeeds very well. For example, there was a very strong level of government support for factory production in Japan from 1946 to 1949, as well as in Taiwan during the 1950s, in South Korea during the 1960s and 1970s, and in Brazil during the 1970s. There were also fast rates of manufacturing output in each one of those economies that were proportionate to the GDP figures.

Mainstream economics has done an impressive job in developing the techniques that allow nations to keep inflation at low levels, and if sound monetary and fiscal policies had been applied in these nations, prices would have surely been in single digits. However, the growth of factory output was a separate phenomenon that depended fundamentally on the level of state support for the sector, and that in its turn was the bottom line for making the economy grow.

The thesis that manufacturing cannot grow without government support was further substantiated in the postwar years, and in particular during the 1960s, by the way in which the different production fields developed in

Japan. The manufacturing field most promoted in the prewar period was shipbuilding, which was also the field that grew the fastest. By the end of the war, about 80 percent of Japan's sea vessels had been destroyed and Tokyo decided to fully reconstruct the fleet and expand it way beyond its former peak size. Because of Japan's archipelago structure, Tokyo again decided that shipbuilding would receive priority, and consequently it continued to supply it with the most subsidies. This coincided with the fact that in the quarter century after the war, ship production grew the fastest. By the end of the 1960s, Japan was the largest producer of ships in the world. This also coalesced with the fact that no other government in the world supplied as many incentives to this field.[28]

Until 1957, production was largely limited to the goods that had been produced up to 1945. That coincided with a policy of government support that concentrated on the traditional factory fields. By then numerous new fields had appeared in other parts of the world and had proven to be in very high demand, but Tokyo decided that since traditional fields had not succeeded in becoming internationally competitive, it was not worthwhile to promote new ones. Japanese capitalists, who were later hailed in the West as being highly entrepreneurial, systematically refused during these years to venture into new fields.

However, when the government decided in the late 1950s that it was time to support new manufacturing fields, suddenly the private sector decided to venture into the new domains. Also worth noting is that of the new fields Tokyo supplied with incentives, it was airplanes that received the least. It is most likely not serendipity that by the late 1960s aerospace was the only field that had failed to become internationally competitive. A relatively weak support for aircraft production correlated with a slow output of planes. In the decades preceding 1945, on the other hand, there had been a very strong promotion of airplanes, and not only did production increase rapidly, but by World War II Japan was fabricating some of the best fighter planes in the world. Again and again, it was made evident that the development of each factory field depended fundamentally on the level of support the state supplied to each.[29]

Japan's nationalism has long historical roots and in the postwar years it survived almost as strongly as before. In the quarter century after the war, Japan's nationalism expressed itself on economic matters in numerous ways, and one of them consisted in prohibiting foreign direct investment (FDI). Tokyo prohibited foreign companies from buying domestic firms or foreign companies from creating production facilities. Many therefore concluded that keeping practically all domestic production in the hands of Japanese contributed significantly to the attainment of fast growth.

However, if that had been an important variable in determining Japan's outstanding performance, then Hong Kong and Singapore should have attained a poor performance during 1950 to 1969 as well as during the whole second half of the twentieth century. In these decades, the majority of investment came from abroad and these city-states granted 100 percent ownership rights to foreigners. The fact, however, is that Hong Kong and Singapore attained faster economic growth from 1950 to 1997 than Japan, and Hong Kong outperformed Japan even during the period 1950 to 1969. Once again, the correlation was missing.[30]

Deceleration and Its Causes

Japan grew so rapidly in the postwar years that by the late 1960s it was rapidly showing signs of soon catching up with the United States and surpassing all other OECD countries. Until the 1950s, most economists and policy makers in developed countries thought that Japan lacked the endowments to ever match North America and Western Europe. By the late 1960s,however, most had changed their minds and had begun to actually see in Japan a nation with the potential to supersede them all.[31]

Tokyo had promoted manufacturing much more enthusiastically than all other OECD countries precisely for this reason. There were also reconstruction motivations and strong balance-of-payments concerns that drove the government to support the sector, but because of the strong nationalism of the country, catching up with the most developed nations was the main motivation. As Japan began to reach this goal, Tokyo was driven to the conclusion that there was no longer a need to subsidize manufacturing as much. The other two main goals propelling support were also largely achieved during this decade. After 1964, for example, Japan started enjoying regular trade surpluses, and foreign exchange reserves grew at a very fast pace.[32]

Reconstruction had also been largely completed by then. Even though since 1952 parity had been attained with the highest GDP peak levels of the early 1940s, the abundant physical scars of the war were still very evident during the early 1960s. The physical destruction acted as a psychological remainder and as a source of pressure on the government that priority had to center on investment and production, and these two fundamentally had to do with manufacturing. However, by the late 1960s most of these physical scars had vanished and the justification for concentrating resources on investment and production began rapidly to decrease.

Therefore, as a result of having largely succeeded in achieving the goals of catching up, of increasing foreign exchange reserves, and of reconstructing the country, Tokyo considerably lowered the level of support for manufacturing in the 1970s.

Even in the mid-nineteenth century, when Japan was developmentally behind the West in gigantic proportions and when its foreign reserves were nonexistent, most policy makers concluded that the promotion of factories could only be justified as long as the country had not caught up. They all thought that the support should not last forever. Over the years, the bulk of the political class remained convinced that once the country had reached parity on military and civilian matters, there was no longer a need to give the sector preferential treatment. By the early 1970s, the same vision of the world continued to prevail among policy makers.[33]

Nobody in Tokyo actually believed that factory production was the key to economic growth, because nobody understood that this sector is particularly endowed with the capacity to generate technonolgy. Precisely because of this nature, it is in constant need of massive investments. Nobody understood that, because of the sector's high investment-absorbing nature and the massive risks that go with it, only when the government reduces those risks by supplying incentives does the private sector decide to venture into it. If the incentives were reduced, the risks would increase and the private sector would venture less; and that is precisely what happened during the 1970s. The factory-promotion efforts of the state were reduced by a large margin, which coalesced with a very pronounced deceleration in the rate of factory output that was paralleled by a considerable deceleration in the rate of GDP. Factory output averaged about 6.2 percent annually and GDP about 5.2 percent

On the other hand, Japan had still not caught up with the United States and there was therefore a significant gap that policy makers wanted to eliminate. Tokyo thus decreed a relatively high level of subsidies. In no other OECD country were the ideological motivations during the 1970s for the promotion of factories as strong as in Japan. No country in Western Europe, Oceania, or Canada was trying to catch up with the United States. They had absolutely nothing against becoming as developed as the United States, but that was not the goal in itself. Nor did they have strong national security concerns or other strong drive that would have pushed them to channel abundant resources into the sector. Tokyo's stronger support relative to all other OECD nations during this decade coalesced also with the fastest rates of factory output and of GDP among this group of nations.

Many people in Japan and abroad had criticized the government for spending so many resources in manufacturing during the quarter century after the war. In their view, the other sectors had been neglected and the living conditions of the population had not improved much. In reality, agriculture, other primary domains, construction, services, and living conditions in general had grown and improved faster than at any other moment in history. However, many people were convinced that if more had been invested in housing, education, health, the primary sector, and the like, living conditions would have improved more.

With these ideas in mind, the government began to transfer resources from manufacturing to the other sectors. It diminished the grants and subsidized lending to the sector and increased this sort of incentives to nonmanufacturing activities. Tokyo began to let commercial banks engage a much larger share of their assets into loans to nonfactory domains, and interest rates stopped being so favorable to the sector. From 1950 to 1969 the share of total bank assets allocated to manufacturing was about twice as high as during the 1970s, and that coincided with factory and GDP rates that were also about twice as fast as those of the 1970s.[34]

The creation of housing-loan banks in the 1970s was illustrative of the transfer of resources to nonmanufacturing domains. It was thought that by investing a larger share of the nation's resources in housing, the accommodation resources of the population would improve faster. This common-sense reasoning proved to be wrong. Construction, and more particularly housing, expanded at a slower pace than during the preceding two decades. More was also invested in agriculture, but this domain's rate of output decelerated massively. While farm output averaged about 4 percent annually from 1950 to 1969, in the 1970s the figure dropped to just 1 percent. The same result occurred with most other domains. A larger share of overall investment in nonfactory domains did not translate into their faster growth. Aside from much lower financial incentives for manufacturing, there was also a reduction of fiscal and nonfinancial incentives to this sector.[35]

Japan was not the only country that experienced a noticeable deceleration of its GDP figures during this decade. Numerous other countries, particularly OECD nations, went through a similar ordeal. Many economists concluded that this situation was the result of the oil shocks that significantly increased the import bill of most countries by a very large margin. It was argued that besides oil, a number of other primary sector goods experienced a rapid rise in prices during the 1970s, which elevated costs of production for OECD countries, reducing their capacity to compete internationally. It was also asserted that the lax monetary policy in developed nations and the accompanying rise in inflation contributed to the slowdown.

Had these really been the variables responsible for the deceleration, the 1980s should have experienced an acceleration of the GDP figures because during this decade the price of oil and of practically all primary goods fell significantly in world markets. On top of that, practically all OECD countries applied tight monetary policies and inflation fell. The fact is, however, that economic growth decelerated further for Japan and for practically all other developed nations. Commodity prices in the 1990s continued to fall, tight monetary policies continued to be applied, and inflation was even lower, but once again the economy of these countries continued to decelerate instead of accelerate.[36]

While this thesis does not correlate with the facts, the one of manufacturing succeeds very well. In practically all OECD countries, there was a noticeable drop in support for manufacturing during the 1970s, which was shadowed by a proportionate drop in factory output. A further decrease in subsidies for this sector characterized the 1980s and a further deceleration in the rate of factory output was experienced. In the 1990s, the trend continued, as practically all OECD governments promoted factory output even less and manufacturing production slowed down some more.

During the 1970s, it was Japan that experienced the steepest deceleration among OECD countries relative to the preceding decade. Many argued that it was the result of Japan's total dependence on imports of oil and most other primary goods. If that had really been the cause of such a phenomenon, then the NICs should have endured a similar or worse outcome because they were even more dependent than Japan on imports for their commodity needs. However, each one of these four economies attained a GDP rate that was about twice as fast as that of Japan. In the case of South Korea and Singapore, there was even a noticeable acceleration in the rate of growth relative to the 1960s.[37]

These economies imported up to the last drop of oil as well as practically all other commodities that they consumed. If impressive rates of growth were attained under those circumstances, then it becomes very evident that the rise in commodity prices had practically nothing to do with Japan's pronounced deceleration. Here again, the thesis of manufacturing is consistent with the facts. There was very strong government support for manufacturing in all of the NICs during the 1970s, which was shadowed by double-digit rates of factory output. Not only was support much stronger than in Japan, but in the case of South Korea and Singapore it was stronger than in the 1960s.[38]

Japan initiated foreign direct investment in the 1970s and relatively large amounts of capital flowed beyond the country's borders. Less direct forms of foreign investment also increased significantly during this decade, and many began to argue that this situation subtracted considerable amounts of funds for domestic investment. The export of capital, according to this argument, was largely responsible for the pronounced deceleration of the rates of growth.

This was not the first time that such an argument had been utilized. British economic deceleration in the late nineteenth century had been explained in a similar way. As in Britain, however, in Japan this argument did not match the facts. In the Japan of the 1970s, there was an abundance of capital that could have been utilized to increase investment by a very large margin. By then the archipelago had the highest level of savings as a share of GDP among OECD countries and had very large foreign exchange reserves, which were growing rapidly. Had more capital been exported, there would still have been an abundance of it left behind and capable of increasing domestic investment by a large margin. The export of capital was actually a reflection of its abundance.

Others stated that Japan's deceleration was largely the result of the large reduction of trade barriers that Tokyo undertook. However, if trade protection had had a positive effect on growth, then Hong Kong and Singapore should have performed poorly not just in the 1970s but also before and after. That was not so. Their average performance was so impressive during the whole twentieth century that they outperformed Japan, and every other country in the world while practicing the most uncompromising free trade regime.[39]

Exports decelerated considerably during the 1970s. After averaging about 17 percent in the preceding decade, they slowed down to just 9 percent. Many who argued that exports were the engine of growth thought that they were corroborated by these figures. The case of the United States during the late nineteenth century, however, demonstrated that it was possible to grow rapidly while exports grew at a slower pace. The case of the Soviet Union from 1920 to 1940 went a step further and demonstrated that it was even possible to grow fast while exports did not grow at all or even contracted. During other periods, numerous other countries experienced rates of economic growth that were much faster than those of exports (Britain during the 1930s). This clearly indicated that some other variable was responsible for the expansion of the economy.[40]

Others argued that the deceleration of the 1970s was to a great extent the result of fiscal imbalances. The 1960s were characterized by budget surpluses, but during the following decade budget deficits became the norm. Most Japanese economists and policy makers had long believed that fiscal imbalances were intrinsically harmful to the economy, and there is much empirical data that supports such an idea. However, the historical evidence of Japan and numerous other countries reveals that the negative effects of budget deficits are only marginal and that the deficits Japan experienced during the 1970s could not have possibly been fundamentally responsible for the pronounced deceleration.

The budget surpluses of the 1960s had averaged about 1 percent of GDP while the deficits of the following decade accounted for just 2 percent. The difference was not large enough to justify a drop of more than 50 percent in the rate of economic activity—not to mention that during the 1950s there were small deficits and the economy nevertheless grew at an impressive pace. Also worth noting is that during the 1980s the deficits shrunk by half to average just about 1 percent of GDP, but the economy did not experience any improvement. Quite on the contrary, the economy decelerated further. In 1990 to 1997 there was more fiscal consolidation and the budget was largely balanced. However, once again the economy did not improve and actually decelerated by a significant margin. The correlation was systematically absent.[41]

Further evidence substantiating the manufacturing thesis is found in the way in which this sector developed during the 1970s. New fields such as computers and microchips were born, once again coinciding with a decisive supply of incentives from Tokyo to these new fields. Practically every new factory field of production during this decade or the preceding ones was a result of government support. Japanese capitalists proved as unwilling as ever to venture on their own into new domains, which was understandable, considering the gigantic investments that were required and the massive risks that they represented. Some of the new fields were promoted largely as a result of the oil shocks. The quadrupling of oil prices in 1974 coincided with the first recession in the postwar years, and many policy makers became convinced that in order to reactivate the economy it was necessary to move into new fields that consumed little oil.[42]

By then the bulk of Japan's production was labor-intensive, which also tended to be oil-intensive. Across-the-board support to manufacturing had begun to decrease in the early 1970s, but since 1974 it decreased considerably more as the oil shock misled policy makers even further. Incentives for labor-intensive manufacturing were reduced by a large margin. There was an increase in support for capital-intensive production, but it far from compensated for the large drop in the other domain. Tokyo thought that it had solved its oil problem, but the fact is that the economy was never again capable of regaining the speed of the postwar years. In the 1980s, the government made further efforts to promote technology-intensive fields, and on top of that the price of oil dropped. However, notwithstanding all of these efforts, the economy decelerated further.

The fact that countries such as South Korea, Taiwan, and Brazil attained impressive GDP figures in the 1970s while producing mostly labor- and oil-intensive goods reinforces further the idea that there was no causality linkage between oil-intensive production and the deceleration. In South Korea and Brazil, economic growth was even faster than in the 1960s when oil was cheap. However, much suggests that if Tokyo had fully transferred the strong support it supplied to labor-intensive fields in the 1960s to capital-intensive domains in the 1970s, then it is highly likely that the overall rate of manufacturing would have remained very high. Had that occurred, Japan would have almost surely retained its double-digit economic growth during the 1970s and the following decades.

Between 1979 and 1980, oil prices once again rose by a large margin and policy makers concluded that the country should distance itself even more from labor-intensive production. There was nothing wrong in this effort, as long as the overall support for manufacturing remained constant or it increased. Unfortunately, that was not the case. Other events, aside from the rise in oil prices, misled Tokyo even more. During the 1970s, Japan had averaged 5 percent GDP growth while the United States had only gotten 3 percent. Under those circumstances, Japan had moved closer to its main goal of catching up developmentally with the United States, and consequently Tokyo concluded that support for manufacturing should decrease further.[43]

Government financial institutions during the 1980s therefore reduced grants and subsidized financing to the producers of factory goods. Tokyo also deregulated commercial banks further so that they could lend a larger share of their assets to nonmanufacturing sectors, and so that interest rates would become more market-determined (which translated into less favorable rates for factory loans). On top of that, fiscal and nonfinancial incentives were also reduced.

Aside from being closer to catch-up, there was the ever-present idea in Tokyo that once at parity with the most developed nations, the private sector would deliver great results on its own. Most policy makers were convinced that, once parity was achieved, support for the sector was intrinsically harmful to the economy. There was also the idea that Japan had reached a stage where it could afford better living conditions for the population, and that such a goal could best be achieved by investing more in nonmanufacturing sectors.

However, as soon as support for the sector decreased, factory output began to grow at a slower pace. As soon as commercial banks were given more liberty, they began to increase the share of their lending to quick-return activities such as real estate and financial speculation. Massive funds went into the construction of condominiums, golf course parks, stock exchange transactions, and currency dealings. A few new manufacturing fields appeared during this decade, but practically all of them were the result of government support. Japanese capitalists once again proved unwilling to venture into new fields and banks refused to risk their funds in new manufacturing domains. They even proved unwilling to lend as much as formerly to well established fields.[44]

During the 1980s resources were again transferred to nonmanufacturing sectors in the belief that in this way these sectors would grow faster, and once again they grew much more slowly than in the first quarter-century after the war. Construction, services, and agriculture grew much more slowly than in the years of impressive factory output. While factory output had grown by about 12 percent during the years 1950 to 1969, in the 1980s the figure was only of about 5.3 percent annually. GDP as well fell from its double digit pace to a 4.0 percent rate. Less support relative to the 1970s coincided with a further economic deceleration. If wealth was created at a slower pace, it was only inevitable that the other sectors would also grow more slowly.

Economic growth was nonetheless much faster than in the other OECD countries and that coalesced with much lower levels of government support for manufacturing in those countries, as well as with much slower rates of factory output. On average these countries minus Japan averaged a rate of production of this sector of about 3.1 percent annually in the 1980s and a GDP growth of about 2.7 percent.[45]

As Tokyo was more strongly motivated than all of these countries, it was inevitable that support was stronger. In spite of the much faster growth, the deceleration of the 1980s caused much alarm in Tokyo and there was much speculation over its causes. Many argued that it was the result of the further trade liberalization that took place in that decade, but there is little evidence of such a causal linkage. An objective analysis of the historical evidence actually indicates that trade liberalization had positive effects on the economy. Hong Kong and Singapore, for example, applied absolutely no tariff or non-tariff barriers during this decade and they nonetheless attained a much faster rate of growth that averaged more than 7 percent per year, almost double the rate of Japan.

During this decade several of the few state-owned companies, such as Nippon Telegraph and Telephone, Japan Airlines, and the National Railways, were privatized. This measure reduced direct state intervention in the economy and was interpreted by some as having negative effects on the economy. It was stated that such a measure was partially responsible for the slowdown. The experience of Japan across time leaves no doubt about the far superior levels of efficiency of private firms compared to state-owned ones. Much suggests that this measure had a positive effect on the economy.

There was also some dismantling of cartels, as pressure on this matter mounted from Washington, which saw the keiretsus as impediments to trade. The mounting bilateral trade deficits with Japan drove the United States to insist on this matter. There is also much in history that strongly suggests that cartels and other competition-hindering practices are harmful to the economy. This measure also seems to have had a positive effect.[46]

The reduction of trade barriers, the privatization efforts, and the reduction of cartels are actually very strong evidence in support of the manufacturing thesis. These measures, which are considered as very important (if not determinant) by most economists for the attainment of growth, began to be undertaken in the 1970s and were pursued with more enthusiasm in the 1980s. If they were as important as most experts claim, then the economy should have improved. However, the exact opposite took place, and the only thing that can consistently explain such a phenomenon is the decreased factory promotion efforts of the state. During the 1990s, the same phenomenon was repeated. There was further liberalization and deregulation, but the economy instead of improving, once again took the opposite direction. It appears as a paradox, as long as it is not taken in consideration that during the last decade of the twentieth century Tokyo decreed even fewer subsidies for manufacturing.

The Crisis of the 1990s

The deceleration of the 1970s, although very pronounced, had not worried Tokyo much because a rate of 5 percent for a nation that had already attained development was nonetheless a strong performance. In addition, it was much faster than that of all other OECD countries. The further economic slowdown of the 1980s was also not seen by policy makers as a sign of economic malaise because a 4 percent annual growth still allowed for unemployment to be kept at very low levels and for real incomes to rise relatively fast. Also, that rate remained much faster than that of all other OECD countries. However, between 1990 to 1997, economic growth averaged only 2.1 percent per year and it was no longer possible to hold unemployment at very low levels. With that rate also, Japan could no longer boast that it was the fastest grower among developed countries. The average rate of growth for the other OECD countries during these years was exactly the same as Japan's.[47]

This time policy makers became really alarmed and many measures were taken to try to reactivate the economy. There was hardly an area of the economy that was not scrutinized and promoted during these years in an effort to reverse the tide. However, all of these efforts failed. Perhaps the most noteworthy of these efforts, because of size, were the fiscal stimulation measures taken by the government. In 1992, Tokyo began to spend gigantic funds in public works, housing, other construction works, agriculture, and small and medium-size firms.[48]

The bulk of the spending went into construction and more particularly into public works. At first, the spending was not that large because it was thought that with just a small amount of funds the economy would recover. Even though the 1992 stimulation package was not huge, it was the biggest dose of Keynesian fiscal stimulus since the end of the war. As this effort failed to deliver positive results, the government decreed a much larger package.[49]

By 1993 the stimulation package had grown to enormous proportions and had become by far the largest among OECD countries. That year, public works as a share of GDP accounted for about 8 percent, while in most other developed nations, including the United States, Germany, and France, they accounted for only about 2 percent. The economy, however, refused to come out of stagnation. In 1994, the massive expenditures in construction, infrastructure, and other domains were continued. They became, by far the largest in the world and in spite of it all, the economy remained stagnant.[50]

Between 1995 to 1997, Tokyo pursued these efforts further, but the economy only barely grew. There were also tax reductions on income and sales taxes were cut in an effort to stimulate consumption. Policy makers thought that this measure would increase demand and therefore production. Output, however, remained stagnant, as did consumption. There were also tax breaks for small and medium-size companies, the vast majority of which operated in nonmanufacturing domains. That also failed to deliver recovery.[51]

Driven mostly by speculation, Japan's stock market rose to stratospheric levels in the 1980s and in the early 1990s it crashed. The depression in the value of stocks drove many in the government (and in particular in the Finance Ministry) to the conclusion that a recovery in the price of stocks was necessary for a recovery of the economy. The Finance Ministry therefore allocated a large amount of funds for stock buying so as to raise the value of these securities. There was eventually some recovery in the price of stocks, but the economy refused to show signs of dynamism.[52]

The plan to prop up the price of stocks was supplemented by the concerted effort of the Central Bank to lower interest rates. The Central Bank, however, was mostly interested in reducing the cost of capital so that investment would pick up again. From the early 1990s, therefore, the Central Bank began to lower the discount rate and by 1993 it was at its lowest level in about forty years. As usual, commercial bank rates shadowed the discount rate and fell down by a very large margin. Capital became cheap and there was an abundance of it, but the economy remained stagnant.[53]

Since no results were obtained and inflation was very low, the Central Bank felt that it could afford to lower interest rates even more. By 1995, the cost of capital was at a historic low. At no other point during the twentieth century had real interest rates (inflation adjusted) been as low. However, notwithstanding the very low cost of capital, investment continued to be very weak. By 1997, interest rates had fallen even more but they still could not reactivate the economy. This was not the first time in the history of Japan that an abundance of cheap capital had failed to drive investment up. In the 1920s after the 1923 earthquake, Tokyo had also adopted an easy-credit policy by lowering interest rates significantly and taking other measures to reactivate the economy, but there was never an investment boom.[54]

In the early 1990s, nonperforming loans among Japanese banks rose to very alarming levels. Reckless lending to construction companies during the 1980s sent property prices out of consumers' reach and in the 1990s those prices crashed. The holes in the loan portfolios of these banks led to a conservative lending policy, and policy makers concluded that in order to reactivate investment, it was necessary to reestablish financial health in the banking system. So in 1993, the government came out with a rescue package for the banks, consisting of funds to improve solvency as well as provide some restructuring of their operations. The effort failed and by 1997 the share of nonperforming loans had actually risen.

The problems of Japanese banks, were, by all means, significant, but the financial position they experienced in the 1990s was massively more favorable than what they endured in the immediate years after the war. In the postwar years, even though their assets were much smaller and their risks much higher, they managed to lend abundantly and investment was impressive. It is evident that the bank's financial troubles of the 1990s were not the main reason for their low lending and the overall low level of investment.

It is as well worth noting that during the 1990s China's banks had a much larger share of nonperforming loans than Japanese banks had. It is estimated that nonperforming loans accounted for about a fifth of the loan portfolio of Chinese banks (as a share of GDP), while in Japan it was only about a tenth. China, however, had one of the fastest rates of economic growth in the world together with an investment boom.[55]

In the early 1990s, there were numerous political scandals dealing mostly with corruption among the ruling party. The Liberal Democratic Party (LDP) had long ruled the country and many argued that it had become corrupt and had lost interest in governing with efficiency because it had taken power for granted. It was thought that new blood with fresh ideas was needed to reactivate the economy, so the voters took things in their hands and ousted the LDP. In the elections of mid-1993, the opposition was given a mandate for the first time in almost half a century. The new government was full of honest intentions and new ideas, but none of them aimed to promote manufacturing, and that coincided with the continuation of a stagnant economy. Eventually, the voters got tired of the socialists and gave power once again to the LDP, which promised to eliminate corruption and liberalize the economy. The new government kept its promises, but the economy continued to languish.[56]

Another measure the government took in an effort to reactivate the economy was to increase its economic ties with Asia, particularly with East Asia. Most in government circles were convinced that by increasing the country's linkages with the fastest-growing region in the world, Japan would also profit from the positive spillover effects of its neighbors. East Asian countries indeed grew at an impressive pace between 1990 and 1997 and they relentlessly bought more and more from Japan, but the Japanese economy refused to grow quickly. Australia and New Zealand were also convinced since the 1980s that being so close to East Asia would bring positive results to them. However, during the 1980s and 1990s, these two countries attained only modest GDP figures.[57]

By the early 1990s, Japan's labor costs, in particular in manufacturing, were the same as or higher than in the majority of OECD countries. It had long been argued by many Western economists, but also by several in Japan, that a main reason for Japan's fast growth was its cheap labor force. By the 1990s, labor was no longer cheap, and as a result many concluded that this argument was correct. In an effort to counter this situation, Tokyo took some measures to encourage foreign direct investment (FDI) in developing countries so that production costs of Japanese enterprises would get lowered. FDI had begun in the 1970s and did not fail to increase thereafter. In the 1990s, it reached unprecedented levels, but in spite of having transferred such a large share of production to countries where labor costs were very low, the economy never managed to recuperate. During that period, the correlation actually went in the opposite direction. As FDI increased, the economy decelerated more.[58]

Part of the strategy to encourage FDI consisted in letting the currency appreciate, for a strong yen would further reduce the costs of setting up operations abroad. The yen did appreciate by a large margin, going from about ¥130 to a dollar in 1990 to about ¥90 to a dollar in 1995. At first, the government viewed the rapid appreciation as a sign of the structural strength of the economy, but as the economy remained stagnant, the position of Tokyo eventually reverted. Eventually, many began to argue that the large outflows of FDI were not reducing the costs of production of Japanese firms, but were actually hollowing out the country's production base. In 1996, therefore, the government took measures to depreciate the currency in the hope that the hollowing-out process would decrease and that exports would be promoted. Exports were seen by many as determinant for growth and in the 1990s they had decelerated. The yen did depreciate, the emigration of Japanese firms also decelerated, and there was a rise in exports, but the economy still did not rise.[59]

The situation in Japan during the 1990s was similar to that in the United States during the 1930s. Both nations took numerous similar measures to reactivate the economy and none of them worked. They both tended to adopt a measure on a certain aspect of the economy and once that measure failed to deliver the desired results, they would frequently adopt the inverse measure, which would also end in failure. Aside from both having adopted similar stimulation policies which failed, there was something even more important that was common to both. The governments of the United States during the 1930s and of Japan during the 1990s were both largely unmotivated toward manufacturing, and consequently they promoted this sector only weakly. In both countries, factory output grew at a slow pace. Between 1990 and 1997 manufacturing production in Japan averaged about 2.4 percent and GDP about 2.1 percent annually. In the United States during the 1930s, the respective figures were 0.8 percent and 0.6 percent.

By the late 1980s, Japan had on average caught up developmentally with the United States. In fields such as automobiles, electronics, computers, and microchips it was on a par. On other fields such as steel and shipbuilding, it was even ahead. It had a massive and growing trade surplus with the United States and the rest of the world, and its foreign exchange reserves were the largest in the world. The goal of catching up with the most developed country in the world had finally materialized and that had been the main motivation pushing Tokyo to supply support to the sector. The motivations dealing with balance of payments and war reconstruction had also been completely satisfied. As a result, policy makers no longer saw the need to supply manufacturers with more incentives than those given by the governments of the other OECD countries.

Since Washington and the governments of other developed countries by then supplied their manufacturing sectors with a modest level of subsidies, Tokyo concluded that in order to maintain its leading position it only had to promote factories in a similar way. Convinced that strong support for the sector, once Japan had caught up developmentally, was intrinsically harmful for the economy, Japanese policy makers decided to offer only the same level of support as the other developed nations. This coincided with almost identical rates of factory output for Japan and the other OECD countries. The United States, for example, the nation that attracted the interest of Tokyo the most, attained a factory rate of production during 1990 and 1997 of about 2.6 percent annually.[60]

By the mid-1990s, most economists and policy makers in Japan had come to believe that once it had reached full development, it was no longer possible for the nation to grow fast. Most thought that the best that Japan could attain was average GDP rates of about 3 percent per year. However, notwithstanding those much lower expectations, the economy failed to achieve even that rate of growth. The idea that, once developed, a nation can no longer grow fast was by then more than a century old. It had first appeared in Britain in the late nineteenth century and it was repeatedly proven wrong as this nation attained faster rates in future periods than the ones attained when this idea was born. Other nations went through similar situations. During the 1930s in the United States, for example, the same was argued and once again it was proven wrong as fast growth was achieved in the following three decades.

Japan itself was the living proof that such an idea had no scientific grounds. By 1949, the country was already developed relative to those times, and yet it attained the most impressive growth figures of all its history in the following decades. Even by the 1990s, there was much evidence that fast growth was still possible for Japan. By the late 1980s, Hong Kong and Singapore had reached a level of development that was almost the same as that of Japan and the other OECD countries. By then, their per-capita GDP in terms of purchasing power parity was similar. If the argument about the limits of development were valid, then these two economies should at best have done a little bit better than OECD nations. The fact is that their performance was far superior. While Hong Kong averaged a GDP growth of about 5 percent annually between 1990 and 1997, Singapore attained the impressive figure of 8 percent. OECD nations averaged only 2 percent. By 1997, GDP per capita in purchasing power parity terms for these two economies was higher than that of practically all OECD countries.[61]

Singapore and Hong Kong not only demonstrated that it was perfectly possible to attain very fast growth while being fully developed, but their example also strongly suggested that they managed to achieve that feat because there was ample support for manufacturing. During the 1990s, the governments of these two territories promoted factory production much more enthusiastically than OECD governments did, and factory output grew at a much faster pace. Of these two East Asian economies, it was the government of Singapore that supplied the sector with the most incentives, and it was also in this Southeast Asian nation where factory output grew the fastest.[62]

By the mid-1990s most policy makers in Japan had come to believe that the reason the country could not achieve the 3 percent GDP rate, that they thought was the best that could be attained, was that so many companies had emigrated abroad. There had indeed been an unprecedented outflow of FDI, but the case of Hong Kong and Singapore demonstrated that this phenomenon was not an obstacle for the attainment of fast growth. Since the 1980s, the emigration of companies from Hong Kong and Singapore to nearby countries was proportionately larger than that of Japan. However, during this decade economic growth averaged more than 7 percent for both, while Japan only got 4 percent. During the 1990s, the outflows of FDI from Hong Kong and Singapore increased by a very large margin. Due to their direct proximity to China and Malaysia respectively, the outflow of enterprises from the two city-states that searched for cheaper production bases was much faster than in Japan. And in spite of it all, they got much faster rates of economic growth. In the case of Singapore, growth was even faster than in the preceding decade.

It is also worth noting that during the 1920s in Japan and the 1930s in the United States, there was hardly any emigration of domestic firms to lower-cost countries. In spite of that, the economy of Japan grew at about the same pace as during the 1990s and that of the United States performed miserably. The empirical data clearly indicates that a poor performance is not the result of emigrating firms, even when it occurs at a very fast pace. The evidence suggests that independent of how fast the emigration occurs, the economy will nevertheless grow rapidly as long as there is strong support for manufacturing.[63]

Technology and Growth in Japan

Further evidence substantiating the thesis of manufacturing was found in the way technology developed in Japan during the second half of the twentieth century. Technology grew the fastest during the first two decades of this period, coinciding with the years when factory output grew the fastest. During these years, the bulk of the technology was imported, but the nation experienced an unprecedented improvement of living conditions. During the 1970s, the rate by which technology grew slowed down considerably, and that correlated with a significant deceleration in the rate of factory output. In the 1980s, the level of support for manufacturing decreased some more and technology once again shadowed this situation by growing at a slower pace. And in the 1990s, Tokyo's factory promotion efforts were even lower, the rate of factory output decelerated further, and technical progress did the same.[64]

It is interesting to note that in the 1970s Japan began to produce the majority of the technology it consumed, and in the decades that followed, the share of domestically created technology grew more and more. In 1973, for example, Japan began to have more new contracts for the export of technology than for its imports and the surplus progressively increased over time.

However, this phenomenon coincided with a noticeable deceleration in the overall rate by which technology was consumed and by a deceleration of the economy. By the 1990s, at a time when the country was producing the bulk of its technology and was among the most innovative in the world, it was attaining the slowest GDP rates of the postwar years. The 1990s were actually the time during the twentieth century when Japan produced by far the largest share of the technology it utilized, and it was also the time when GDP figures were at their lowest. By the 1990s, Japan had one of the highest per-capita registrations of patents in the world. It was similar to the case of Switzerland, which by then had the world's highest per-capita output of patents, and in spite of such a positive achievement economic growth was extremely low. From 1990 to 1997 Switzerland's GDP averaged less than 1 percent annually.[65]

The research and development expenditures also presented a similar paradoxical phenomenon. During the 1950s, for example Japan's R&D expenditures as a share of GDP averaged only 1 percent and the economy grew by about 9 percent annually. Between 1990 and 1997, on the other hand, R&D accounted for about 3 percent of GDP, but the economy only expanded by about 2 percent per year. The historical evidence clearly indicates no correlation between R&D and growth, nor between the capacity to self-produce the bulk of the technology utilized and growth.[66]

For a long time, numerous policy makers in many countries have believed that the capacity to self-produce the bulk of the technology that a nation consumes is important for the attainment of growth. The evidence, however, suggests otherwise.

When a nation finds itself lagging technologically with respect to the most advanced countries in the world, it becomes inevitable that the bulk of the technology utilized is imported. How fast technology grows is fundamental for a nation, not if it manages to self-create the technology it consumes. If it imports everything but its growth is very fast, a nation will improve its living conditions very rapidly. That was the case of Japan and more still of Hong Kong during the 1950s. History suggests that for this to occur, strong support for manufacturing must be decreed.[67]

However, if technology grows slowly even though the bulk is self-created, living conditions will improve slowly. That is precisely what took place in Japan and in practically all other OECD countries during the 1990s. That also coincided with weak government support for manufacturing in practically all OECD countries. As a nation catches up developmentally with the most advanced ones, it is almost inevitably driven to produce the bulk of the technology it consumes or at least a large share of it.

That, however, does not mean that once it reaches that stage a nation cannot attain fast technological growth and fast economic growth. The case of the United States during World War II clearly demonstrated that a nation could very well be at the very top of world economic and technological development and still attain an impressive growth of technology and GDP. In this case, practically all technology was created domestically.

In the case of Hong Kong during the 1990s, the bulk of the technology that was consumed was not created domestically. However, this economy demonstrated that even though it attained a similar level of economic and technological development as OECD countries, it could still grow technologically very rapidly. During these years it continued to import technology at a very fast pace and by the mid-1990s, this city-state was among the most technologically advanced in the world.[68]

The common denominator in all of these cases was a strong level of government support for manufacturing. The factory promotion efforts of Washington during World War II were the strongest during the twentieth century and that was shadowed by the fastest growth of technology during this century. During the 1990s, support for the sector in Hong Kong was very strong and rates of manufacturing output were also fast.

Technology in Japan during the second half of the twentieth century did not develop homogeneously. While some fields developed at an impressive pace, others had a much slower pace. The differing speeds by which each field grew coincided with the differing levels of subsidies that Tokyo supplied. Shipbuilding and steel were the fields more enthusiastically promoted during the whole period and they were also the ones that developed technologically the most. Other fields such as automobiles, electronics, computers, microchips, and machine tools were slightly less strongly supported and their technological development (although very fast) was slightly less pronounced.

Aerospace technology, on the other hand, advanced very slowly and that coincided with very weak promotion efforts from Tokyo. By the 1990s, Japan continued to be incapable of producing airplanes in a competitive way. Until 1945, however, Japan had perhaps the best fighter plane technology in the world, which correlated with massive government support in the preceding decades to that field. During the second half of the twentieth century, Tokyo decreed practically no support for the production of military planes and only a few incentives for the production of civilian planes. As a result of the war, Japan was not allowed to invest abundantly in military manufacturing but it could have invested massively in civilian airplane production.[69]

The thesis that manufacturing is the prime creator of technology was also corroborated by the fact that technology systematically materialized as a factory good. Technology in Japan during this half-century period was at first mostly imported, and it came fundamentally in the form of machinery and equipment. It came also in the form of patents, which ended up almost always transformed into a manufactured good. The imported technology also came in the form of factory goods of all sorts, which were dismantled so as to be able to reverse-engineer on them and copy their numerous parts. The technology that was domestically created appeared likewise in a similar form. The bulk of patents were directly linked to a manufactured good. Inventions such as the video cassette recorder and the Walkman were among the most outstanding discoveries, and it comes without saying that they were manufactured goods.

A long-term analysis of the development of technology reveals as well a tight correlation with manufacturing. Technology had been almost totally stagnant throughout Japan's history up to the mid-nineteenth century, and during all of that time government support for the sector was practically nonexistent. Then, during the second half of the nineteenth century, there were relatively strong factory promotion efforts, and all of a sudden technology grew rapidly. During the first half of the twentieth century, the state considerably raised the level of subsidization to this sector and technology grew at a much faster pace. During the second half of the twentieth century, Tokyo promoted factory production in a still more intensive way and technology expanded at a still faster pace.

Relative to other countries the correlation was also very strong. Until the nineteenth century support for manufacturing was stronger in most of the future OECD countries than in Japan, and it was Japan which was among the slowest in technological development. During the first half of the twentieth century, however, Tokyo promoted the sector more decisively than any other country in the world, and it also had the fastest pace of technological development in the world. Then, in the second half of the twentieth century, Japanese policy makers promoted factory production more enthusiastically than any other OECD government, and it was Japan that again had the fastest growth of technical progress among this group of nations. The NICs, however, showed a faster technological development than Japan during this period, which coincided with stronger support of the sector and a faster rate of factory output.

The thesis of manufacturing is further substantiated by the way productivity behaved. Since productivity is fundamentally determined by technology, it follows that whatever is responsible for the generation of technology must also be the cause of productivity. If manufacturing is the cause of technology, then it must also be the cause of productivity. It would thus be logical to expect a correlation between the fluctuations of this sector and those of productivity, and the empirical data of the second half of the twentieth century reveal precisely that. Overall productivity in Japan grew the fastest during 1950 to 1969, coinciding with the period in which factory output grew at the fastest pace. In the following three decades rates of productivity progressively decelerated, and that was paralleled by a progressive decrease in government support for the sector.[70]

The long-term development of productivity also shadowed that of manufacturing. During the second half of the nineteenth century, productivity grew for the first time at a rate that was no longer stagnant and that coalesced with the first time in Japan's history in which there was a relatively strong support for manufacturing. The 1900 to 1949 period witnessed a much faster growth of productivity and that was accompanied by a simultaneous much faster rate of growth of factory output. During the second half of the twentieth century, Tokyo supplied a still higher level of subsidies for the sector, which was paralleled with an unprecedented growth of productivity. Cross-country comparisons reveal also a similar correlation. During 1950 to 1997, it was Japan that attained the fastest rates of productivity among OECD countries. Productivity averaged about 5 percent annually, while in practically no other developed country did it average much more than 3 percent per year. That ran parallel to the fact that no other OECD government supplied manufacturing with as much support as Tokyo did.[71]

The strong linkages between manufacturing and technology were also expressed in the way research and development took form. The bulk of the R&D done in Japan during the second half of the twentieth century was carried out in the laboratories of manufacturing companies, and the vast majority that was done in independent laboratories and universities was intended ultimately to conceive a manufactured good. Also, the majority of R&D expenditures were the doing of the state. Tsukuba Science City, for example, which is Japan's largest effort to put under one roof as much R&D as possible, since its birth in the early 1970s divided private and state investment in relatively equal shares. By the mid-1990s, about half of the research institutes were government-owned and only about half of the researchers labored in private laboratories.[72]

Although the majority of the R&D during the second half of the century was done by private companies, a very large share of the costs of these firms was absorbed by the state. Tokyo supplied numerous grants and other incentives for R&D to private manufacturers. Between the incentives it supplied to the private sector and the direct expenditures the state made, the majority of the funds for this purpose ended up coming from the coffers of the state. It is also logical to presume that if the government had not paid for that large share of R&D and also had not supplied an abundant amount of other incentives to manufacturers, the private sector would have invested practically nothing in R&D. As the bulk of Japan's history demonstrated, without government support for the sector there was practically no growth of manufacturing. If the sector did not develop, there was no growth of technology. Under those circumstances, there is no private-sector investment in R&D. Not by chance is it that to the mid-nineteenth century, R&D expenditures in Japan were nonexistent.

The Nonmanufacturing Sectors

The way the other sectors of the economy developed during the 1950 to 1997 period corroborated further the manufacturing–technology thesis. On the one hand, the fluctuations of these sectors coincided with the differing levels of state support for factories. Agriculture, for example grew the fastest during 1950 and 1969, averaging about 4 percent per year. Never before had agriculture grown so fast and never before had the state promoted factory output as decisively.

The unprecedented growth of agricultural productivity during these years as well correlated with an unprecedented mechanization of the domain. The production and utilization of farm machinery, as well as fertilizers, pesticides, and irrigation equipment, grew at a rate never seen before. Much indicates that it was these factory goods that were fundamentally responsible for the fast growth of output and productivity, because technology was concentrated on these goods. Since technology was added to the land at a very fast pace, productivity inevitably had to increase rapidly.[73]

During 1970 to 1997, the rate of agricultural output and productivity decelerated by a very large margin and that coincided with a very significant reduction of subsidies for manufacturing and with a very large deceleration in the rate of factory output. The other sectors behaved similarly. Construction also grew by far the fastest between 1950 and 1969, even though only a very small share of the nation's resources was allocated to it. Although an increasing share of total investment flowed into construction in the following decades, the growth of this sector progressively slowed down.[74]

Services also grew the fastest when Tokyo strongly promoted factory production. During 1950 to 1969 services grew by about 10.3 percent annually. In the 1970s, the factory promotion efforts of the state decreased by a large margin and services also experienced a considerable deceleration and averaged about 5.0 percent per year. In the following decade, fewer subsidies for the factories coincided with a growth of services of just 3.2 percent. In 1990 to 1997, Japanese policy makers promoted factories considerably less, there was a proportionate deceleration in the rate of this sector, and services experienced a similar slowdown. Services expanded by just 2 percent annually.[75]

What the empirical data strongly suggests is that the primary sector, construction and services, is fundamentally dependent for its growth on the growth of the manufacturing sector. The preceding history also reveals a strong correlation. Primary production, construction, and services had experienced practically no growth up to the mid-nineteenth century, and then during the second part of this century the government for the first time promoted manufacturing in a relative decisive way. Suddenly, all the other sectors began to grow at a relatively fast pace. In the following fifty years, the state transferred an even larger share of the overall resources of the nation to the factories, and all the other sectors grew at a much faster pace. Then during 1950 to 1997 Tokyo allocated a still higher level of subsidies to manufacturing and once again the other sectors followed and grew at a faster pace.

During 1950 to 1997, services grew at a very fast pace. They grew so fast that their share of GDP expanded considerably. While in 1950 they accounted for about 40 percent of GDP, by 1997 they had grown to represent approximately 60 percent. During this period, in practically all other OECD countries, services grew rapidly and became also by far the largest sector of the economy. This development of events convinced many economists that services had become fundamental for the attainment of economic growth. Since there was no long-term correlation between services and GDP prior to the mid-twentieth century, it was argued that services acted only as an important determinant of growth once a nation had reached a relatively high level of development.[76]

However, even under this form the argument was incapable of adding up consistently with the facts. On the one hand, during the whole of the second half of the twentieth century, services attained extremely low rates of productivity in Japan and in the rest of OECD countries. The historical records also show, that in the preceding periods, services performed miserably on the productivity front. A sector that performs so badly on the productivity front is highly unlikely to have the capacity to act as a pulling force on the rest of the economy. Manufacturing, on the other hand, attained throughout the whole of history the fastest rates of productivity. In Japan and in the other OECD countries during the second half of the twentieth century, productivity in manufacturing grew on average more than twice as fast as in services. Only manufacturing, which grew faster than GDP and which attained a strong rate of productivity, had the capacity to act as a propeller.[77]

Also, as the share of services in GDP expanded, the economy decelerated more and more. If services were so helpful for the attainment of growth one would expect that as their size in the economy expanded, the economy would improve. However, exactly the opposite took place. During the 1990s, at a time when services accounted for by far the largest share of GDP in the history of Japan and the rest of OECD countries, the economy was at its worst. In Japan, the rate of economic growth was the lowest in the twentieth century, and in practically all other OECD nations GDP figures were the lowest during the second half of the century. It was argued that services were very useful for creating jobs, but the fact is that the unemployment and underemployment figures for OECD countries during the 1990s were among the worst in the twentieth century.

The desire of numerous countries to experience a considerable development of their service sectors is by all means a goal worth pursuing, but history ultimately demonstrates that the best way to achieve that goal is by promoting manufacturing as much as possible. Since there is much evidence that manufacturing is the sole creator of technology, it is inevitable that the faster this sector grows, the larger the development possibilities for services and the other sectors, for it is technology that allows all sectors to grow.

The way the proliferation of services in Japan during 1950 to 1997 was directly bonded to factory goods further substantiates this idea. It was, for example, switch boards, telephones, fax machines, telex machines, copper cables, fiber optic cables, cellular phones, and satellites that made possible a revolution in telecommunication services. It was airplanes, jet engines, supertankers, double-hulled ships, high-speed trains, automobiles, and trucks that delivered a revolution in transport services. Without these goods, the rapid development of telecommunication and transport services would have been impossible.

Educational services also experienced a large transformation and a significant upgrading, fundamentally as a result of factory goods such as computers, calculators, laboratory equipment, and books. By the 1990s, Japan had one of the most developed educational systems in the world, and the vast advances on this front would have been impossible without these educational factory goods and the abundance of wealth that the manufacturing sector delivered. Had the economy not grown so rapidly in the previous decades, Japan would never have been able to afford to educate its population so well; and much evidence suggests that fast growth was only possible thanks to the decisive factory promotion efforts of the state.[78]

Health services also experienced revolutionary progress, as mortality and morbidity dropped significantly. While in 1950 the average life expectancy in Japan was just 54, by 1997 it had risen to 80, the highest in the world. The extensive development of health services would have been impossible without the unprecedented production of pharmaceuticals and medical equipment. Without antibiotics, anti-arthritics, analgesics, anti-hypertensives, tranquilizers, and vasodilators, as well as x-ray machines, operating instruments, ultrasound equipment, magnetic resonance machines, and laser equipment, the impressive developments of medical services are unthinkable. It was these factory goods that made the advances possible because they were the depositories of technology.[79]

During 1950 to 1997, Japan not only attained a rate of economic growth that was much faster than any preceding half-century period, it also grew much faster than any other OECD country. Its rates of growth were about twice as fast as the average of the other OECD nations. There was much speculation over the causes of such a phenomenon and none of the proposed explanations added up consistently with the facts.[80]

The one subject that has most attracted the attention of analysts has been culture. Numerous social scientists and economists have argued that the fundamental factor explaining the far superior GDP figures of Japan relative to the West was its culture. It was asserted that the country's culture drove the Japanese to work harder, to study more, and to save a larger share of their income than in the West. It was supposedly culture that enabled the creation of a system in which people, not profit, took priority. Culture was also what supposedly drove the government, business, and labor into collaborative arrangements so as to devise the policies that were best for the whole of society and not just for a part of it.[81]

Japanese cultural traits such as trust, security, and cooperation apparently delivered superior business practices than in the West (long-term relationships in distribution and subcontracting). Lifetime employment and a harmonious relationship between business and labor, according to this line of argument, was also culturally determined and highly positive for the economy.[82]

Culture did, by all means, play a role in molding the behavior of the Japanese, and it is quite evident that the aspects that analysts have noted are valuable assets. However, if culture in its numerous aspects had been a determinant for growth, then Japan should have outperformed the West since the Middle Ages, if not earlier, because culture is something that is very constant over the long term. The fact is that only since the early twentieth century did Japan begin to grow faster than the West. Before the mid-nineteenth century, that same culture systematically failed to deliver even a very slow pace of growth. Many, therefore, argued that Japan's success lay not just in its culture, but also in having combined it with the best from the West.[83]

Impressed by Japan's superior performance, the West began in the late twentieth century to imitate Japan in several of those characteristics that analysts asserted were pivotal for the archipelago's success. That was particularly so with business practices. To a large extent, the West began to absorb the best from Japan, but in spite of that, its economic performance did not improve. It actually deteriorated. If Japan had attained impressive growth largely as a result of having combined its culture with the best from the West, it would be logical to expect that the West would attain similar results by absorbing the best from Japan. That was not so—not to mention that several of those aspects were not very different from some that several OECD countries had practiced for a relatively long period of time. Job security and harmonious business–labor relations, for example, were endorsed in Scandinavian nations, Germany, and Switzerland during the second half of the twentieth century, but these countries grew considerably more slowly than Japan. This argument evidently could not explain Japan's performance.

It also could not explain why there had been such a significant growth difference between the first and the second half of the twentieth century. Many argued that the faster growth of the second half was the result of the intelligent political leadership of the Liberal Democratic Party, the high professionalism of the country's technocrats, and the radical change from military to civilian production. To that was added the numerous economic, political, and social reforms the Americans instituted after the end of World War II. There is no doubt that Japan benefited from all of this, but if these factors had been determinant of the fast growth, the country should have continued to grow rapidly during the 1990s.[84]

The fact that the GDP figures were slower in the 1990s than during any other decade of the twentieth century very clearly demonstrated that these factors were not the determinant ones. The weak performance of the 1990s also demonstrated that culture, in whichever form, was not the cause of the fast growth.

While none of these variables, whether presented individually or as a group, managed to blend coherently with the facts, the manufacturing variable did. Levels of government support for manufacturing consistently correlated with the GDP figures in the centuries prior to the mid-nineteenth century as well as in the following periods.[85]

Only strong government support for manufacturing during the second half of the twentieth century is capable of explaining why, in spite of the high trade barriers, it was still possible to attain fast growth. After much trade liberalization since the 1970s, several studies estimated that by the early 1990s Japanese trade barriers (tariff and nontariff) were still about five times higher than those of the United States. To that it had to be added that cartels were also more numerous than in the U.S. However, notwithstanding those policy errors, Japan attained a rate of growth that was more than twice as fast as that of the United States during the whole period. That phenomenon becomes understandable only when it is considered that Tokyo promoted factories much more decisively than Washington did during the 1950 to 1997 period and that factory output grew on average more than twice as fast.[86]

The thesis of manufacturing finds itself also reinforced by the way in which investment and savings behaved. The generation of technology is by far the most resource-absorbing activity in the world. If manufacturing is the prime creator of technology, then vast investments are inevitably needed in order to materialize it, and if it is the prime creator of wealth, then savings should shadow its development because the possibilities for savings are always larger when wealth is created at a faster pace.

The historical data show precisely that. During the 1930s, for example, Tokyo supported manufacturing enthusiastically, and investment and savings were high. In the 1940s, factory output grew much more slowly and the share of investment and savings from GDP decreased. In the 1950s, factories were promoted in a considerably more decisive way and the level of investment and savings rose. During the 1960s, the government supplied an even higher dose of subsidies to the sector and that coalesced with a still larger share of investment and savings as a share of GDP. In the 1970s, Tokyo's promotion efforts for factory production diminished and the level of investment and savings did likewise. During the 1980s, the correlation was once again observed. Factory output grew at a slower pace and the share of investment and savings became smaller. In the 1990s, a further decrease in the factory-promotion efforts of the state was paralleled by a further decrease in the share of investment and savings.[87]

The way exports performed during the second half of the twentieth century substantiates further the thesis of manufacturing. If this sector is the fundamental generator of wealth, then the development of exports should be largely dependent on the development of manufacturing because only when wealth gets created is it that exports can grow. The faster the growth of this sector, the more possibilities for a faster growth of exports, and vice versa. The empirical data of this period reveals precisely such a correlation. There was strong support for manufacturing during the 1950s and exports grew by about 14 percent annually. In the 1960s, Tokyo promoted the sector some more and exports rose to average about 17 percent per year. In the 1970s, factory output slowed down significantly and exports shadowed by averaging only 9 percent. During the 1980s factory output decelerated some more, and exports decelerated to 7 percent. Between 1990 and 1997, Tokyo considerably reduced the incentives to manufacturers, and exports averaged only 4 percent per year.[88]

It is also worth noting that exports during the second half of the twentieth century increased at an unprecedented pace, which coincided with unprecedented government support for the sector and with unprecedented rates of factory output. Throughout history, the long-term development of exports systematically correlated with the differing levels of government support

support for manufacturing. Considering that factory goods are the easiest to export and the ones that earn the most hard currency, this situation was only inevitable.

Living conditions improved significantly in Japan during the second half of the twentieth century. Never before had they improved by as much and by the end of the century they were above those of the immense majority of nations in the world. Morbidity and mortality plummeted, levels of education rose impressively, housing improved by leaps and bounds, violence decreased greatly, and the rule of law, for the first time in the history, became firmly entrenched. Environmental protection, for the first time in history, was seriously endorsed, and women's rights made major progress.

If manufacturing is the prime creator of wealth, as history strongly suggests, then it is understandable why the largest improvement in living conditions in the history of Japan coincided with the strongest promotion of factories.

Much also suggests that the improvements in living conditions could have been considerably larger if Japanese policy makers would have promoted manufacturing more. If Tokyo would have continued to support the sector during the last three decades of the twentieth century as it did during the preceding two, it is almost a fact that the economy would have continued to grow at a double-digit pace.

2
China in Modern Times

China During the 1950s

The second half of the twentieth century was the first time in China's millenary history in which the country succeeded in attaining a fast rate of economic growth. Its growth was without parallel and living conditions improved in an unprecedented way. After having grown by much less than 1 percent annually during all the preceding fifty-year periods, the country suddenly grew by about 5 percent. The official figures were much higher for the whole 1950 to 1997 period, averaging about 9 percent per year (see tables in appendix). However, since market forces were so distorted and the quality of the goods was so low, it is necessary to adjust those figures so as to make them compatible with those of capitalist economies with highly competitive markets.

Even the adjusted figure of 5 percent was about ten times faster than the rate of the first half of the twentieth century, and it was also faster than the GDP rates attained by the majority of developing and developed countries. The massive acceleration in China's rate of growth was a dramatic slap in the face to the numerous analysts who had long asserted that China was culturally predisposed to remain forever in stagnation. It was also a strong piece of evidence in support of the thesis that economic growth is fundamentally an endogenously led phenomenon. During most of this period, and in particular during the initial decades, China imported few resources from abroad, and in spite of having also accumulated so little wealth up to the mid-twentieth century, managed to accelerate its rate of growth exponentially.

The events that took place in China during this period, in particular on matters referring to policy making, presented a striking paradox to those who interpreted reality from the perspective of orthodox ideas. These interpretations assert that the worst economic policy that a government can undertake is one in which market forces become totally distorted. There is no doubt that distorting market forces reduces efficiency, but if it were determinant for economic growth, then China should have performed even worse than during the 1900 to 1949 period.

During the first half of the twentieth century China was a country fundamentally driven by market forces, and economic growth averaged only about 0.5 percent annually. In the second half of the century, however, the country was governed by Communists, and as declared enemies of capitalism they got to the task of distorting market forces on a massive scale. If the premise of orthodox analysts were valid, then the performance of the economy should have significantly deteriorated, but the fact is that GDP figures accelerated impressively.

In 1979, Peking began to liberalize the economy, and over time market forces took an increasingly more active role. However, even by 1997 the private sector accounted for only about a tenth of GDP and China was one of the most regulated economies in the world. The late twentieth century was therefore a period in which market forces were still massively distorted, yet China attained one of the fastest rates of economic growth in the world. The extreme distortion of markets very blatantly demonstrated that the main agent determining economic growth was not the one that most analysts believed in. The historical data reveal that the agent with the highest potential for acting as a propeller of the economy was manufacturing.[1]

During 1950 to 1997, a gigantic increase in government support for manufacturing coalesced with a double-digit pace of factory output. A completely new class of policy makers appeared, who were strongly motivated to promote factory output. However, among the new rulers in Peking there was never a clear understanding about the causality linkages between this sector and growth. Their support was strictly of an ideological nature. It was the socialist autarchy goals, the national security concerns, the balance-of-payments considerations, and the desire to catch up with the most advanced nations that inadvertently drove them to promote the sector.[2]

By October 1949, all of mainland China was under the control of Mao Tse-tung and his comrades. By then, the level of development of China was lagging terribly behind numerous countries, and because the country was so economically and militarily weak, over the preceding century it had been repeatedly abused and humiliated by foreign powers. The Communists were fixated with the idea of avoiding future abuses from abroad and were therefore prepared to do whatever was necessary to make the country economically and militarily strong. China's economic weakness was expressed in every aspect of everyday life, but nowhere was it more evident than in its incapacity to produce factory goods such as tractors, trains, metals, machine tools, and consumer goods. Nowhere was the country's military weakness more evident than in its incapacity to produce weapons. The leaders therefore decided to allocate a very large share of the nation's resources for the production of civilian and military manufactures.

Because of their communist vision of the world, the new leadership was not interested in trading or maintaining other forms of economic linkages with the capitalist world, so it opted for a policy of autarchy. Had the new leadership not believed in communism, their strong desire to make the country economically and militarily strong would have still driven them to support manufacturing significantly. However, because of their desire to become economically self-sufficient, they were driven to promote factory production even more so as to produce everything domestically.[3]

Even before the Communists took over power, they had become fully convinced that there was a need to make massive investments in armaments so as to avoid a repetition of the Japanese occupation (1931-1945), which had ended up costing about five million Chinese lives. Before their triumph over the Nationalists, they had also become convinced that the possibilities of a large-scale war against a foreign nation would be higher than before, precisely because of their anticapitalist convictions. The Korean War debuted only one year after they took over power and a number of circumstances drove Peking to take part in it. Chinese forces were ill-equipped and had to confront the most technologically advanced weapons in the world. This event pressed the matter even more and forced the government to make larger allocations than originally planned for the production of weapons. Very large investments were therefore made during the 1950s for the sake of enlarging and modernizing the country's armed forces.[4]

Defense expenditure as a share of GDP during the first decade of communist rule averaged about 23 percent, which far exceeded the approximately 2 percent of the preceding decades. Such a gigantic "waste" of resources for the production of goods that could not be consumed should have, according to common sense, delivered an even worse economic performance than during the preceding decades. Considering also that the new economic system greatly conspired against efficiency, it becomes almost impossible to see how could the economy not have experienced a gigantic contraction. The fact, however, is that the economy accelerated its pace exponentially.[5]

By the time the Communists took over power, they had cordial relations with the Soviet Union and they immediately solicited Moscow's help so as to materialize their large development programs. Moscow provided loans, machinery, equipment, patents, and engineers for the creation of numerous large factories in fields such as iron, steel, machine tools, and tractors. Although the Soviets contributed significantly to the development of China's manufacturing during the 1950s, the majority of the investment in the sector resulted from the utilization of domestic resources. The very rapid factory output that took place during this decade demonstrated that a fast growth of the sector was not only fundamentally an endogenously led phenomenon but also that it could occur under the worst of circumstances.[6]

The Communists began to distort market forces almost as soon as they came to power. In no time, all the enterprises and assets of the country had been nationalized, and prices, capital, and labor were regulated very tightly. It is also worth noting that during the first three years of communist rule, Soviet help was largely absent. Importing resources from capitalist countries was practically impossible as a result of China's participation in the Korean War and the Allied embargo that followed. There was also a scarcity of domestic resources as a result of the country's very low level of development. In addition, the country was emerging from the economic chaos that reigned in the 1940s and there was a lack of governing experience. However, in spite of all this, factory output grew at an impressive pace, and that coincided with an impressive rate of economic growth.

If fast GDP growth was possible under these terrible circumstances, then it becomes evident that a variable different from those of orthodox interpretations was ultimately responsible for the positive performance of the economy. It becomes evident that such a variable must possess such tremendous growth-generating powers that even under the worst growth-deterring circumstances it still manages to deliver impressive results. The events of these years also demonstrated the endogenous nature of economic growth—showing how even if a nation found itself at the lowest possible levels of development with no flow of resources from abroad, it was still possible to mobilize the existing domestic resources into manufacturing and attain fast growth.

Although Soviet aid began to arrive in large amounts in 1953, during the rest of the 1950s the majority of investment was the result of the utilization of domestic resources. The help China received came in the form of low-quality goods and on top of that, market forces were distorted even further. By the late 1950s, market forces had been almost completely suppressed. However, during the rest of the 1950s there was relatively fast economic growth that once again coincided with decisive support of factory production, as well as with a fast rate of manufacturing output.[7]

Manufacturing during the 1950s expanded on average by about 17 percent annually and GDP grew in real terms by about 4.6 percent. The official growth figure was about 13 percent but, once it was filtered to match the quality and cost-efficiency levels of highly competitive capitalist economies, the figure shrank massively. This was the heavy price that China had to pay for its abusive distortion of market forces. A highly competitive economy would have extracted from such a rate of factory output a much higher rate of GDP. Hong Kong, for example, during the 1950s attained manufacturing output of about 16 percent annually and GDP averaged about 12 percent.[8]

China committed a major policy error by distorting market forces, but such an encompassing distortion was actually the ultimate proof that the bottom line of economic growth resided in government support for manufacturing. Only if it is assumed that factory output is the prime generator of technology, and therefore wealth, is it possible to make sense out of the empirical information. The fact that China also spent extensive amounts on armaments during this decade makes it even harder to understand why growth accelerated so much, unless once again it is assumed that the faster the output of factory goods, the faster the creation of wealth.

It is argued that growth was fast because the production capacity of the country was at low levels as a result of the contraction of the 1940s, and it was therefore only a matter of reactivating it. If reactivation was so easy, why is it, then, that the Nationalists failed to achieve the same feat after the end of World War II? By then GDP had already contracted significantly and they were incapable of reactivating the economy. GDP continued to contract until 1949.[9]

This situation coincided with a decrease in state support for the sector between 1946 and 1949. There was indeed much reactivation in the early 1950s, but by 1952 the economy had already reached the same level of the early 1940s. From then on, growth no longer had anything to do with bringing back into production the existing installed capacity. As the immediate post–World War II years demonstrated, bringing back into production the installed capacity was possible only when there was support for the sector, and a continuation of economic growth once the production base had reached its limit was also possible only if support was continued.[10]

In their factory-promotion efforts, the Communists did not utilize the capitalist methods of fiscal, financial, and nonfinancial incentives, but rather decreed very large direct allocations. The need to produce armaments and the belief that producer goods were pivotal for development drove Chinese policy makers to decree a stronger level of support for heavy manufacturing than for its lighter counterpart. Light manufacturing was associated with consumer goods and therefore with a domain that hampered investment. Mao was not only convinced of that, but also of the idea that steel was the key link for the rest of the sector. It was similar to Lenin's fixation with electricity and Stalin's with machine tools.

However, it is worth noting that the much stronger support for heavy manufacturing was followed by a much faster growth of this domain; and within heavy manufacturing, it was steel that grew the fastest. As in capitalist economies, growth in centrally planned economies was the strongest among the manufacturing fields that received the most support from the state.[11]

The fast growth of China during the 1950s astonished many analysts with orthodox visions of the world. Many had been convinced that China would never grow quickly even under capitalism, and as China became communist they predicted total economic ruin. Many therefore argued that the fast growth was the result of the accumulated transfers of capital and technology from foreigners since the mid-nineteenth century. Such an argument obviously forgot that for about a century such transfers had systematically failed to deliver fast growth. Why would those transfers all of a sudden begin to deliver what they had failed to do during the preceding hundred years? The largest transfer had taken place in Manchuria, but even there it was very small by the time the Communists took hold of power. At the end of World War II the Soviets took to the USSR the bulk of the machinery and equipment that the Japanese had left behind. This argument obviously failed to explain the facts consistently.

Others asserted that the fast growth was the result of a hard-working and well-educated population with abundant commercial experience. Once again, the inevitable question was, why did all those qualities so systematically failed to deliver fast growth during the preceding decades? It must also be wondered why those qualities would deliver positive results under a centrally planned system and fail in a capitalist economy. According to most analysts, it should have been the other way around. The same goes for the preceding argument. Why did the transfers of capital and technology from advanced capitalist countries fail for so long to deliver fast growth in a market economy and why did the accumulated transfers deliver fast growth under a communist system? Both of these arguments failed totally to match the facts.[12]

During the 1950s, China experienced by far the fastest rate of population growth to that date. Population expanded by about 2.3 percent annually, compared with the much lower 0.2 percent rate of the preceding fifty years. Many therefore argued that fast population growth created additional demand and allowed for a much higher level of output. The population-growth argument had failed to bond coherently with the empirical data in many countries, and in the case of China many inconsistencies were also observed.

If the big acceleration in the rate of population growth during the 1950s was responsible for the big acceleration in the GDP figures, then the economic figures of the following decade should have been similar. During the 1960s, the rate of population was almost the same (1.9 percent) as that of the preceding decade. The fact however is that GDP rates decelerated by about 60 percent. Also worth noting is that in the 1990 to 1997 period, the population grew about half as fast as during the 1950s, and in spite of that, real

GDP grew about twice as fast as during the 1950s. The fact that population grew at a much slower pace than GDP during the 1950s makes it also highly improbable that this variable could have had a pulling effect on the rest of the economy. While population grew by about 2.3 percent annually, the economy did so by about 4.6 percent.[13]

Although none of these explanations adds up consistently with the facts, the thesis of manufacturing does very well. There was, for example, an extremely low level of support for manufacturing until 1949. That is why—notwithstanding the market-driven economy that prevailed, the significant transfers of capital and technology, and a hard working population—China failed to attain fast growth. In the 1950s, there was a very strong promotion of factories, which is why—despite the massive distortion of market forces, the absence of foreign trade, and the huge investment in weapons—fast economic growth was still attained. In the 1960s, the manufacturing thesis was substantiated further.

The Great Leap Forward and Agriculture

There was an unprecedented mobilization of resources during the 1950s in order to promote manufacturing production, and even though it was one of the largest by international standards, it was nonetheless far from being as large as possible. Mao was never enthusiastic about manufacturing, and only because of very strong ideological motivations did he acquiesce to decisively promote factories. Throughout his life, he remained deeply convinced that the key to economic growth resided in agriculture. That is why the very first policy that Mao decreed in 1949 dealt with land redistribution.[14]

Mao was convinced that the accumulation of capital depended on the development of agriculture. According to him, only an increased purchasing power of the peasant majority could create a demand for nonagricultural goods and services. Mao was not the only one who thought like that; most in the Party hierarchy agreed. Many analysts have asserted that such an ideology was the result of the peasant roots of Mao and his comrades, as well as their low levels of education. However, such ideas by then were conventional wisdom among development specialists throughout the world. For them, increased agricultural output and increased peasant income were the bottom line for rapid economic growth; and for that to happen, governments had to concentrate on promoting this domain.[15]

This idea had actually been around for centuries and was also widely diffused throughout the world. That is the reason that during most of history, in just about every corner of the world, the fundamental (if not the only) developmental policy of practically all governments was the promotion of agriculture. However, in every single corner of the world, the policy systematically failed to deliver the desired results. In China, this was the only policy pursued for millennia up to 1949, and during all that time the GDP and agricultural figures moved at a terribly slow pace. Mao and his comrades could have very easily decreed more support for manufacturing during the 1950s, but they were convinced that the lower the share of overall investment that flowed into agriculture, the longer it would take China to overcome poverty.

From the early 1950s, the Party leadership hoped that national security concerns would diminish so that they could allocate more resources to agriculture. By the later part of that decade, these concerns had largely diminished, as the Korean War was completely over. The restrictions on armaments the U.S. imposed on Japan also made it highly improbable that Tokyo could again decide to invade China. On top of that, the Communists had succeeded greatly increasing the country's arsenals. The country had also made major advances in becoming autarkic as a vast array of new goods had made their appearance.

Peking therefore concluded that it no longer needed to produce as many weapons and civilian factory goods, and that it could therefore transfer resources from manufacturing to agriculture. One thing that drove policy makers to be even more inclined to increase the allocation of resources to agriculture was the fast population growth of those years. From 1950 the population grew much faster than ever before and by 1957, the country had more than 600 million people. The last large-scale famines had taken place in the late 1940s and by the late 1950s, malnutrition was still endemic. The leadership was obsessed with the food supply, and the large population figures drove them to conclude that a major famine would occur soon unless they invested more in agriculture.[16]

In 1958, therefore, the Great Leap Forward was initiated. This program aimed to increase food production by a very large margin and resources were allocated to agriculture in much larger amounts. The end result, however, was exactly the opposite of the one intended. Agricultural output not only failed to expand greatly, it failed to expand even slowly. It was not even capable of maintaining the same level of production of 1957 and the next year it contracted. Worst still was that the contraction was on a large scale. Convinced that the country needed some time to adjust, the new policy was continued and promoted even more the next year, but in 1959 agriculture once again contracted significantly. More support for agriculture the next year coincided again with a further contraction of output. Between 1958 to 1960, agricultural production contracted by about 20 percent.

Many argued that the drop in farm production was the result of natural catastrophes, including drought and flood, that were felt during those years, but these natural calamities took place in 1960. There is no doubt that they owned part of the blame, but the fact that the contraction started in 1958 makes it very evident that something else was the culprit in the food crisis (at least between 1958 and 1959). It was also argued that the contraction was the result of the withdrawal of Soviet aid, which destabilized the Chinese economy. China had become, to a certain extent, dependent on Soviet assistance and surely such an abrupt termination of the aid program had destabilizing effects, but once again there is the problem of timing. Aid was withdrawn until August 1960 and the contraction of agriculture started much earlier. The cause of the crisis obviously resided somewhere else.[17]

While these arguments fail to explain the crisis, the thesis of manufacturing succeeds very well. There was a decrease in the support of this sector as the government transferred resources to agriculture between 1958 and 1960, which correlated not just with the food crisis, but with a contraction of factory output and GDP. The Communists were hoping to feed the masses better and precisely the opposite took place. About twenty million Chinese died from hunger during the period of 1959 to 1961.[18]

The same error that had taken place for thousands of years was repeated. Large investments in agriculture never delivered a large food supply. For a long time, however, a correlation between large investments in manufacturing and a fast growth of the food supply had existed. The fastest growth of agriculture in the first half of the twentieth century, for example, took place during the 1930s, and it was during this decade that factories were promoted the most. During the years 1950 to 1957, massive investments in manufacturing were made and agriculture expanded at an impressive pace, compared with anything previously experienced. Then, during the Great Leap Forward, investment in agriculture increased considerably and its share of total investment rose to account for almost one-half. During the 1990s, on the other hand, Peking neglected agriculture and concentrated mostly on manufacturing. Investment in agriculture as a share of the total averaged only about 1 percent, but farm output expanded by about 3 percent per year (one of the fastest rates in China's history).[19]

If it is assumed (as history very strongly suggests) that manufacturing is the fundamental creator of wealth, then this apparent paradox becomes resolved. As Peking largely reduced the resources it allocated to manufacturing in the late 1950s, the overall technological level of the country contracted, as there was a contraction in the goods that embody technology. As a result, fewer tractors were produced than before, as well as less fertilizer, irrigation equipment, and pesticides. Without technology, agriculture cannot develop, and without these farm-factory goods, Chinese farmers were left without the means to apply technology to their work.

Khrushchev's withdrawal of aid in 1960, because of Mao's hostility to his policy of détente with the West, did have a negative effect on manufacturing and therefore on agriculture. Since the bulk of the aid was for the development of factories, its cessation paralyzed the numerous production establishments the Soviets had installed. However, had Mao at least sustained the strong support for manufacturing of 1950 to 1957, the withdrawal of Soviet aid and the natural catastrophes would have delivered, at worst, just a slowdown in 1960. Between 1980 and 1997, at a time when a very strong factory-promotion effort was undertaken, there were also numerous floods, droughts, frosts, and pests that hampered and destabilized China's agriculture, but they never generated a food crisis. Quite to the contrary, agriculture grew at an unprecedented pace, which coincided with a double-digit growth of factory output. By the 1990s, for example, China was the largest producer of chemical fertilizers in the world and a major world producer of farm machinery and pesticides.[20]

There was a reversion of policies in the early 1960s and Peking began to increase its support for manufacturing. Immediately, agricultural output and GDP began to increase. However, during the rest of the 1960s, support for factories was inferior to that of the preceding decade. Since there were no major national security concerns forcing large investments in armaments, since the goal of being autarkic was closer to getting realized than before, and since the desire to catch up was not that strong, support was lower than in the 1950s. It was not as if catching up with Japan and the West was irrelevant for Mao and the Party leadership, but they were almost totally obsessed with agriculture and the food supply. They just could not get out of their heads the idea that the key to growth resided in investing as much as possible in agriculture.[21]

Although there were no major national security concerns such as the Korean War, the tensions with the USSR that had erupted since 1960 forced much larger investments in armaments than what Peking otherwise would have made. The ideological motivations of autarky and catching up were less pronounced than in the 1950s, but they were still present; and it was because of these motivations that manufacturing received support. The compounded effect of these ideological motivations ended up delivering a level of support to military and civilian manufacturing in the 1960s that was considerably lower than in the preceding decade. That was paralleled by a considerable deceleration in the rate of factory output and GDP. Factory output averaged about 10 percent annually and real GDP about 2.2 percent. The official growth figure was about 5 percent, but the massive distortion of market forces delivered an untold amount of goods that were of such a bad quality that they actually could not be consumed.[22]

Mao and most of the Party leadership lamented the investments they had to make in armaments and most other fields of manufacturing. The few among the leadership who gave priority to catching up with the most advanced nations and wanted therefore more investment in manufacturing were sidelined. Deng Xiao-ping was among those few, and he was purged during the Cultural Revolution. The Cultural Revolution, which began in 1966, also contributed to diminishing government support for the sector as the massive political agitation that took place frequently paralyzed or slowed down factory production. This political phenomenon regularly rerouted the leadership into nonmanufacturing areas, and it also tended to persecute those who were supportive of the sector.[23]

During the 1960s, therefore, Mao continued to favor agriculture at the expense of manufacturing, and once again the goal of rapidly increasing the food supply was not attained. Agriculture grew at a much slower pace than during 1950 to 1957. The goal of quickly increasing the food supply was praiseworthy, but it was the means to achieve it that was obviously failing.

Much suggests that if Peking had concentrated on allocating as many resources as possible to food-related factory production, a much faster rate of food output and everything else would have been attained. Logic suggests that the most rational policy would have been one that gave strong support to fields such as farm machinery, fertilizers, irrigation equipment, pesticides, and food processing.

Better results would have also been achieved if the economy had not been distorted. Under a highly competitive system, any given level of support for manufacturing would have delivered a much larger output of food and everything else. Japan for example, with a 13 percent rate of manufacturing output during the 1960s got an 11 percent rate of GDP and a 4 percent annual growth of agriculture. China on the other hand manufactured at a 10 percent annual rate, real GDP averaged 2.2 percent, and farm output was only 1.5 percent.[24]

Until Mao's death in 1976, support for manufacturing continued to be relatively low because of the priority status for investment that he assigned to agriculture, because of relatively low national security concerns, and because of the political chaos that the Cultural Revolution continued to cause. Social unrest caused by ideological struggles in the mid-1970s, for example, brought manufacturing production to a standstill. During 1970 to 1976, factories were only modestly promoted, which correlated with a modest rate of manufacturing-output and a slow GDP rate. As during the previous decades, it was heavy manufacturing that received the most support and it was also this domain that expanded the fastest.[25]

The fiasco of the Great Leap Forward made many in the Party start to lose faith in Mao's policies. As the economy attained a modest performance in the 1960s and a mediocre showing between 1970 and 1976, the attitude of the Party progressively turned into one of repudiation of his policies. Mao's designated successor was Hua Guo-feng, who had been chosen precisely because he seemed to endorse all of Mao's policies. However, as soon as he took power, the policy of giving priority to agriculture was terminated. Giving priority to heavy manufacturing was continued but there was an across-the-board large increase in investment in factories. Factory output immediately rose to double-digit rates from 1977 to 1978 and economic growth suddenly accelerated its pace by a very large margin.

The Hua years were also characterized by a partial renunciation of the policy of economic autarky, and foreign trade with the rest of the world increased. A large share of the machinery and equipment utilized for the new factories was imported from Japan and other capitalist nations. The desire of Hua and his comrades to catch up developmentally with the most advanced countries was much stronger than Mao's and that was fundamentally what drove them to endorse such policies.[26]

In December 1978, Deng Xiao-ping gained the upper hand in the Politburo, although Hua continued to have a strong voice in the Party until 1982. Deng had very strong desires to make China, in a relatively short time, a nation as advanced as the OECD countries. For him and his followers, priority had to be assigned to manufacturing because large production of factory goods was the main characteristic of advanced nations. But he also believed that the economy needed to be significantly liberalized because the most developed nations had all capitalist economies. In 1979, therefore, there was again strong support for manufacturing, and this once again coalesced with fast growth.[27]

All in all, the factory-promotion efforts of Peking during the 1970s fluctuated erratically from one extreme to the other and on average it was slightly stronger than during the preceding decade. The small increase in support coincided with a slightly faster rate of factory output and with a proportionate acceleration of the GDP figures. Manufacturing averaged about 11 percent annually and real GDP about 3 percent. The official figure was about 6 percent but, in spite of the small liberalization at the end of the decade, China remained a tightly centrally planned economy during the 1970s. The price for such a major policy error was inefficiency on a grand scale.[28]

Even though real economic growth during 1950 to 1979 averaged only 3.3 percent per year, it was a rate far superior to anything previously experienced. It was inevitable to ask how China could have attained such a growth for so long while distorting market forces almost completely. The only thing that can consistently explain such an apparent paradox was the very strong level of support that Peking supplied to the manufacturing sector, which delivered an average factory rate of about 12.6 percent per year.

By the end of the 1970s, living conditions had improved greatly in relation to the progress made during the first half of the twentieth century, and much seems to indicate that manufacturing was to be thanked for those progresses. A nation that prior to 1950 could not produce tractors, trains, or watches was by the late 1970s capable of producing not only those goods, but also computers, earth satellites, oral contraceptives, and nuclear weapons.[29]

Economic Reform Under Deng

Deng and his followers were in power during all of the 1980s and their strong desire to catch up with the most developed nations drove them to assign investment priority to technology-intensive fields. Since factory goods are particularly endowed with the capacity to embody technology, resources ended up being abundantly allocated to this sector. Another sign of the repudiation of Maoist policies was the emphasis the new rulers gave to light manufacturing. The reformers were convinced that the favoritism to heavy manufacturing of the preceding decades was a wasteful allocation of resources that depressed consumption.[30]

The reformers thought that a proportionate distribution between heavy and light manufacturing was the best for the nation. To correct the imbalance, therefore, they promoted the second domain in a much more decisive way. The case of Hong Kong clearly demonstrated that the ideas of the Maoists were unfounded, but it also showed that having a balance was irrelevant. The colony attained the fastest rates of economic growth in the world during the second half of the twentieth century while producing fundamentally light goods. Heavy manufacturing accounted for only a very small share of total factory output during the whole period.[31]

Other countries throughout history went through situations in which the output of the two factory domains was largely equal and in spite of that no growth was experienced. The United States during the 1930s is a clear example. What history very strongly suggests is that there is no causality linkage between the distribution of the two domains and economic growth. Whether in balance or not, growth has systematically materialized whenever the overall rate of factory output expanded. The faster the rate of its expansion, the faster the GDP figures.

To catch up as soon as possible, China needed to import as much technology as possible. Since China had few foreign exchange reserves and technology came fundamentally in the form of factory goods, Deng and his followers aimed to substitute as many manufactured imports as possible by producing them domestically. The scarcity of foreign exchange also drove them to export as much as possible. Since the most exportable of goods are manufactures, this motivation delivered further support for the sector.

National security concerns were not a motivation during the 1980s as external threats diminished by a large margin. In consequence, defense as a share of GDP dropped from about 15 percent in 1980 to about 6 percent ten years later. There was actually a considerable reduction in support for military manufacturing.[32]

Despite the large reduction in the share of investments allocated to the production of armaments, the developmental catch-up motivation and the balance-of-payments pressures were so strong that they drove the government to increase by a very large margin the level of support for civilian manufacturing. As a result, overall support for the sector was much higher than before. That was shadowed by a much faster rate of factory output than in the 1970s, which correlated with a much faster rate of GDP. Manufacturing expanded during the 1980s by an average annual rate of about 16 percent and real GDP grew by about 7 percent. The official growth figure was about 10 percent but in spite of significant liberalization during those years, China remained with a very distorted economy.[33]

By 1989, less than 2 percent of the economy was in private hands. Although market forces were allowed significantly to sculpt the development of the rest of the economy that remained in state hands, the liberalization was only great when compared to the total market distortion of the Mao years. Because of the massive distortions that were still present, waste and inefficiency were still experienced on a wide scale and China was incapable of extracting as much from its manufacturing output as were the most efficient market economies. Hong Kong, for example, with a rate of factory output about half as rapid as that of China managed to attain a GDP rate that was actually faster. During the 1980s, manufacturing grew by about 8.8 percent annually in the colony and the economy expanded by about 7.5 percent.

The liberalization that took place nonetheless reduced the wastage by a significant amount and the units of factory output needed to deliver a unit of GDP diminished. In the 1970s, the ratio was about 3.3 to 1 and in the worst of the Mao years (1960s) the ratio was about 4.7 to 1. During the 1980s, however, the ratio dropped to just 2.3 to 1.

During the 1980s, the government continued to mostly allocate resources directly to manufacturing. However, it also began to utilize capitalist mechanisms. Peking began to supply (in particular to foreigners) fiscal, financial, and nonfinancial incentives. In 1979, it created four special economic zones (SEZs), in 1984 it opened fourteen coastal cities, and in 1988 it opened Hainan Island to foreign investment. Manufacturing was not the only activity desired from foreigners and it was not the only one that was supplied with incentives, but because Deng was fixated on acquiring technology, and technology found its materialization in factory goods, priority was given to foreign direct investment (FDI) in this sector. It was manufacturers, therefore, that received the most attractive incentives.[34]

To be able to equip its factories with superior technology, Peking borrowed from international organizations, from commercial banks in advanced capitalist countries, and from capital markets. As a result, its external debt as a share of GDP went from about 1 percent in 1979 to about 13 percent in 1989. Not all of the large borrowing was for manufacturing, but the vast majority flowed into that sector.

With the exception of a few foreign and Chinese private companies, practically all factories during the 1980s were in state hands. The majority were owned and managed by the central government, although provincial and local authorities were in control of numerous firms. All of these factories, but in particular the ones in the hands of provincial and local authorities, were increasingly allowed to take responsibility for profits and losses. This situation opened the way for financial failure, so to protect them, Peking erected very high trade barriers.[35]

This was evidently a policy error. These enterprises were clearly without the capacity to compete, but what they needed in order to withstand the competition from superior imports was not trade protection but higher levels of fiscal, financial, and nonfinancial incentives. Without trade barriers and at the level of support that the government supplied, most of them would have gone bankrupt. However, if no trade barriers had been applied and simultaneously a much higher dose of incentives had been offered, it is highly likely that the majority not only would have survived, but their levels of efficiency and quality would have massively improved. A much higher level of incentives also would have allowed the ones that survived to expand capacity considerably. In this way, they would have been able to absorb the workers made redundant by rationalization and bankruptcies.[36]

The government could have borrowed much more and supplied those factories with larger financial resources so that they could have acquired better capital goods and other inputs for production. Even though it borrowed in large amounts, it was not enough to supply all the existing factories and the numerous new ones with inputs from advanced capitalist countries. A much larger foreign borrowing was needed to supply all factories with superior machinery, but a large foreign debt was politically unacceptable.

Peking could have also mobilized more the domestic resources and allocated a much larger share of overall investment to manufacturing, but there were still strong sympathies among the leadership toward agriculture. Even Deng was convinced that investing in agriculture was very important for economic development. He was not obsessed with agriculture as Mao was, but he and most of his followers were persuaded that this domain deserved a significant share of investment.[37]

That is actually the reason why liberalization started first in agriculture and during this decade went further in this domain. Had the large investments in agriculture, in the other primary sector activities, in construction and in services, been mostly transferred to manufacturing, then the trade barriers would have not been needed and Chinese factory goods would have managed to compete successfully against imports. The case of Hong Kong and Singapore in the second half of the twentieth century very clearly demonstrated that free trade is no obstacle for a very fast factory output. Their example in the early twentieth century also made it evident that free trade does not inhibit the growth of manufacturing even when an economy has no previous experience in producing those goods.

It was also a policy error not to have privatized the untold amount of enterprises that, even after the liberalization they underwent, experienced much lower levels of productivity than private ones. Levels of efficiency ran parallel to how market-driven those factories were. The ones run by the central government had the worst performance; those managed by provincial and local authorities that were more responsive to the market attained higher levels of efficiency; and those that were fully in private hands had by far the best levels of quality and cost-effectiveness. Had all the factories been privatized, economic growth surely would have been much faster.

It was not just the factories that were in state hands that needed to be transferred to the private sector, but practically all other enterprises. All the land of the country, for example, remained during the 1980s the property of the state. Farmers negotiated lease periods and conditions with local governments. If enterprises in all other sectors of the economy had been privatized, the performance of the economy would have also been better.

Because of their monopolistic practices, state firms also tend to generate higher prices and therefore higher inflation. Budget deficits also tend to generate inflation, and during the 1980s China experienced budget deficits in every single year. The bulk of government expenditure was utilized to finance state firms and a very large share of those funds was wasted in inefficiency. It is obvious that if those firms had been privatized and monopolies dissolved, the same strong support for manufacturing could have been supplied without having to run budget deficits. Since the funds would have been more efficiently utilized, less would have been needed, in turn delivering downward pressure on prices.[38]

The state firms and the trade barriers were major policy errors, but they simultaneously acted as the ultimate proof that the bottom line of economic growth resides in manufacturing. How else could a nation with about 99 percent of its enterprises in state hands and with gigantic trade barriers have attained a real economic growth of about 7 percent annually during a whole decade? So decisive was the promotion of the sector that during the 1980s about three-fourths of government expenditure and about 70 percent of state bank loans went to finance factories.[39]

The financial structure of China during this decade was also indicative of the causality linkage between manufacturing and growth. Until the mid-1980s, banks were practically nonexistent as the Central Bank was responsible for almost all commercial banking activity. Then the Central Bank was relieved of those activities and four large specialized banks as well as eight smaller generalized banks were created to take care of commercial lending. By 1989, they accounted for practically all bank lending and every single one of those banks was in state hands. Nongovernment financing in the form of small trust and investment companies had grown rapidly since 1979, but by the end of the 1980s they only accounted for a very small fraction of overall lending. As for the Central Bank, by the end of the decade it was still extremely far away from operating like its counterparts in capitalist countries. Interest rates were fixed by the Central Bank with very little notice of market signals, and commercial banks were not allowed to deviate even a little from the fixed rate.[40]

China's banks were among the most inefficient in the world and among the least market-driven. Added to that, not a single foreign bank was allowed to set up operations in China. Allowing them in would have surely improved the performance of the whole financial system but since Peking could not see a direct linkage with technology, it concluded that it did not need them.

As a service, banking is indeed not a technology-intensive activity, but banks can play an important role in the promotion of factories. Peking could have opted for a policy that combined aspects from Japan and Hong Kong. In the 1950s and 1960s, Tokyo regulated private banks so that they would lend the majority of their assets to manufacturing at preferential rates, and Hong Kong since those years allowed foreign banks to operate in its territory. If Peking had allowed foreign banks to operate in China on the condition that they would lend fundamentally to manufacturing, it would have mobilized even more funds into the sector and improved financial productivity.

With such an inefficient financial system, it is hard to see how China could have attained one of the fastest rates of economic growth in the world. The structure of the country's financial system is actually one more piece of evidence in support of the manufacturing thesis. On the one hand, the system allocated the bulk of its resources to this sector, supplying it with the basic ingredient for growth, and on the other hand it was highly inefficient. Only if it is assumed that manufacturing is the prime creator of wealth can the terrible inefficiencies of the financial system become compatible with the fast GDP figures.[41]

The development of FDI during the 1980s is another piece of solid evidence in support of the manufacturing thesis. China promoted it abundantly relative to its total exclusion during the Mao years. However, compared with numerous developing and developed countries, it attracted little. In per capita terms, it attracted only a very small fraction of what nations such as Singapore, Hong Kong, Mexico, Malaysia, Indonesia, the United States, or Britain did. As a share of GDP, it was also much smaller than in these and several other countries. Even in pure volume terms, it was so small that even Singapore managed to attract more during this decade.[42]

Many tried to explain China's fast growth by arguing that it had been largely aided by the large inflow of FDI. The fact is that it was large only when compared to its absolute absence in the preceding decades. By international standards, it was small and the vast majority of nations that received far larger per-capita net inflows of FDI had much slower rates of economic growth. In addition, these countries had the benefit of possessing capitalist economies. Countries such as Mexico, which received much larger amounts of FDI, averaged an economic growth of not even 1 percent during the 1980s. Even by 1989, the accumulated stock of FDI in China accounted for about only 1 percent of GDP. It was impossible, therefore, to consistently argue that FDI propelled the Chinese economy or that it had significantly assisted it during this decade. It is obvious that some other factor was responsible for the growth and the only factor that intertwined logically with the facts was manufacturing.[43]

During this decade, FDI was mostly in factories and it concentrated almost exclusively in the coastal provinces, but even there the overall foreign participation in the economy was small. However, in part because of FDI, manufacturing output in these provinces was faster than in the interior and this coincided with a faster rate of economic growth.

Nonetheless, there was one province where FDI accounted for a considerable share of total output. Guangdong province was the main beneficiary of Hong Kong investors, who were the main source of FDI during the 1980s. Largely because of the FDI, Guangdong attained the fastest rates of factory output in China and it was also this province that attained the fastest GDP figures. It is nonetheless worth noting that the majority of the factory output in this province came from Chinese companies. The only place where FDI accounted for the majority of output was in the Special Economic Zones (SEZs) and, together with the rapid growth of investment from Chinese companies, overall factory output was much faster than in the rest of Guangdong. Once again, this was paralleled by a much faster rate of economic growth in the SEZs than in Guangdong.[44]

The correlation was very constant and straightforward. It was in the coastal provinces where the promotion of manufacturing was the strongest and it was these provinces that had the fastest GDP figures. Within the coastal provinces, it was Guangdong where support for the sector in its different forms was more decisive and it was this province that had the fastest rates of economic growth. Relative to Guangdong, the SEZs supplied an even more attractive incentive package to manufacturers, and attained an even faster rate of economic growth.

Promoting FDI in manufacturing was by all means a very positive undertaking, but during the 1980s it flowed in too small quantities for it to affect the economy considerably. However, it could have flowed in much larger amounts had Peking decided to supply to foreigners a much more attractive package of incentives. Unfortunately, communist ideology and nationalist sentiments hindered the leadership from attracting more investment in factories.

FDI is actually a better alternative to borrowing and importing the technology, because the foreign companies are already experts in managing that technology and the risks of financial failure are therefore lower. Many argued that if Peking had decreed a more favorable incentive package, FDI would have not increased by much because FDI is mostly supply led. However, the 1990s demonstrated that vastly larger flows were possible, and the fact is that they materialized immediately after the government supplied larger possibilities for profit to foreign firms.[45]

Understanding China's Development

Many thought that they could explain China's fast growth by presenting education as the propelling agent. China had indeed made massive educational progresses during the first three decades of communist rule and there is no doubt that a more qualified labor force is more productive. However, even with a workforce that is the best educated in the world, a nation is not supposed to attain growth if it distorts market forces in such a wide-ranging way as China did during the 1980s.

According to several studies, by the 1990s Japan had the workforce with the highest levels of education in the world and on top of that it was a capitalist economy where hardly a state company existed. However, GDP averaged only 2 percent. China, on the other hand, with a much inferior educational system, in the 1980s averaged 7 percent. It is also worth noting that even though literacy and math skills improved considerably during the years 1950 to 1979, most of what the Chinese learned was unscientific communist ideology. The Cultural Revolution pushed this educational nonsense to its extremes as universities and schools were closed for a year, as college training was shortened, as only politically subservient and ideologically correct people could enter universities, as scientific research came almost to a halt, and as intellectuals were persecuted.[46]

If education was the driving economic force, than GDP figures should not have varied so much from the 1970s to the 1980s. In the Deng years, education became considerably less ideological, but there were no major changes in quality and quantity that could have justified such a large growth differential between the two decades. From almost every perspective, it is impossible to find a correlation between education and growth.

Others argued that China's fast growth was the result of the culturally determined workaholic attitude of the population. A cultural factor was helpful in overcoming the difficulty of explaining the massive market distortions, but if culture had such powerful growth qualities, why is it, then, that the same culture attained miserable results prior to 1950? During the early twentieth century, for example, the average Chinese worked about ninety-five hours per week, and in spite of that, economic growth was almost completely stagnant. To that it has to be added that capitalism then prevailed and market distortions were minimal compared with the 1980s.[47]

During the 1980s, on the other hand, the average Chinese worked only about sixty hours per week and economic growth was one of the fastest in the world. The argument of culture, is not consistent through time. China's culture is one of the oldest of the world and for thousands of years the Chinese worked much longer hours than during the 1980s. During all of that time, however, economic growth was terribly slow. It must also be taken in consideration that during the Mao years Chinese culture (especially in the form of Confucianism) was attacked and suppressed by the authorities, and in spite of that, economic growth was much faster than ever before.[48]

China's foreign trade and in particular its exports grew very rapidly during the 1980s, and many came to conclude that exports were the engine of economic growth. Exports grew by about 12 percent per year and their share of GDP went from about 6 percent in 1980 to about 17 percent in 1989. By international measurements, a share of 17 percent is not small, but considering the massive market distortions that prevailed during this decade, exports needed to have grown much faster so that such a thesis could have had a chance of being logically acceptable. In a market economy, a 12 percent export growth could have perhaps justified such an argument but in an economy where about 99 percent of the enterprises were in state hands, only a much faster rate of exports could have had a chance of compensating for such deficiencies. Not to mention that cross country studies of market economies clearly indicate an absence of a consistent correlation between exports and growth.[49]

Further evidence substantiating the thesis of manufacturing is found on matters referring to corruption. Corruption has negative effects on the economy for it represents an additional cost on production. During the 1980s, there was a massive increase in corruption in China compared with the preceding decade. Considering the weight that most analysts assign to corruption, it would have been logical to expect that the performance of the economy would have deteriorated. The nation, however, attained a rate of economic growth that was more than twice as fast as that of the 1970s.

What this data actually reveals is that there must have been a factor among the many that intervened during those years that possessed such strong growth-generating powers that in spite of the large rise in corruption, it still managed to deliver fast growth. Since the factors that mainstream economics designates as being responsible for growth were absent due to the massive market distortions, and the ones presented by social scientists also did not add up, the only one left is manufacturing.[50]

Contrary to the other ones, this one did add up consistently. In cross-country comparisons, manufacturing was the only factor that could also explain why, in spite of very high levels of corruption, several countries in the region attained fast rates of economic growth. Japan, for example, during the 1950s and 1960s had very high levels of corruption and it nonetheless attained double-digit growth. In the 1990s, however, at a time when corruption had decreased by a very large margin, it was only capable of attaining a GDP rate about one-fifth that of the postwar decades. Taiwan and Hong Kong also had much higher levels of corruption in the 1950s than in the 1990s, and in spite of that it was in the 1950s when the economy grew faster. The same goes for South Korea and Singapore, in comparing the 1960s with the 1990s.[51]

A similar phenomenon has also been observed in countries that are culturally very different from those of East Asia. In the United States of the late nineteenth century, for example, government corruption was endemic and much higher than at any point during the twentieth century. However, it was at this moment that the country attained its fastest rates of economic growth in all of its history.

All of these apparent paradoxes are no longer a riddle once they get analyzed from the perspective of manufacturing. Support for this sector, for example, was much stronger in China during the 1980s than during the preceding decade. In Japan, the state gave massive subsidies to factory producers during the 1950 to 1969 period and during the 1990s it supplied only a few. In Taiwan and Hong Kong, support and factory output were much higher in the 1950s than in the 1990s. In South Korea and Singapore, the same applies in comparing the 1960s and the 1990s, and the late nineteenth century in the United States was the moment in all of the country's history when Washington subsidized manufacturers in the most decisive way.

Also conspiring against fast growth in China during the 1980s was the distorted exchange rate system. On the one hand, it was fixed and largely overvalued. Not only was it not responsive to market signals, but it also acted as a hindrance on exports because of its overvaluation. To make matters worse there were two distorted exchange rates. It is important to notice that in spite of this incoherent exchange rate system, fast economic growth was attained as well as fast export growth. This situation once again reinforces the idea that there must have been a factor so decisive for the generation of growth that could supersede the inherent deficiencies of such a deformed exchange rate system.[52]

Most economists and social scientists consider that infrastructure is a very important factor needed for development. In the China of the 1980s, however, infrastructure was terribly deficient. Electricity was nonexistent in large parts of the country and even in the most advanced coastal regions, brownouts occurred regularly. The rail system was decrepit and inefficient and its coverage of the country was not dense. Telecommunications were scarce and obsolete, and the road system was even worse than in sub-Saharan Africa. Investment in infrastructure grew relatively fast but it expanded only half as fast as GDP, which ended up delivering even more congestion. In several other countries of East Asia, a similar phenomenon took place. During the 1980s, investment in infrastructure in the region grew by about 4 percent per year while GDP expanded by about 8 percent.[53]

With such a deficient infrastructure system it is hard to understand how China was capable of attaining a much faster rate of growth than numerous developed and middle income countries that had far superior infrastructure systems, and which had the advantage of possessing market economies. If infrastructure were very important for growth, the course of events should have been considerably different. The confluence of events once again suggests that a variable different from the ones that have been considered by orthodox interpretations was the one fundamentally responsible for the fast growth of China.

In the 1990s, the policies of the preceding decade were pushed a step further. During 1990 to 1997, Deng remained the most important politician in China and he succeeded in placing in top government positions people who were as fixated as he was with the idea of catching up developmentally with the most advanced nations of the world. These were years in which Deng and his followers were more in control of power than before and it was also a time when those who believed that investment priority should be allocated to agriculture became an even smaller minority. Therefore, the desire among the political class to acquire the best technology of the world as fast as possible expanded during the 1990s, and as a result more efforts were made to import technology so as to increase the technology-producing capacity of the country. Since the bondage between technology and factory goods is very strong, government support for manufacturing was increased.

Direct and indirect support was increased by means of budgetary allocations and by means of more fiscal, financial, and nonfinancial incentives. The bulk of the budget continued to be utilized to subsidize factories. Direct government grants, however, diminished relative to the 1980s, but overall funding increased as state banks lent to manufacturing much larger amounts. Peking not only enlarged by a wide margin the financial assets of the existing state banks, but also created new ones. In 1994, for example, three new state banks were created, with the main function of supplying funds to factories. Although the private sector expanded rapidly during the 1990s, the bulk of factory output continued to come out of state companies. These state factories, however, were considerably liberalized, and as this process progressed, a myriad of tax exemptions and deductions was introduced that more than compensated for the decrease of direct subsidies the government had formerly supplied.[54]

To finance the larger inflow of technology that the leadership wanted, more foreign borrowing was undertaken and more incentives were supplied to foreign manufacturers. Foreign debt as a share of GDP rose to about 16 percent by 1997 and higher profit possibilities were offered to foreign firms. That translated into a larger importation of machinery and equipment and a much larger inflow of FDI, the majority of which was in manufacturing. The increased support for manufacturing in the 1990s coincided with a faster rate of factory output as well as with a faster rate of GDP (factory output expanded by about 18 percent annually and real GDP by about 8 percent). The official growth figures were about 11 percent, but once again they needed to be discounted because of the still large market distortions and their accompanying inefficiencies.

The fact that manufacturing grew faster than in any other country between 1990 and 1997 drove China to position itself as the third largest producer of factory goods in the world. That coincided with the fact that it also became the third largest economy in the world. Illustrating this fast growth was steel production. In 1987, China was way behind Japan and the United

States in steel production but by 1997 it had become the largest producer in the world. Steel also demonstrated the endogenous nature of growth because by the end of the 1990s practically all of China's steel production was fabricated by Chinese firms and consumed in China.[55]

Catching-up technologically was not the only motivation propelling government support for the sector. There were also balance of payments concerns. However, the determining motivation was the desire to catch up developmentally. National security concerns diminished further and investment in armaments as a share of GDP actually decreased. However, the desire to catch up was so much stronger than in the 1980s, that it more than compensated for the drop in the production of weapons. Even the military were utilized to promote the catch-up goal. It had already started in the 1980s, but during 1990 to 1997 the Armed Forces became even more involved in the production of civilian factory goods.[56]

The correlation between manufacturing and growth was appreciated in practically every aspect of the economy. It was, for example, in the coastal provinces once again where support for the sector was more enthusiastic and it was there that the economy expanded the fastest. Among those provinces, it was in Guangdong where a combination of the efforts of central, provincial, and local governments delivered the most incentives to manufacturers, and it was this 180,000-square-kilometer province that got the fastest GDP figures. However, the factory-promotion efforts in the four mainland SEZs were even more decisive than in Guangdong and the SEZs got once again a faster economic growth than Guangdong.

Guangdong and the SEZs are very important examples of the almost unlimited growth possibilities of manufacturing. By 1997, only thirteen countries in the world had a larger population than that of Guangdong, which had a population of about 70 million. It is important to notice that by 1979 this province was not even the most developed in China; nine others were more advanced. However, from 1980 to 1997, it attained a real economic growth of about 14 percent per year, while factory output averaged about 24 percent annually. In the SEZs, which by 1997 had a population of about three million, economic growth in real terms during 1980 to 1997 was about 26 percent per year while manufacturing grew by about 35 percent annually. By the mid-1990s, Mongolia, Gambia, Congo, Lesotho, Macedonia, Namibia, Letonia, Panama, Botswana, Stonia, Trinidad and Tobago, Gabon, Oman, Slovenia, Kuwait, the United Arab Emirates, and Singapore had smaller populations than the SEZs.[57]

What Guangdong and the SEZs seem to demonstrate is that the limits of fast economic growth over sustained periods of time are not the approximate 10 percent that Japan and the NICs attained at different moments during the second half of the twentieth century. Guangdong and the SEZs demonstrated that much faster rates are possible for large and small countries, and that the key for such an achievement resides in a very strong level of government support for manufacturing.

These two economic entities as well demonstrated that double-digit factory growth was not in need of trade protection. From the start of the reforms, Guangdong had the most liberal trade regime among all the thirty Chinese provinces, and the SEZs practiced almost free trade. The whole regional structure of China actually presented a pattern in which the higher the protection, the slower the growth. It was in the provinces of the interior where factory and GDP grew at the slowest rate, and it was there that trade barriers were the highest. Factory output in the interior grew slower because it was less promoted, but the data clearly reveals that trade protection offered no aid to manufacturers.

The fact that efficiency correlated with the degree of trade openness strongly suggests that protectionism actually harmed the cause of manufacturing. In the SEZs, with almost free trade, only 1.3 units of manufacturing were needed to produce a unit of GDP. In the interior, on the other hand, more than three units of factory output were needed to generate a unit of growth. Also, a major justification for the trade protection was the fear of trade deficits. However, despite Guangdong's very liberal trade policies, it consistently experienced trade surpluses. With an annual average export growth of about 35 percent during 1980 to 1997, it was only inevitable that it ran surpluses. Practically all of those exports were factory goods.[58]

The case of Guangdong and the SEZs also demonstrated that market distortions in no way assist the economy and, on the contrary, are harmful for growth. Guangdong was the province where state companies were less numerous, where monopolies and other competition-hindering practices were less present, and where the private sector was given more liberties. The SEZs were actually the only regions in China where the private sector accounted for the bulk of output, where competition was cutthroat and where regulation was only barely felt, and this was the place where economic growth was by far the fastest.

Between 1990 and 1997, China made much progress in liberalizing the economy across-the-board throughout the whole country. However, in spite of significant liberalization, the economy remained greatly distorted. On average during these years, the private sector accounted only for about 7 percent of GDP. The immense majority of enterprises were owned and operated by central, provincial, or local governments. There was also much trade liberalization during this period but, relative to most other countries in the world, trade barriers were still very high. Tariffs were very high and nontariff barriers were even higher. About half of China's imports were subject to nontariff barriers. The duty system was not even homogeneous throughout the country, varying from province to province, and import quotas went often unpublished.[59]

Although regulations on business were significantly reduced, they were still much higher than in most developing and developed countries of the world. The banking system experienced some restructuring along market lines, but during the 1990s it was still light years away from resembling that of a capitalist country. The Central Bank only barely oriented its decisions because of market signals, the immense majority of banks were state-owned and were among the most inefficient in the world, nonperforming loans were very large, foreign banks were not allowed to operate in China, and domestic capital markets were small and primitive. By 1997, the four biggest state banks accounted for about 90 percent of all bank lending. On top of that, from 1990 to 1993, the exchange rate continued to be highly distorted and overvalued (by 1993 the Yuan was overvalued by about 30 percent). During those years, a two-tier exchange rate continued to operate. From early 1994, however, it was unified and allowed to float.[60]

During the 1990s, the government tried to become more fiscally sound in its finances but, in spite of its efforts, budget deficits were once again the norm. During these years, the central government ran budget deficits averaging about 4 percent of GDP. Worse still was that those deficits were run mostly for the sake of financing state enterprises that were known for their very high levels of inefficiency. Those deficits also contributed to accelerate inflation, and prices rose faster than during the 1980s. Inflation averaged about 9 percent annually during 1990 to 1997, while in the preceding decade the figure had only been about 6 percent.[61]

During these years, there was also a rise in corruption compared with the 1980s. Corruption became terribly high not just by Chinese standards, but also relative to other developing countries. Many foreign businessmen with ample experience in Africa and Latin America asserted that corruption in China was even worse. In 1994 PERC, a Hong Kong private consulting firm, conducted a survey and rated Asian countries by their levels of corruption. The survey concluded that China had the highest levels in the region. By the mid-1990s, it had them even higher than India, Indonesia, and the Philippines.[62]

Although there were large investments in infrastructure in the 1990s, they continued to grow at a much slower pace than the economy, and the end result was even more congestion in roads, railways, ports, and airports. The supply of electricity continued to be largely insufficient and telecommunications were capable of meeting only a fraction of demand. Despite progress, by the mid-1990s China still had fewer kilometers of roads per capita than Latin America,

India, and even sub-Saharan Africa. In general, infrastructure was still largely deficient even by the standards of the least developed nations.[63]

Between 1990 and 1997, China was one of the most distorted economies in the world. Only nations such as North Korea, Cuba, and Vietnam were more distorted. On an index of global economic freedom prepared by the Washington based Heritage Foundation, countries were ranked on the basis of how government intervention restricted the economy. Among the large number of countries analyzed, China came out in the last position as having the least economic freedom.

How, then, was it possible that while state companies accounted for the immense majority of output, trade barriers were terribly high, regulations were abundant, the financial system was in state hands, and there was a constant fiscal imbalance, fast economic growth was attained?[64]

If orthodox interpretations of the economy were right, then China should have, at best, experienced stagnation. If it is added that inflation accelerated, corruption worsened, and infrastructure congestion rose, then it becomes even harder to explain the fast growth. Worse still is that it was not just fast growth, but actually one of the fastest in the world. It is beyond any doubt that all of these aspects were policy errors and that the economy would have performed better if there had been whole-scale privatization and liberalization, balanced budgets, an independent Central Bank, low inflation, no corruption, and a better infrastructure.

However, this huge number of policy errors actually was very clear evidence that some other factor had been overlooked that had impressive growth-generating powers. The powers of this variable were so obviously strong that it was still capable of driving the whole economy at a fast pace even under these circumstances. The abundance of market distortions were actually the ultimate proof that manufacturing is the bottom line of economic growth, and the decisive support of this sector was made very evident by the fact that factories absorbed about three-fourths of overall bank lending.

It was argued that FDI was fundamentally responsible for the fast growth, but even large inflows of FDI are not supposed to deliver fast growth when an economy is so massively distorted. During the 1990s, in terms of volume, China became the largest recipient of FDI among non-OECD countries in the world. However, in per capita terms, it continued to be largely overtaken by numerous counties in Asia and Latin America that had similar or much slower rates of growth. On top of that, those nations had market economies and during the 1990s they liberalized them even more. Nevertheless, Latin American countries, for example, attained much slower GDP figures than China. Latin American growth averaged only a little over 3 percent annually.

Also worth noting is that the difference between the inflows of FDI during the 1980s and 1990s was enormous, but the GDP figures were not. In the 1980s, China received on average about US$1.5 billion per year in FDI, while during 1990 to 1997 the amount rose to about US$24 billion. However, real economic growth averaged about 7 percent during the 1980s and 8 percent during the 1990s. If FDI had played a decisive role in growth, then the difference in GDP figures should have been much larger. Not in China or in other countries is it possible to find a consistent correlation. FDI did contribute to growth, but it is evident that it was not a major determinant.[65]

It was also argued that Confucianism and Chinese culture were behind the fast growth of the 1990s because they drove the population to work long hours, study hard, and save a lot. It is worth noting that in the 1990s the average per capita number of hours worked per week fell noticeably from the preceding decade. The economy, however, instead of doing likewise, experienced a faster pace of growth.[66]

Others argued that fast growth was the result of the fast growth of exports. Exports grew by about 16 percent annually between 1990 and 1997, and not only was that a much faster rate than that of GDP, it was also faster than the export figures of the 1980s. However, from the perspective of traditional economic thinking, fast export growth was supposedly impossible in an economy where more than nine-tenths of output came from state-owned companies and where market distortions in every aspect of the economy were extensive.

Still others asserted that fast growth resulted from the improved agricultural productivity resulting from liberalization. During the Deng years, the rate of agricultural output accelerated considerably and it was this domain that was most liberalized. The fact that agriculture still accounted for a large share of GDP and that the large majority of the population lived in rural areas drove many to the conclusion that China's growth was agriculture-led.

The problem with this argument was that in spite of the acceleration, farm output systematically grew at a slower pace than GDP. Even during the 1980s, when farm output grew the fastest, its rate was slower than GDP. For agriculture to have a propelling effect on the rest of the economy, one would have expected a faster rate than that of GDP. The official GDP figure for the 1980s was about 10 percent and that of agriculture was about 6 percent. Even the adjusted GDP figure to capitalist standards was still faster (7 percent). Since inefficiency plagued the whole Chinese economy, it is also logical to assume that if agriculture is measured by strict capitalist standards, then the real figures become smaller. Under those circumstances, the gap between real GDP and real agricultural output widens even more.

It is also worth noting that in the 1990s the gap widened still more as the GDP figures accelerated and those of agriculture decelerated. If there had been a causality linkage between agriculture and growth, the faster GDP figures of the 1990s should have been accompanied by a faster rate of farm output. The fact, however, is that the official growth figure was about 11 percent annually and the average rate of farm output was only 4 percent. Even with the discounted GDP figure of 8 percent, agriculture was still lagging considerably behind. Not to mention that during the whole 1980 to 1997 period, agricultural productivity also was considerably behind that of manufacturing. For agriculture to have acted as the engine of growth, it needed to have attained a rate of output faster than that of GDP or to have at least attained a rate of productivity growth faster than that of any other sector. It did not attain either of the two.

The whole second half of the twentieth century also supplied abundant evidence clearly suggesting it was highly unlikely that agriculture could have had a propelling effect on the rest of the economy. According to the official figures, the average rate of farm output during the 1950 to 1997 period, was about 4 percent, while the official GDP figures were about 9 percent. Once the GDP figures are adjusted to capitalist measurements (5 percent), they still come out higher than the farm figures. It is obvious that agriculture also needs to be adjusted because the inefficiencies of the system rendered a large share of the output as not consumable. Under these circumstances, therefore, the gap becomes even wider.[67]

Technology and Wealth Creation

The way technology developed during the 1990s as well as during the whole second half of the twentieth century was illustrative of a linkage between manufacturing and growth that was not merely fortuitous.

The only possible way the massive market distortions that China experienced during the 1950 to 1997 years could be made to add up with the empirical evidence would be to assume that technology is the source of wealth and that manufacturing is the prime generator of technology. It would also have to be assumed that the prime mechanism for the attainment of manufacturing growth is government support for this sector. Under those circumstances, it becomes fully understandable why since 1950 Peking, in spite of practicing central planning, managed nonetheless to attain a much faster rate of growth than ever before.

If such an assumption is right, it would be logical to find a correlation between manufacturing and technology. The historical data reveals precisely that. There is a very strong correlation between the two during the whole period. During the 1950s, for example, the government supported manufacturing for the first time in China's history in a very decisive way, and technology grew rapidly for the first time. During this decade the bulk of the technology was imported from the Soviet Union. In the following decade, technology was mostly produced domestically, but the rate of technology slowed down considerably, which coincided with a considerable decrease in the level of support Peking gave to factory production.

During the 1970s, technology developed slightly faster and that correlated with a slight increase in the government's promotion efforts with respect to manufacturing. Once again, the majority of the technology was created domestically, but relative to numerous East Asian countries that mostly imported technology, China's development was slow. In the 1980s, Peking began to promote factory production in a much more enthusiastic way and all of a sudden technology began to grow at a much faster pace. This time most of it was imported and practically all of it came from capitalist countries. In the 1990s, factory production grew even faster, reflecting the stronger promotion efforts of the state. This was paralleled by a growth of technology even faster than in the preceding decade. Once again, the bulk of the technology was imported from capitalist countries.

Also substantiating the manufacturing-technology thesis is the fact that technology systematically made its appearance in the form of a manufactured good. Most frequently, it appeared in the form of machinery and equipment, in particular when it was imported in large amounts as during the 1950s, the 1980s, and the 1990s. However, independent of whether imported or created domestically, or imported from a communist or a capitalist country, the bottom line was that factory goods were at the core of technology. Domestic innovation also found its materialization in manufactured goods such as China's first hydrogen bomb (1967) and the first earth satellite (1970). The technology that was imported not in the form of a factory good, but in the form of a patent ended up almost always transformed into a manufacture.[68]

Since productivity is fundamentally determined by technology, it is therefore logical to presume that if manufacturing is responsible for the creation of technology, then the historical data should reveal a correlation between this sector and productivity. History reveals such a parallel. During the first half of the twentieth century, for example, manufacturing was only barely promoted and productivity grew at a very slow pace. Then in the 1950s, factory production was abundantly promoted and productivity grew at an unprecedented pace. During the following two decades, support for manufacturing was lower, factory output decelerated, and the rate of productivity slowed down. In the 1980s and 1990s, factory output accelerated by a very large margin and productivity did likewise. Also worth noting is that during the whole 1950 to 1997 period, productivity in manufacturing was faster than in all the other sectors. During the 1980s, for example, output per worker in manufacturing was almost four times faster than in agriculture.[69]

The creation of technology is the most resource-absorbing activity in the world and precisely for that reason it is extremely investment-intensive. If technology is fundamentally generated by manufacturing, there should be a correlation between the fluctuations of this sector and those of investment. The empirical data reveals that such a parallelism exists. Investment as a share of GDP averaged about 2 percent during the first half of the twentieth century, coinciding with very weak government support for the sector and a weak pace of factory output. In the 1950s, Peking allocated a very large share of the nation's resources to this sector, which correlated with an unprecedented overall level of investment averaging about 25 percent of GDP. In the 1960s, support decreased considerably, the rate of factory output decelerated significantly, and investment as a share of GDP dropped to about 8 percent. During the 1970s, Peking allocated more resources to manufacturing, and investment immediately rose to average about 12 percent of GDP.[70]

During the 1980s, policy makers decided to increase by a very large margin the share of total resources allocated to manufacturing and immediately the overall level of investment rose by a large margin, averaging about 30 percent of GDP. In the 1990s, the factory-promotion efforts of the state were still larger and investment averaged about 33 percent. The investment-intensive nature of this sector was also appreciated in the fact that it systematically absorbed a much higher share of overall investment relative to its share of GDP. During 1990 to 1997, for example, manufacturing accounted for about 38 percent of GDP, but it absorbed almost fourth-fifths of overall investment.[71]

If manufacturing is the prime creator of wealth, it would be rational to expect to find a long-term correlation between this sector and savings, because the possibilities for savings appear mostly when there is wealth creation. The faster the creation of wealth, the larger the potential for savings to increase, and vice versa. During the first half of the twentieth century, for example the rate of factory output was very slow and savings were almost nonexistent. As a share of GDP, they averaged about 3 percent. In the 1950s, however, the state mobilized resources into manufacturing to an unprecedented extent, and all of a sudden savings rose to average about 30 percent of GDP. During the following two decades, the rate of factory output decelerated considerably and savings also dropped, to average about 18 percent.[72]

In the 1980s, the promotion efforts of the government increased considerably and even though consumerism became acceptable, savings nonetheless rose significantly to about 37 percent of GDP. In the 1990s, the consumerist trend continued to much larger proportions. The authorities no longer condemned as capitalist spiritual pollution the growing inclination of the population to consume. Notwithstanding this trend, savings rose to even higher levels and that coincided with the allocation of a larger share of the nation's resources to the manufacturing sector. The faster rate of factory output that such an allocation delivered correlated with a larger share of savings, which averaged about 41 percent.[73]

A long-term cross-country analysis also reveals a strong correlation between manufacturing, investment, and savings. During the second half of the twentieth century East Asia was the region that promoted factory production most enthusiastically, and it was also this region that attained by far the largest share of investment and savings from GDP. During the first half of the century, however, government support for the sector in East Asia (with the exception of Japan) was weak and way below that of other regions such as Europe, North America, and Oceania. On this occasion, however, the share of investment and savings from GDP was small, and inferior to that of Europe, North America, and Oceania. Most likely, it was not just serendipity that during this period the majority of governments of the future OECD countries promoted the sector more.[74]

The evidence substantiating the manufacturing thesis comes from all directions. The way each sector of the economy developed during the whole 1950 to 1997 period was illustrative of the above. Not only did the output rate of agriculture and the other primary activities, construction and services correlate with the differing levels of government support for manufacturing, but the possibility of progress for each sector and domain showed a direct dependence on factory goods.

Agriculture, for example, grew very slowly in China during the first half of the twentieth century (about 0.4 percent annually) even though the bulk of the nation's resources were allocated to it. Then in the 1950s, when Peking for the first time transferred abundant resources from the other sectors to the factories, farm output rapidly grew for the first time in history (5 percent annually). During the following two decades, support for manufacturing dropped significantly and agricultural production decelerated to an average rate of about 2.3 percent. In the 1980 to 1997 period, Peking once again changed track and increased the share from the nation's total resources that it allocated to manufacturing. Farm output all of a sudden grew much faster, averaging about 4.7 percent annually.

The other primary activities showed a similar parallel. It was precisely when resources were mobilized in larger amounts toward factories that mining, fishing, and forestry experienced their fastest growth. When it is taken into consideration how determinant for primary production are factory goods such as farm machinery, mining equipment, fishing boats, and forestry equipment, it becomes then understandable why a faster rate of factory output systematically coalesced with a faster production of primary goods. If it is assumed that manufacturing is the prime generator of technology, then it becomes even clearer why primary production is so dependent on these goods.[75]

The same goes for trade and in particular exports. During the first half of the twentieth century, even though China was largely open to international trade exports grew at a very slow pace (about 2 percent annually), which ran parallel to very weak support for manufacturing. During the first three decades of communist rule, on the other hand, even though the country was almost totally closed to foreign trade, exports grew at a much faster pace and averaged about 6 percent per year. This correlated with an unprecedented promotion of factory production. Then during the 1980 to 1997 period, Peking allocated a much larger share of the nation's resources to the sector, and exports immediately began to grow at a much faster pace, averaging about 14 percent per year.[76]

The fact that during the whole second half of the twentieth century the fastest growing exports were manufactures goes a step further toward demonstrating that exports were not the cause of growth, but were actually one of the many effects of manufacturing. When it is considered how determinant factory goods are for exports, it becomes understandable why export growth is only possible when there is first a production of manufactured goods. Strong support of this sector delivers not only more factory goods, which are the fundamental reason for the expansion of export possibilities, but also a larger output of primary goods, which enlarges the export possibilities of primary goods. By 1970, manufactured exports accounted for about half of total exports and by 1997 they had grown to account for about 86 percent. In spite of the large manufacturing share, however, by the 1990s China exported by far the largest amount of primary goods in all of its history.[77]

The very fast growth of exports during 1980 to 1997 suggested also that producing factory goods in a competitive way was possible for all nations. Most analysts in the late 1970s predicted that China would never become a major exporter of factory goods to developed capitalist countries because of its historical inability to achieve such a feat. Had China dismantled central planning completely, most would have still argued that because China had never been successful on that front, it was condemned to remain forever in the same condition. Since Peking decided to liberalize the economy only a little, most analysts stated that the chances that it would one day be able to export abundant factory goods was almost nonexistent.[78]

However, in no time it became a major world exporter of textiles, clothes, toys, and footwear and soon afterward it achieved the same feat with tools, steel, auto parts, electronics, microwave ovens, personal computers, and numerous other goods. By the 1990s, China's exports had succeeded so much that they had become one of the main reasons for the trade deficits of North America and Western Europe. During the 1990s, China ran bilateral trade surpluses with all OECD countries including Japan, as well as with all of the NICs. Exports as a share of GDP went from about 7 percent in 1980 to about 32 percent in 1997.[79]

The same was said about numerous other countries that throughout history were classified as incapable of ever becoming competitive manufacturing exporters. However, as soon as the governments of those countries opted for a policy of strong support for the sector, everything rapidly changed and a fast growth of exports of this type of goods was attained.

The development of construction during the second half of the twentieth century took a very similar path as the one that the primary sector and exports did. Construction during the first half of the century grew very slowly. Then in the 1950s, it developed at a pace far superior to anything previously experienced. In the following two decades, construction slowed down and during the years 1980 to 1997, it greatly accelerated. This all was directly parallel to the differing levels of support for manufacturing. By taking into consideration the high concentration of technology that construction manufactures such as shovels, bars, cement, metals, bricks, cranes, earth movers, and other building machines embody, it becomes understandable why this sector is so dependent on factory goods for its development.

Services also followed a similar pattern of development. During the first half of the twentieth century, services grew at a very slow pace. Suddenly in the 1950s, they grew rapidly compared with anything previously experienced, and so did factory output. In the following two decades, a much smaller share of total resources was allocated to manufacturing and the pace of services decelerated. Then, during 1980 to 1997, factory output accelerated considerably and services also grew much faster.

What the history of China very strongly suggests is that the development of services is fundamentally dependent on the share of resources allocated to manufacturing. The larger the share flowing into factories, the faster the growth of services. This apparent paradox nonetheless becomes understandable once the massive technology-embodying characteristics of these goods are taken in consideration. The vast majority of services are not even possible without factory goods.[80]

China made impressive progress during the second half of the twentieth century. Living conditions probably improved more than during all of its preceding history. Morbidity and mortality dropped significantly, education rose, housing improved, violence decreased, the rule of law made major progress, respect for women's rights leaped forward, and environmental protection also made large advances.

By the 1990s, for example, China spent about 0.6 percent of GDP on environmental programs. Up to the mid-twentieth century, however, not even a tiny fraction of that share was spent on protecting the environment—understandable considering that until then, famine, epidemics, poverty, and war had constantly absorbed the minds of policy makers. There was no time to ponder other matters. Not only was there no motivation to think about protecting flora and fauna, there were no means to finance environmental programs.[81]

During the second half of the twentieth century, the practice of binding and deforming the feet of women was abolished as was the practice of taking multiple wives. As long as the main preoccupation of policy makers and the population was to avoid falling prey to starvation and malnutrition and as long as people lived in ignorance, it was practically impossible for the male population to understand that such mistreatment was unacceptable.

It was the rapid creation of wealth during the 1950 to 1997 years that put an end to hunger and ignorance. It was also wealth that permitted China, for the first time in its history, to afford to take measures to protect the environment and to recognize the inalienable rights of women.[82]

For centuries, plague, cholera, poliomyelitis, tuberculosis, schistosomiasis, and many other diseases constantly ravaged the Chinese population. Then, during the second half of the twentieth century, all of a sudden China acquired the technology to control these diseases.

Throughout all of the country's history until the mid-twentieth century, China's rulers attempted to improve rapidly living conditions. However, they systematically failed and the only moment when they succeeded was when they decided to allocate a very large share of the nation's resources to manufacturing. If this sector is the main creator of technology, as history very strongly suggests, then it is perfectly understandable why strong support for manufacturing correlated so tightly with a rapid pace of wealth creation.[83]

Although support was massively stronger during the 1950 to 1997 period than before, it was far from being as strong as it could have been. Much suggests that if no market distortions would have taken place during this period the 14 percent rate of manufacturing output that was attained would have delivered a real economic growth of about 11 percent, instead of the 5 percent that was achieved. Much also suggests that factory output could have been 20 percent per year or even higher if many more resources would have been allocated to this sector. Had that occurred the economy would have also grown much faster. Overall levels of development would have been much higher by the late 1990s, and democracy and environmental protection would have made more progress.

3

The Newly Industrialized Economies of East Asia

The Case of Hong Kong and Taiwan

Hong Kong and the Motivation to Promote Factories

The fastest growing economy in East Asia during the second half of the twentieth century was Hong Kong's. Not only did it grow faster than any other economy in the region, but it grew faster than any other nation in the world. Economic growth averaged almost 9 percent per year and living conditions improved at a breathtaking pace. From a vast shantytown in 1949 where electricity was rare, running water was scarce, and sewers were almost nonexistent, Hong Kong by the turn of the century became one of the wealthiest places in the world.[1]

The colony attained this spectacular rate of growth while simultaneously practicing probably the most liberal economic policies in the world. By the end of the twentieth century, liberal economic policies were adopted by most nations of the world. Governments were so impressed by the collapse of communism that practically all nonliberal economic policies became largely sidelined. However, notwithstanding the triumph of liberal economics, numerous policy makers in developed, middle-income, and developing countries remained distrustful of the policies that this school of thought demanded. Their distrust resulted from the fact that on numerous occasions the application of these policies did not deliver the desired results.

The case of Hong Kong was the ultimate proof that at least the vast majority of liberal economic policies were capable of delivering very positive results. It was evident that this school of thought had not uncovered all of the variables that intervene in the growth process, but it was also evident that it was correct about the vast majority of what it asserted.

Hong Kong demonstrated both ideas simultaneously. The fact that free trade was practiced, that government expenditure as a share of GDP was very small, that regulations on business were minimal, and that budgetary surpluses prevailed was suggestive that liberal economic policies were conducive to fast economic growth. The fact that the exchange rate was largely determined by the market, that a responsible monetary policy was practiced, that expenditures on social welfare were minimal, that foreign firms were allowed to invest without restrictions, and that state companies were nonexistent reinforced such an idea.[2]

On the other hand, the fact that during the first half of the twentieth century, the rate of GDP in the colony was considerably slower while it practiced the exact same liberal policies was also suggestive that the bottom line of economic growth resided somewhere else. Growth during the 1900 to 1949 period was just one-third as fast as during the following fifty years. If liberal policies had been fundamentally responsible for the colony's fast growth in the second half of the century, then rates of growth should not have varied much in the preceding period. The fact that there was a very large difference clearly indicates that a nonidentified variable was at work.

The fact that economic growth during the second half of the nineteenth century was even slower while the same liberal policies were also practiced further reinforces this idea. From 1850 to 1899, GDP averaged only one-ninth of the figure a century later. At that time, market forces were actually more unrestrained (government expenditure as a share of GDP was but a fraction of what it was a century later, regulations were fewer, and so on). On top of that it was the exact same British colonial government that ruled the territory. If liberal policies were really the engine of Hong Kong's growth, the second half of the nineteenth century should have actually attained a faster rate of GDP than the 1950 to 1997 period. The fact that it attained an extremely slow pace of growth very clearly indicates that liberal policies were not the bottom line of growth.

Much seems to indicate that the variable at the core of growth was manufacturing. Unlike the liberal economic thesis, the theory of manufacturing succeeds in adding up with the facts. During the second half of the nineteenth century, for example, there was only a very low level of government support for manufacturing and the rate of output of this sector averaged only 1.4 percent annually, reflected by a GDP rate of just 1.0 percent. In the following fifty years, a change in the politics of China and the region drove the colonial government to increase substantially the subsidies to the producers of factory goods, and this sector's rate of output accelerated considerably averaging about 4.0 percent per year. GDP immediately increased its pace to about 3.5 percent. During the 1950 to 1997 period, a new and radical change in the politics of China pressured the territory's government to elevate by a large margin the financial, nonfinancial, and fiscal incentives to factory producers. This coalesced with a rate of manufacturing output that averaged about 10.8 percent annually and a rate of GDP that averaged about 8.8 percent. (See tables in appendix.)

It was a number of exogenous factors that pressured the British authorities to change the course of economic policies, but it was not exogenous means that made the growth possible. It was the change in domestic policies that delivered the growth, not the resources that came from abroad. What came from abroad was the pressure to change policies.

As soon as World War II ended, it became clear that the Communists had a good chance of taking over power in China, and thousands of people fearing such a development began to emigrate. Between 1945 and 1949, about 1.3 million people fled to Hong Kong, more than trebling the colony's population.[3]

Even before the arrival of this massive influx of people, the territory was enduring high levels of unemployment. Such a mass emigration threatened the political stability of the colony because the employment possibilities were practically zero. Poverty and unemployment on such a scale risked becoming explosive, in particular at a time when communist ideology was on the rise. Since the colony was resource-poor the only possibility the authorities could think of as a way for generating employment was manufacturing. Since this sector was also the only one that offered vast export possibilities, the government decided to concentrate its emergency efforts on promoting factories. Exports became a sort of national security necessity in order to pay for the vast import of foodstuffs and other primary-sector essentials needed for the newcomers.

The colonial authorities therefore began in the late 1940s to offer abundant incentives to factory producers. The government supplied land for the factory site at very subsidized prices—a major subsidy considering the very high land prices in the colony. As the owner of all of the territory's land, the government was capable of abundantly supplying this resource. The authorities also constructed multistory flatted factories and leased the facilities at very subsidized prices to entrepreneurs, to reduce their costs of production even more. Numerous grants and subsidized lending were also supplied for the acquisition of machinery, equipment, and other vital production inputs. In addition, corporate income tax was set at very low levels (12.5 percent) and utilities were supplied at below market prices.

Manufacturers were burdened with neither social security costs nor minimum wage decrees that would raise their costs of production. Just about everything was done to increase the opportunities for profit of factory producers in order to induce a much higher level of investment in that sector.[4]

As soon as this massive promotion effort began, factory output also began to grow at an impressive pace, and the economy did likewise. There was much emigration of textile producers from Guangdong and Shanghai beginning in the late 1940s, although it was more pronounced after the Communists took over power. These emigrants brought their machinery, their capital, and their management expertise. This obviously contributed noticeably to the increased factory production and the improvement of the economy, but it accounted only for a minority share of the overall factory production of those years. The most impressive rates of factory output occurred in the late 1940s even though the bulk of the textile emigration took place in the early 1950s. Also worth noting is that by the late 1950s, factory output had considerably decelerated, and then in the early 1960s it accelerated suddenly again by a large margin. That occurred even though at that moment there was no new influx of machinery and capital from the new wave of emigrants.

This absence of a correlation suggests that the large emigration of textile producers during those years was not the fundamental variable explaining the fast growth of manufacturing and of GDP. The differing levels of government support for the sector did, however, correlate with the differing rates of factory output. Confronted with an overwhelming amount of immigrants in the late 1940s, the British authorities decreed a huge increase in support for the sector, and factory output averaged almost 30 percent in each year (1948 and 1949). The economy shadowed and grew by about 23 percent annually. Since the unemployment problem rapidly decreased by the late 1950s, the level of incentives to factory producers was significantly lowered, which coincided with a rate of factory output that was only like one-third as fast as a decade earlier.

The large-scale famine created by the Great Leap Forward in the early 1960s sent a new wave of Chinese emigrants into Hong Kong. None of them, however, emigrated with machinery, capital, or any type of assets. Practically all of them were peasants who at best knew how to read and write. It had been originally argued that the Shanghaise were the most entrepreneurial and hard-working of all Chinese and the large number of them who fled to the colony around 1950 contributed enormously to the rapid growth. However, during the second wave of immigrants in the early 1960s there was hardly a Shanghaise among them. The bulk came from the surrounding regions in Guangdong. Factory output and the economy nonetheless accelerated, which coincided with a considerable increase in the level of government support for manufacturing in an effort to rapidly create employment for the thousands of newcomers.[5]

The pressing circumstances of the 1950s therefore forced the colonial authorities to supply an extremely attractive package of incentives to manufacturers, which was reflected by a rate of factory output that averaged about 16 percent annually and a rate of GDP that averaged about 12 percent per year.

In the 1960s, the new wave of immigrants was smaller than the one the colony confronted in the preceding decade. As a result, the authorities were under less pressure to create employment, so they decreed a level of support for the sector not as strong as before. The resource-poor structure of the colony with its overcrowded situation continued also to drive the authorities to see exports as vital, for they provided the foreign exchange needed to import primary goods; but the need to generate vast exports was also less pressing than in the 1950s. As a result fiscal, financial and nonfinancial incentives were somewhat lower, and this was paralleled by a decelerated pace of factory output, which averaged about 14 percent per year. Once again the GDP figures moved in tandem and averaged about 10 percent.[6]

During the 1970s, population, employment, and balance-of-payments pressures were lower, although they still remained strong. As a result, the government decided to promote factories less enthusiastically. Incentives were lowered somewhat and this sector's rate of production averaged about 11 percent, which was reflected by a GDP rate of about 9 percent per year.

The performance of Hong Kong during this decade was the most blatant proof that the argument that justified the significant economic deceleration of OECD countries on the large increase in the price of oil and other commodities was false. Hong Kong not only imported up to the last drop of oil, but also up to the last primary good it consumed. During the 1970s, primary sector production as a share of GDP accounted only for about 1 percent. Hong Kong was therefore hit in a more direct way than any OECD country, and perhaps any other economy in the world, by the rise in commodity prices. That, however, did not inhibit the colony from growing at the still impressive pace of 9 percent.[7]

There was a significant rise in inflation in OECD countries during this decade and it was also asserted that higher prices had also played an important role in the considerable economic deceleration of developed nations. Hong Kong also experienced a noticeable acceleration in the rate of inflation. After having averaged just a little more than 2 percent annually during the 1960s, it went on to average almost 8 percent in the following decade. That, however, did not inhibit an impressive rate of economic growth. Inflation in OECD countries had been almost as high (9 percent) and yet economic growth averaged only 3 percent during the 1970s. GDP figures did decelerate in the colony, but it is highly unlikely that even that was the result of inflation. Inflation in South Korea, for example, became also higher (18 percent), but instead of decelerating, the economy went from an average of about 8 percent in the 1960s to 10 percent in the 1970s.[8]

The absence of a correlation was evident in all directions. Hong Kong demonstrated that it was possible to attain fast growth even if inflation rose considerably and South Korea demonstrated that it was possible to even accelerate an already fast GDP rate while an already fast pace of inflation accelerated even more. The quick inflation clearly had negative effects on the economy, but much suggests that they were not directly related to growth.

The 1980s gave further evidence that there was practically no causality linkage between inflation and growth. During this decade, prices in Hong Kong dropped a little and averaged about 7 percent annually. However, the economy of the colony did not grow faster than formerly nor did it even maintain the same pace of the 1970s. It decelerated further and averaged about 7.5 percent. While the inflation-growth correlation was again missing, the one with manufacturing was as strong as before. During this decade, government support for the sector decreased some more and factory output averaged about 8.1 percent annually.[9]

During the 1990s, the parallelism continued. Support for the sector was lowered more, which coalesced with a further deceleration of the rate of factory output and of GDP. The population, employment, natural resource, and balance-of-payments pressures were much lower than before and as a result the factory promotion efforts of the state were also much lower. Some of these motivations had almost ceased to exist, but the factory promotion efforts of the recent past had so directly coincided with fast growth that the authorities suspected that factory promotion was at least partially responsible for the positive performance. While the empirical data strongly suggested a causality linkage, the beliefs of the vast majority of policy makers were of a liberal nature and unsympathetic to government intervention, for manufacturing or for anything else. This mixed situation ended by delivering a level of promotion that was much lower than in the 1950s, but was nonetheless stronger than in most nations. This correlated with a rate of factory output and economic growth that was faster than the one attained by most nations, but much slower than the one attained during the 1950s.

There was large-scale emigration of labor-intensive manufacturing to China during the 1980s, which took even larger proportions during the 1990s. Worries about a hollowed-out manufacturing base that would cripple the export capacity of the colony also drove the authorities to promote capital-intensive manufacturing. Since technology-intensive production is less likely to emigrate, relatively large incentives were offered to would-be investors in this domain during the 1990s. Numerous government bodies dispensed the subsidies for the production of technology-intensive goods as well as for export promotion reasons.[10]

There was, for example, the Hong Kong Applied Research and Development Fund, which since its founding in 1992 began to offer financial assistance in the form of equity injections or subsidized loans of up to 50 percent of the cost of research and development expenditures. Although the equity injections ultimately translated into a direct partial state ownership of the firm, majority ownership systematically remained in private hands. There was also the Hong Kong Industrial Technology Center that offered subsidized accommodation to technology-based firms as well as grants. This center supplied ready-made state-of-the-art factory sites at very subsidized rents, as well as grants for the acquisition of high-tech machinery and equipment. The New Technology and Training Scheme that was also founded in 1992 concentrated, on the other hand, on offering grants to factory producers in new technology domains so that they could better cover their training costs. There were even specialized technology promotion schemes like the Hong Kong Plastic Technology Center, which provided technical advice and financial assistance to plastic manufacturers.

Between 1990 and 1997, there were also a number of other government schemes to promote manufacturing that were not technology intensive. These schemes pursued mostly to promote traditional exports. The government, for example, offered export credit insurance across-the-board to practically all manufacturers who exported. The Hong Kong Productivity Council offered grants to firms intending to improve their manufacturing quality and efficiency. The grants were mostly utilized to acquire state-of-the-art machinery and equipment. The Hong Kong Industrial States Corporation supplied multistory flatted factory facilities at very subsidized rents.

Although the bulk of the assistance was supplied across-the-board, there were nonetheless some manufacturing fields that were promoted more enthusiastically. Textiles and apparel received by far the most subsidies because of their very large export share. By the late 1980s, these two fields alone accounted for about 40 percent of exports. Footwear was another field that received preference because of its large export share. Other schemes such as the Additional Funding for Industrial Support Scheme provided grants and subsidized lending for diverse factory fields.[11]

Although considerably lower than during the 1950s, the level of subsidization the colonial authorities gave to the manufacturing sector during the 1990s was nonetheless very strong by international standards. That coincided with a rate of factory output and GDP that was much slower than before, but faster than that attained by many other economies in the world.

Manufacturing output between 1990 and 1997, averaged about 5.2 percent annually and GDP about 5.0 percent.[12]

Technology and Education

There is much to suggest that the strong correlation between manufacturing and growth during the whole second half of the twentieth century was not due to chance. Technology is the bottom line for the creation of wealth, and the fact that technology during this period was so directly intertwined with manufacturing and so bonded to factory goods suggests that this sector was fundamentally behind the generation of technology. If that was really the case, the empirical evidence should express itself in the form of a correlation. The existing data reveals precisely that.

The development of technology, for example, experienced a massive acceleration during the 1950s relative to the preceding decade, which coincided with a massive increase in the factory promotion efforts of the colonial authorities. Support and the rate of factory output slowed down a little in the following decade and there was a simultaneous slowdown in the importation of technology. In the 1970s, the rate of factory output decelerated some more, as did the rate at which technology was imported and consumed. In the 1980s, the trend continued and a lower promotion of the sector correlated with a slower pace of technical progress.

The 1990 to 1997 period was the first time the British authorities worked to promote the creation of technology, but it was actually the moment during the second half of the twentieth century when technology grew at the slowest pace. That was paralleled by the slowest rates of factory output during the whole period.

A long-term analysis of Hong Kong's history reveals a similar correlation. The second half of the nineteenth century experienced unprecedented technological development, which coincided with the first time in the territory's history when there was government support for manufacturing. It was nonetheless a very slow pace of development, and by international standards support of the sector was also very weak. During the 1900 to 1949 period, however, the factory promotion efforts by the colonial authorities rose by a large margin and there was a concomitant large acceleration in the pace at which technology was imported and consumed. Machinery and equipment, methods of production, patents, and consumer technology goods were imported at a much faster pace than in the preceding fifty years. In the second half of the twentieth century the rate of manufacturing production accelerated much more, as did the rate by which technology grew.

A very peculiar aspect about the way technology developed in Hong Kong gave further credence to the idea that manufacturing was the means by which technology expanded. From the start of colonial rule until the late twentieth century, practically all of the technology that Hong Kong utilized was imported. By the late twentieth century, several East Asian economies had become net technology exporters or at least major technology producers. Hong Kong, however, even by the 1990s continued to import the bulk of the technology it consumed. By then, the territory was more economically developed than most OECD countries, and in spite of its lack of inventive capacity, it managed to attain a much better economic performance.[13]

The typical development of events for most nations that over time caught up economically with the most advanced countries, was first a period in which they imported practically all of the technology they consumed. Then, as these countries grew faster than the most advanced ones and narrowed the developmental gap, they increasingly began to produce their own technology. Finally, by the time they attained the same standards of living or even surpassed the most advanced nations, they were creating most of the technology they consumed.

That was the case of nations such as the United States, Germany, Japan, most other OECD countries, and even Taiwan, South Korea, and Singapore. The case of Hong Kong was different. It never moved on to the second, less still to the third stage, even though it developmentally surpassed most developed countries. During 1990 to 1997, at a time when the most advanced OECD countries created the large majority of the technology they consumed and spent about 2.7 percent of their GDPs on R&D, the colony spent only 0.08 percent. The other NICs, on the other hand, had largely increased their R&D expenditures, which had come to account for about 1.8 percent of GDP.[14]

That insignificant share of R&D expenditures was a mirror of the almost total dependence on technology imports. Many analysts have long been convinced that a developed nation needs to create at least a very large share of the technology it consumes if it is to continue to grow. The case of Hong Kong, however, demonstrated that it was not necessarily so. It even turned the whole idea upside down because while investing an infinitesimal amount in R&D it managed to attain a much faster rate of growth than OECD countries. While the economy during the 1990s averaged about 5 percent annually, in OECD countries the figure was just 2 percent. Japan, which was the largest spender in R&D among developed nations (3.0 percent from GDP) and which amazed the world with its innovations during these years, averaged an economic growth of about 2 percent. Hong Kong, on the other hand, which spent a much smaller share in R&D (0.08 percent) than even India, China, Thailand, or Indonesia, grew more than twice as fast. It is worth noting that during 1990 to 1997, measured in purchasing power parity terms Japan had on average a slightly lower GDP per capita than Hong Kong.[15]

While the thesis about attaining development and being obliged to be inventive in order to continue to grow does not match the empirical evidence, the theory of manufacturing does. Government support for this sector was much stronger in Hong Kong than in Japan and the rest of the OECD countries, and rates of factory output were also much faster. This was perhaps one of the most potent pieces of evidence indicating a direct causality linkage between this sector and technology. How else could the impressive development of technology during the second half of the twentieth century, but more still during the 1990s, be explained? By 1996, Hong Kong's R&D expenditures as a share of GDP was only 0.1 percent. Even though insignificant, it had grown considerably since the 1950s; at that time it was zero.[16]

The constantly growing expenditures in R&D had actually coincided with progressively decelerating rates of GDP. In the 1950s, GDP averaged about 12 percent per year and by the 1990s the figure was just 5 percent. The decreasing levels of government support and the decelerating rates of factory output could explain such a phenomenon.

How could Hong Kong have grown so fast during the whole 1950 to 1997 period without creating technology? How did it manage to grow economically and technologically faster than everybody else while spending absolutely nothing on R&D?

All of this apparently contradictory information becomes understandable if it is departed from the premise that manufacturing is the prime creator of technology. Since the Hong Kong government promoted the sector more enthusiastically during the second half of the twentieth century than any other government in the world, technology was inevitably driven to develop at a faster pace. Even during the 1990s, when it was as developed as the most advanced countries, it continued to promote the sector more decisively than OECD countries, and the economy was inevitably driven to absorb technology at a faster pace.[17]

Education has long been considered by economists and social scientists as a very important component of development, and numerous studies attempting to explain the fast growth of East Asia have claimed that the region's impressive results were largely the result of abundant investments in education. Even though the vast majority of the successful economies in the region invested abundantly in education, Hong Kong did not. The colonial authorities systematically neglected education and invested little in it.[18]

By the 1990s, even though the colony had reached a level of development that was at parity with the most advanced OECD countries, it spent only 3 percent of its GDP on education, while the average for the most developed OECD countries was about 7 percent. Investment was so low that even by 1996, about 85 percent of primary school students attended classes part-time because there were not enough schools, and buildings had to be shared. By then, the colony spent in primary education only one-third (per capita) of what Japan did.[19]

The low investments in education were experienced at all levels. The per capita investments that countries with similar levels of development made in secondary and university education were also much higher than in the colony. During the 1990s, for example, Hong Kong had about one university for every 800,000 people, while in the United States the figure was one university for every 80,000. Not only did Americans have ten times more access to higher education, but the quality of those universities was also superior to those in Hong Kong. If education were so important for growth, the colony should have attained a much slower rate of growth than Japan, the United States, and most other OECD countries. The fact, however, is that the territory vastly outperformed practically all OECD countries during these years.[20]

During the entire second half of the twentieth century, Hong Kong underinvested in education and not only did it outperform OECD countries, which during that period were by far the largest per capita investors in education in the world, but also the three other NICs. The colony attained a much faster GDP growth (9 percent) than OECD countries (4 percent), and the other NICs (8 percent), which also invested considerably more in education. Investing in education is surely a positive undertaking, but an absence of a correlation is very evident. This apparently paradoxical situation was actually one more piece of evidence in support of the manufacturing-growth thesis. Only if it is assumed that this sector is the prime creator of technology does such a phenomenon become logically explainable.

Most analysts believe that the development of technology is largely dependent on the educational level of a nation. If such a causality linkage existed, the colony would have attained a slower pace of technological growth during the second half of the century than OECD nations and the other NICs did. The fact is that the utilization and consumption of technology grew at a much faster pace.

The empirical evidence strongly suggests that the best way to promote technology is by allocating as large a share of overall resources to this sector, and that is precisely what the colonial authorities did. During the 1950 to 1997 period, support for the sector was stronger than in any other place in the world. Since the promotion of factories is not dependent on levels of education, but on supplying abundant incentives to the sector, it becomes therefore clear why, in spite of the low investments in education, the colony still managed to attain a spectacular growth of technology.

The Nonmanufacturing Sectors and Small Government

Other aspects of the economic history of Hong Kong also suggest a causality linkage between this sector and wealth creation. The fact that all primary-sector activities accounted by 1950 for only a very small share of the economy invalidates the argument that there was a transfer of resources from this sector to the others, which allowed for much larger investments in nonprimary activities. By the mid-twentieth century, the sum of agriculture, forestry, fishing, mining, and quarrying accounted only for about 9 percent of GDP.

The colony was by then a developing economy and many have argued that for fast growth to occur under those circumstances, it was necessary that there be a large transfer of resources from a large primary sector to the rest of the economy. That was not the case in Hong Kong, nor did such a situation occur in the preceding decades. Since the mid-nineteenth century, Britain never used the colony as a supplier of raw materials but as a trading post and as a naval bunkering station. Part of the reason was that only a tenth of the land was suitable for agriculture. Besides that the whole territory was just one thousand square kilometers.[21]

There was never a transfer of resources from the primary sector to the other sectors, and in spite of that the economy grew relatively fast during the first half of the twentieth century, and then at the fastest pace in the world during the second half. This situation, as well as that of other nations strongly suggests that developing countries do not need to invest abundantly in primary production so as to extract a surplus that allows for the financing of the other sectors. This situation demonstrated that even impressive growth was possible while starting at a very low level of development, as long as factories were abundantly promoted. The systemic failure of numerous developing countries to attain fast growth during the past half century, despite having a large primary sector and concentrating investment in this sector, gives further credence to this idea.

The development of events in the territory between 1950 to 1997 also demonstrated that other generally accepted beliefs were not consistent with the empirical data. Numerous governments throughout the world tried during this period to reduce income disparities, believing that the best way to achieve that goal was by means of redistribution policies. Minimum wage legislation and numerous other social welfare measures were implemented so as to transfer more resources to the share of the population with the lowest incomes.

These efforts, however, rarely delivered the desired results. Most developed, middle-income and developing countries experienced practically no improvement in income distribution during this half-century period. In several cases, it even worsened. In Hong Kong, however, practically nothing was undertaken to improve income distribution. Minimum wage legislation was never enacted, there was no protection against unemployment, and overall expenditures on social welfare during the whole period accounted for just 3 percent of government expenditure.[22]

However, income distribution improved considerably and actually more than in most other nations of the world. From having a terribly uneven distribution of wealth in the late 1940s, the colony attained by the end of the century a distribution similar to that of OECD countries. By then the territory had the highest number of billionaires per capita in the world, but the distribution of wealth had nonetheless become more equitable. More important still than distributing wealth was the fact that real incomes for the whole population had risen dramatically. Had income been distributed in exactly equal parts in 1950, then the whole population would have been poor. The poor just would have been less poor because the overall wealth of the colony was very low. By 1997, there were still strong income cleavages, but poverty as defined by the World Bank had been completely eliminated.[23]

The case of Hong Kong clearly demonstrated that eliminating poverty had practically nothing to do with minimum wage regulations and social welfare measures. That becomes more evident when it is considered that during the whole period government expenditure as a share of GDP was among the lowest in the world. In numerous OECD countries, social welfare expenditure accounted for as much as a third of the budget and overall government expenditure for about a third of GDP.[24]

In Hong Kong, however, public spending as a share of GDP averaged about 14 percent and social welfare expenditures accounted for only a fraction of the budget during the second half of the twentieth century. Seen from this perspective, the efforts of social engineering to eliminate poverty and improve income inequality were almost nonexistent. In spite of that, poverty was eliminated and income distribution was significantly improved. The case of Hong Kong demonstrated that the best way to rapidly improve the living conditions of the masses was by attaining very fast rates of economic growth, and much suggested that the means to achieve that goal was by supporting manufacturing as enthusiastically as possible.

During the whole period, it was manufacturing that systematically paid the highest wages—an inevitable phenomenon, considering the technology-creating nature of this sector. Since factory production grew at an unprecedented pace and employment in this sector grew faster than ever, it was inevitable that as a larger share of the population labored in the best-paying sector, wage inequality decreased. It is interesting to note that at a time when manufacturing grew the fastest (1950–1969), wage inequality decreased at the fastest pace, even though social welfare expenditures were much smaller than in the 1990s. In 1961, for example, nearly half of Hong Kong's work force labored in manufacturing while by 1997 the figure had dropped to just 15 percent.[25]

Liberal economists are right when they claim that social welfare expenditures are not conducive to growth, as well as when they assert that labor market distortions such as minimum wages and other regulations do not achieve the goals they pursue. Cross-country comparisons, however, reveal that these measures are not the main reason an economy is hindered from attaining fast growth. During the second half of the twentieth century, sub-Saharan Africa spent very little on social welfare as a share of GDP and labor regulations were few or went unobserved. Economic growth was nonetheless slow, which coincided with weak government support for manufacturing. While in Hong Kong the majority of state expenditure concentrated on promoting factories, in sub-Saharan nations it concentrated on promoting agriculture and other primary activities.

At the same time, Hong Kong is the ultimate proof that the bottom line for growth is not an absence of labor regulations and low social welfare expenditures. During the first half of the twentieth century, these were much lower and GDP figures were nonetheless only one-third as high as during the second half. During the second half of the nineteenth century, these distortions were totally absent, and in spite of that, economic growth was extremely slow. A long-term correlation was impossible to observe.

Most mainstream economists are also correct when they claim that a large government is by no means conducive to growth. Over half a century, Hong Kong attained the fastest GDP figures the world has ever seen, while government expenditure accounted for a very small share of GDP (14 percent). Even by the 1990s, at a time when it was as developed as OECD countries, it spent only 17 percent, while OECD countries spent 45 percent. The colony's expenditure was so small that even by the 1990s it was smaller than that of numerous developing countries such as India and Indonesia.[26]

During the 1950 to 1997 years state expenditure as a share of GDP rose more and more in OECD countries, paralleled by a progressive deceleration of the economy. Such a correlation seems to suggest a causality linkage and many economists came to such a conclusion. Even the colony showed a similar pattern of behavior because during this period the progressive enlargement of the territory's government expenditure was accompanied by a deceleration of the GDP figures. However, during the second half of the nineteenth century Hong Kong attained a very slow pace of economic growth while government expenditure accounted for only a fraction of what was spent between 1950 and 1997. This situation clearly indicates that small government was not the cause of the fast growth.

While this thesis fails to coincide with the facts, the one of manufacturing does. Support for manufacturing in OECD nations and in Hong Kong progressively decreased during the second half of the twentieth century, while during the 1850 to 1899 period the factory promotion efforts of the colonial authorities were very weak.

The situation of Hong Kong during the second half of the twentieth century was similar to that of the United States during the second half of the nineteenth century. In both cases, government expenditure as a share of GDP was very low and the bulk of it was utilized to promote factory production. In the United States, a federal budget that accounted for about 5 percent of GDP was mostly used to assist the producers of trains and other heavy manufactured goods, and in Hong Kong a budget of about 14 percent was mostly used to aid the producers of textiles and other light factory goods. In the United States, such an effort delivered an economic growth of about 5 percent annually and in Hong Kong GDP averaged about 9 percent. History suggests that governments do not need to spend a lot, but that they do need to concentrate their expenditures on manufacturing in order to attain fast growth.

During the whole 1950 to 1997 period, the colony's budget was almost always in surplus. Most mainstream economists have long argued that the maintenance of healthy public finances is important in order to attain a positive economic performance. The case of Hong Kong substantiates such an idea, although there is reason to believe that to a large extent the surpluses were the result of the fast economic growth and not the cause of it.[27]

Services grew very fast during the entire second half of the twentieth century. On average, they actually grew faster than GDP and even faster than manufacturing. They grew so fast that by the 1990s they accounted for about four-fifths of the whole economy. This development of events drove many analysts to conclude that services were the determinant variable propelling economic growth during this period. However, that is highly unlikely considering that services systematically attained a much inferior productivity performance than manufacturing. Also, as the share of services increased over time, the economy decelerated. As with OECD countries, the expansion of services actually correlated with a deterioration of the GDP figures.

The argument that services were the propeller of the economy was mostly utilized to explain the performance of the colony in the late twentieth century, because by then Hong Kong had attained a very high level of development and developed economies were supposedly driven by the service sector. When analyzing the 1990s and comparing the performance of the colony with the other NICs, it became evident that such an idea carried little weight. During these years Hong Kong attained the slowest economic growth of the four, even though it was the one that had by far the fastest service growth.

During 1990 to 1997, services grew about twice as fast as in South Korea, Taiwan, and Singapore, but GDP growth in the colony averaged about 5 percent while in the other three NICs it averaged about 7 percent. By then, Hong Kong had also a much larger service sector than the other NICs, but it was the one that performed the least well.

When all that is taken into consideration, it becomes evident that services could not have possibly been the determinant variable behind growth. When it is recognized that support for manufacturing was stronger and factory output was faster in the other NICs, it becomes clear that the manufacturing thesis is the only one that makes sense.[28]

The rapid growth of services seems rather to have been a by-product of the fast economic growth of the second half of the twentieth century, which in turn seems to have been the result of the strong support that manufacturing received. The historical evidence substantiates such a view. During the 1900 to 1949 period, for example, the authorities utilized the colony mostly as a service economy and in spite of that, services grew at a much slower pace than in the following half-century. In the second half of the nineteenth century, it was utilized exclusively for service purposes, yet services grew at a still slower pace. This all correlated with the differing levels of government support for manufacturing.

The same goes for construction. The rate of growth of this sector progressively accelerated over time, coinciding with the progressive increase in the factory promotion efforts of the authorities. It was during the 1950 to 1997 period that construction grew the fastest and it was then that factories were more enthusiastically promoted.

The strong dependence of construction and services on factory goods is also suggestive of them being a by-product of manufacturing. For example, the construction marvels that the territory achieved by the end of the century were only possible as a result of cranes, earth movers, trucks, cement, steel, glass, and dredgers. These marvels included two airports built in the sea, the longest bridge in the world, and numerous buildings constructed on reclaimed land from the sea. By the 1990s, more than half of the world's fleet of dredgers were in Hong Kong. These factory goods made all that impressive construction possible because it was these goods that were the depositories of technology.[29]

Exports seem to have been also a by-product of manufacturing. During the whole half century, exports grew faster than GDP and many concluded that exports were the driving force of economic growth. However, the correlation within this time period was not very consistent. During the 1960s, for example, exports averaged about 13 percent annually and GDP expanded by about 10 percent, while during 1990 to 1997, exports expanded at a faster pace (about 15 percent), but growth did not follow. GDP averaged only 5 percent per year. The correlation with manufacturing, however, is very consistent. Factory output was twice as fast in the 1960s than in the 1990s. The fact that exports grew very slowly in the second half of the nineteenth century, much faster in the following fifty years, and even faster during the 1950 to 1997 period further substantiates the view that they were a by-product of manufacturing.[30]

Many analysts, impressed by the entrepreneurial and workaholic spirit of the colony's population during the second half of the twentieth century, concluded that the fast growth was largely the result of the vast immigration since the late 1940s of the most business-driven people in China.

Because Hong Kong was a business center, Chinese capitalists fled in large numbers to it. If this had been the propelling agent or at least a major variable, then Taiwan should have attained a much slower rate of growth because most of the immigrants to this island were soldiers and politicians. The fact is that Taiwan was the second fastest growing economy in the world during the latter half of the century. While Hong Kong averaged a GDP rate of about 8.8 percent, Taiwan's rate was about 8.4 percent. The fact is that Taipei also promoted factories in probably the second most aggressive manner in the world during that period.[31]

Others asserted that the colony's impressive results were largely due to the absence of democratic institutions. Since interest groups were not allowed to lobby for benefits, a rational allocation of resources was not hampered. If that were true, then economic growth should have been as fast during the first half of the twentieth century, because at that time the same undemocratic institutions were in place. The fact is that economic growth was much slower. During the 1850 to 1999 period, the population was even less free to protest, demonstrate, and lobby for its interests, and GDP figures were even slower and almost stagnant. It was just impossible to present a consistent correlation. With manufacturing, however, the correlation was always present.[32]

Despite the impressive growth, there is much to suggest that economic growth in the territory could have been much faster during this half-century period. There is no evidence that the colony could not have maintained the 12 percent rate of growth of the 1950s up to the 1990s. A logical analysis of the empirical evidence leads to the conclusion that such a rate could have been easily maintained if factory producers had continued to be supplied with as many incentives as during the 1950s. If it was possible to channel as large a share of the territory's resources into the sector during the 1950s when the colony was poor, then it was even easier to do the same in the following decades. Budget surpluses were experienced after the late 1950s and government debt was nonexistent during most of this period. There were no logical grounds for not continuing such a policy.

However, the colonial authorities never supported the sector because of the belief that manufacturing was the prime generator of technology and the fundamental creator of wealth. It was a number of ideological motivations plus a confluence of historical circumstances that forced policy makers into the implementation of factory promotion policies. As those circumstances changed and became less pressing, in the government's vision of the world there was no longer a need to supply as much support to the sector.

Taiwan During the Early Years of Fast Growth

Taiwan was the second fastest growing economy in the world during the last half of the twentieth century. With the exception of Hong Kong, no other economy in the world attained a GDP rate as fast as that of the island. It was an impressive rate, not just by the standards of the rest of the world but also by the standards of the island itself. Economic growth during the 1950 to 1997 period was almost six times faster than during the first half of the century. While GDP averaged only about 1.4 percent in the first half, the figure averaged about 8.4 percent annually during the second half. (See tables in appendix.)

This drastic change of fortune in the performance of the island coalesced with a drastic increase in the level of government support for manufacturing. The support that the Japanese and Peking had supplied to this sector during the first half of the century was relatively weak, but a number of political circumstances during the 1950 to 1997 period pressured the Nationalist government so much that it felt forced to endorse a very enthusiastic factory-promotion policy. Factory output grew at an impressive pace and averaged about 11.8 percent per year, which was about six times faster than the rate attained during the first half of the century (2.0 percent).[33]

The case of Taiwan presented a paradox to mainstream economics because during the whole period market forces were greatly distorted. State companies accounted for a very large share of total output, trade barriers were high, cartels and monopolies were numerous, and the Central Bank was largely an appendix of the Finance Ministry. In addition, the financial system was mostly in state hands, the few private banks were heavily regulated, the exchange rate was fixed, and there were numerous controls on prices and labor. The large distortions that Taiwan endured between 1950 and 1997 strongly suggested that the fundamental variable explaining the fast growth was not any of those that have been regularly presented by mainstream economics.

However, if instead of presuming that free markets are essential for growth, it was assumed that government support for manufacturing is the bottom line of growth, there is no longer a paradox.

Taiwan, however, attained during this period a rate of manufacturing output that was slightly faster than that of Hong Kong, although it had a GDP rate that was slightly slower. The empirical data clearly shows that it was Hong Kong that maximized its manufacturing investments the most. This information suggests that even though mainstream economics has not identified the fundamental variable causing economic growth, it has nonetheless successfully identified all the other variables responsible for growth. It is logical to presume that if Taipei had practiced liberal policies while supplying as much support to manufacturing as it did, economic growth would have been faster.

The historical data very strongly substantiates such a thesis. As World War II came to an end, Japan had to relinquish possession over Taiwan and the Allies devolved the island to China. The Nationalist government of Chiang Kai-shek was by then convinced that China's best possibilities for development resided in agriculture. As during the previous decades in which he had ruled the country, practically no support for manufacturing was decreed. As Taiwan became again a province of China, these same policies were applied to the island and they coincided with recession. The economic situation became so critical that rebellion broke loose in 1947 and the government ordered a crackdown that resulted in the death of thousands.[34]

After their defeat in 1949, Taiwan became the place of immigration of most Nationalists. About two million mainlanders flowed to the island of six million and increased the population by about one-third. The defeat shocked Chiang and the Kuomintang leadership, which had long believed that the Communists would never win. It was clear from the start that economic policies had to be radically changed. The defeat forced the KMT to review its former policies. The fact was that giving investment priority to agriculture had miserably failed during the preceding decades. With the mass immigration, the island became overpopulated in the eyes of the Nationalists, and natural resources per capita were seen as scarce. This gave further credence to the idea that by concentrating on primary sector activities the country would not overcome underdevelopment.[35]

Since nobody in government circles believed that services could present an alternative for development, promoting manufacturing was seen as the only remaining possibility. There were also very large national security concerns that pressed the matter further in that direction. When Chiang and his followers fled to the island, they were convinced that it was only a temporary retreat, and that once they had regained military strength they would return to the mainland and defeat the Communists. Since from every perspective their position had become militarily disadvantageous, the only way to compensate for the Communists' larger armed forces was by possessing an abundance of weapons that were technologically superior. Taipei therefore decided to allocate a large share of the island's resources for the fabrication of armaments. During the 1950s, Taiwan's defense expenditure as a share of GDP was over 10 percent.[36]

While the Nationalists dreamed of invading the mainland, the Communists did likewise, and they assured that at the first opportunity they would overrun the island so as to finally unify the country under their rule. The threat that the Communists would launch an invasion and that they could blockade the island's foreign trade, drove Chiang to the conclusion that there was also a need to produce as much domestically as possible so that the economy would not be paralyzed in case of war. Since by 1949 Taiwan imported the bulk of the manufactured goods that it consumed, a large-scale effort to substitute those imports was undertaken.[37]

In order to attain the superior armament technology that it needed to compensate for the mainland's numerical superiority, Taipei needed to import much technology. The imports were expensive and the island possessed only a small level of wealth. Since its foreign exchange reserves were low, it attempted to attract large amounts of foreign direct investment (FDI) in the fields of manufacturing related to armaments and in numerous strictly civilian factory fields.

These balance-of-payments considerations also drove the government to institute a policy of export promotion. By 1949, primary-sector exports accounted for more than nine-tenths of total exports and were capable of generating only a small amount of foreign exchange. Since in Taiwan primary exports had never grown rapidly, the government decided that the best chance of quickly increasing exports was to aggressively promote manufacturing.

Therefore, as soon as he set foot on the island, Chiang ordered a large-scale mobilization of resources so as to channel a large share of them into manufacturing. The strong motivations from the shock of the defeat, the low level of natural resources in per capita terms, the pressing national security concerns, and the balance-of-payments worries forced the government to procure a very strong level of subsidies to the sector.

The bulk of the immigrants in 1949 were soldiers and politicians, but there were also some businessmen. Although the majority of the influx of manufacturing went to Hong Kong, a noticeable amount went to Taiwan, the bulk of it was in textiles. This situation drove the KMT to believe that the best civilian factory promotion policy was to concentrate on textiles. From their arrival the Nationalists promoted this field more than any other. The core of a loosely formulated plan for industrial development in 1951 was anchored on textiles.[38]

Abundant incentives were supplied to private factory producers, and numerous state firms in this sector were created. Lavish fiscal, financial, and nonfinancial incentives were offered to domestic and foreign capitalists. There were ample tax exemptions and tax reductions; grants and subsidized loans were supplied in large amounts; and land, ready-made factories, and utilities were made available at very subsidized prices. There were also guaranteed government purchases at prices that assured a profit as well as the supply of raw materials at below market prices. Other efforts to promote the sector were trade protection and the acceptance of cartels.[39]

A very large share of the budget was utilized to create state-owned factories as well as supply grants and other incentives to private manufacturers. The government also created several banks and supplied them with an abundance of funds. These state banks were ordered to lend the bulk of their assets to manufacturing at subsidized rates and with longer than normal periods of maturity. Commercial banks in their turn were heavily regulated so that they would allocate a large share of their loans to the factory sector. The Central Bank was also managed to serve the interests of the sector and a fixed exchange rate at an artificially low level was set so as to promote exports.[40]

Many have argued that Taiwan's fast factory output during the 1950s was largely the result of the immigration of textile makers and their machines from the mainland, but the fact is that the majority of textile output during this decade came from newly created state firms. Even the output that came from the factories of the immigrants was largely the result of the incentives. It is highly unlikely that without the grants and the other incentives they would have bought as much new machinery and equipment as they did.

In most of the other manufacturing fields, the participation of the state was even more pronounced. Fuels, chemicals, plastics, electricity, metals, processed food, and glass were produced mostly by public companies in the 1950s. The bulk of the private sector's output in all of these fields was the direct result of the abundant incentives the government offered. Most other manufacturing fields also experienced a similar development as the direct and indirect support of the state was very clearly appreciated. The rapid development of every single field is practically unthinkable without the assistance they received from Taipei.

In the 1950s, public enterprises accounted for over half of manufacturing output. The government was in a hurry to increase factory production, which is why it created so many state firms. The KMT leadership was convinced that this was the only way to achieve that goal. They were wrong. The case of Hong Kong very clearly demonstrated that there was not a need to create a single state company to generate a fast factory output. As a matter of fact, manufacturing grew much faster in the colony during this decade than in Taiwan. Hong Kong demonstrated that if the government offered massive incentives to the private sector, capitalists would invest extensively in manufacturing. It all depended on the level of support.[41]

High trade barriers and high levels of cartels also characterized the 1950s. Even monopolies were created in relatively large numbers. As with state firms, the goal of the state was to facilitate production to manufacturers and Taipei thought that by reducing competition it would increase the incentives for investment and accelerate the pace of output. Here again these policies failed to deliver the desired results as these practices hampered productivity and reduced efficiency. During the 1950s, Hong Kong attained a faster rate of factory output and GDP than did Taiwan, while applying absolutely no trade barriers and being largely free of cartels. Hong Kong also attained higher levels of efficiency.

State banks were numerous during this decade, were characterized by low levels of efficiency, and accounted for the bulk of bank assets. Throughout history, Chinese bankers had always proven reluctant to lend to manufacturing. That led Chiang to believe that only if the state had direct control of financial assets could the manufacturing sector receive ample capital. The case of Japan, however, demonstrated that higher levels of financial efficiency could be attained if the bulk of assets was in the hands of private banks, and that it was possible to simultaneously allocate the majority of those assets to manufacturing by means of incentives and regulation. The case of Hong Kong demonstrated that even higher levels of banking efficiency could be attained if foreign private banks were allowed in and if competition was fostered.[42]

In spite of the numerous policy errors, there was fast economic growth that averaged about 8 percent annually, and that coincided with a very strong level of government support for manufacturing as well as with a rate of factory output that averaged about 12 percent per year. Had Taipei not distorted market forces so much, it is evident that the same factory rate would have delivered a faster pace of economic growth. Policy makers also could have increased considerably the level of support to the sector so that factory production would have grown at even a faster pace. In Hong Kong, policy makers did just that and the sector grew by 16 percent annually.

There were strong political, national security, and balance-of-payments concerns pushing the government to support manufacturing. However, Chiang and most in the KMT leadership had for so long believed that investing in agriculture was the best for development that it was hard for them to disassociate completely from such an idea. Had they not allocated a large share of resources to agriculture and other primary activities, and would have used those resources to promote factories, manufacturing would almost surely have grown much faster and the economy as well.[43]

Many argued that Taiwan owed its fast growth during this decade to the large flows of American aid. Aid was indeed considerable and it accounted (economic and military) for about 15 percent of GDP. Washington was so determined to prevent the island from falling into communist hands that the U.S. Navy was even assigned to patrol the seas separating it from the mainland in order to curtail an invasion. There was a great amount of aid, but it is worth noting that American aid to Germany after World War II was only 2 percent of GDP, and was terminated in 1952, yet economic growth in the Federal Republic of Germany during the 1950s was exactly the same as that of Taiwan. The aid Taiwan received was seven times larger in per capita terms than what Germany received and it had nothing to reconstruct. If aid was fundamental or essential for growth, the island should have attained a much faster rate of growth than Germany.

Also worth noting is that economic aid was significantly reduced in the early 1960s and terminated in 1965. Economic growth, however, was faster in the 1960s than during the preceding decade. In the 1970s, when the island received no economic aid and considerably less military aid, GDP figures were also faster than in the 1950s.[44]

Much evidence suggests that the aid did contribute to the fast growth but in a marginal way, and only because part of it was utilized to finance factories producing civilian and military goods. This was indeed one way by the government to procure foreign exchange and which allowed it to import state-of-the-art technology. The sophisticated machinery and equipment that was utilized to produce military uniforms, for example, was also utilized to fabricate civilian garments. Several of the machines that were imported to make light and heavy artillery weapons were also utilized to make metals and machine tools. To acquire foreign exchange and superior technology, the government also promoted FDI and borrowed from commercial banks from OECD countries.[45]

There are many who argued that the island's fast growth during the 1950s was largely the result of the developed agriculture and infrastructure the Japanese left behind as well as the large amounts of gold and other valuables that the Nationalists took with them as they left the mainland. If that had been true, at least some growth should have been observed between 1945 and 1949, because by then the developed agriculture and infrastructure were already there. The fact is that the economy contracted. Japan, which had been massively destroyed, nonetheless experienced fast growth during those same years. This all ran in tandem with an absence of support for manufacturing in Taiwan and with very enthusiastic promotion efforts in Japan. As for the gold that was hoarded from China, it was only a small amount relative to the vast amounts of imports the island undertook during the 1950s. It is highly unlikely that these variables played a determinant role in growth.[46]

Historical Overview

During the 1960s, national security concerns were somewhat lower as China began to relinquish the idea of invading the island, although fighter-plane clashes over the South China Sea ended only in 1969. There were also fewer balance-of-payments concerns as trade deficits began to turn into surpluses and foreign exchange reserves began to rise. However, there were still strong pressures from these two fronts.

More important still was that the idea about the priority status of agricultural investment for the attainment of development had been significantly diluted among the leadership, because the 1950s was a decade when it was not applied and fast growth was nonetheless observed. On top of that, agriculture as a share of GDP had significantly shrunk, which subtracted further weight to the agricultural lobby.

In the 1960s, Chiang began to abandon the idea of returning to the mainland and defeating the Communists militarily. Since that seemed no longer possible, his still very strong rivalry with Peking was substituted by the desire to outperform China economically and technologically. Since technology and economic prowess were fundamentally associated with factory goods, manufacturing was seen as the only alternative for achieving that goal.

As a result therefore of strong national security concerns, balance-of-payments worries, the growing suspicion that growth was associated with factory production, and above all the increased desire to outperform China, the government increased the level of support to the sector.

During the 1950s, Taipei had mostly promoted textiles because it was convinced that this was the factory field that the Chinese were better qualified to produce. The other fields that it promoted were mostly those that the Japanese had established during their colonial occupation, and because they were already there, they were also seen as goods that could be produced by the island's population. When the strong factory-promotion policy was launched in the early 1950s, most foreign and domestic economists argued that it was highly unlikely that Taiwan would ever succeed in producing even these goods in large amounts. Since the island or any other region of China had never managed to produce factory goods in large amounts and in an internationally competitive way, they concluded that Taiwan was condemned to forever fail on that front.

This vision of things was rapidly proven wrong as the island succeeded during the 1950s in producing these goods in large amounts and in an internationally competitive way. With the new desire to demonstrate that Taiwan could outperform China on all fronts, the KMT leadership decided in the early 1960s to develop numerous new manufacturing fields. Until the 1950s, light manufacturing had predominated, and by then China had significantly developed its heavy manufacturing domain. Producing large amounts of heavy goods therefore became a political necessity. In addition, there was a growing belief among the KMT leadership that the long-term growth of the economy was impossible without the development of the heavy factory domain. It was believed that capital goods were the key for the overall development of manufacturing.[47]

As Taipei launched this new effort, numerous foreign and domestic analysts predicted that it would turn into a failure and that Taiwan would probably never succeed in producing these goods even in relatively small amounts. Once again, these people were proven wrong as in just a few years the island not only managed to produce vast amounts of these goods but also succeeded in producing them in a internationally competitive way. This situation coincided with a very strong level of government support for heavy manufacturing.

Since much of the support for light manufacturing was sustained and that for heavy manufacturing was increased significantly, the sector as a whole was more decisively promoted in the 1960s than before, which correlated with a faster rate of factory output, averaging about 14 percent annually. GDP reflected this, and was also faster than in the 1950s, averaging about 10 percent per year.

Not a single one of the numerous state factories was privatized and several more were actually created during the 1960s. However, in part as a result of the American disapproval of state companies and in part because of the lower efficiency of government firms, Taipei began to foster more the private sector. Fiscal, financial, and nonfinancial incentives were therefore significantly increased. Unfortunately, the government continued to believe that factory production was also promoted with trade protection and cartels. The cases of Hong Kong and Singapore during that same decade once again clearly demonstrated that these competition-hindering practices were of no assistance to the sector and that, on the contrary, they were actually harmful.[48]

A very positive way in which the government's efforts stimulated factory production was by absorbing a large share of the research and development (R&D) costs. Since the 1950s, but even more in the 1960s, the government established R&D organizations to promote technological upgrading. These organizations were ordered to develop new technology and to supply it to manufacturers at subsidized prices.

Another positive undertaking of Taipei was investing significantly in the educational fields directly related to manufacturing. As in Japan, the strong desire to elevate the educational level of the population rapidly was not acted on indiscriminately. There was no across-the-board promotion of all educational fields. The efforts were fundamentally concentrated in factory-related fields such as mathematics, engineering, and natural sciences.

During the Vietnam War, the United States bought a considerable amount of goods from Taiwan and many argued that this war-induced demand plus the still large military aid that Washington supplied was largely responsible for the fast growth. That surely helped, but it is evident that it was not the bottom line. The fact that during the 1970s no war-induced demand came into being and the island nonetheless attained the same double-digit economic growth as in the 1960s clearly points in that direction.

Also worth noting is that Taiwan would have not been capable of selling to the United States had it not developed a large production capacity, and that would have not occurred had it not seriously promoted manufacturing. The contribution of the war seems to have been in driving Taipei to increase more its factory-promotion efforts. The American military aid seems to have operated in a similar way. The money was utilized to acquire American machinery so as to fabricate armaments and related goods. It had a positive effect, but only because it was utilized to develop the already approved factory-promotion policy. Its positive effect was also felt as the technology utilized to fabricate weapons was soon transferred to civilian uses.

During the 1970s, Taiwan's import bill rose considerably as the price of oil and other commodities increased by a large margin. The island's currency also experienced an effective devaluation in 1971 and 1973 as it was pegged to the U.S. dollar, which was devalued in those years. The devaluation therefore raised even more the import bill and elevated more costs of production. On top of that, since the early 1960s, exports had been experiencing rising protection in OECD countries, wages were rising rapidly and outpricing labor-intensive production, and competition was becoming stronger as South Korea and Singapore began to export the same goods. To make matters worse, there was political de-recognition of the United Nations, Japan, and several other countries in favor of China which caused considerable emigration and capital flight. Most of the immigrants belonged to the intellectual and business elite. Worse still was that inflation rose by a large margin. From having averaged just 2 percent annually during the 1960s, prices jumped to 10 percent in the 1970s.[49]

With so many negative factors interplaying in the island's economy, it would have been only natural for GDP figures to decelerate considerably or perhaps even contract. The fact is that economic growth remained as strong as during the preceding decade and that there was a continuation of a very strong factory-promotion policy. During the 1970s, there were still strong national security concerns and above all the very entrenched desire to outperform China economically and technologically. Since agriculture had shrunk further during the 1960s, the idea that this domain could act as an economic propeller became somewhat obsolete and thus the people within the KMT who favored manufacturing were confronted with a lower level of opposition. These ideological motivations therefore drove the government to enthusiastically support civilian and militarily manufacturing.

By then, the island was producing not only light and heavy artillery pieces but also helicopters, radar equipment, and fighter planes. The quadrupling of the price of oil in 1973 and 1974 shocked the Taiwanese economy and, after having grown by 10 percent annually between 1970 and 1973, the economy tumbled to just 1 percent in 1974. However, because of its strong motivations, the government decided to respond with an even stronger level of support for manufacturing than before and the sector's rate of output averaged 11 percent in the remainder of the decade. Priority was given to electrical goods, electronics, machine tools, petrochemicals, and computers. All in all factory output averaged about 14 percent per year during the decade and GDP averaged about 10 percent.

During this decade, the majority of the budget continued to be utilized to promote factory production. The majority of bank assets were in state hands and these banks were again ordered to lend mostly to manufacturing at subsidized rates. Private banks continued to be heavily regulated so that they would mostly concentrate on the sector. During the 1970s, about 70 percent of bank loans overall were used to finance manufacturing.[50]

Fiscal and nonfinancial incentives continued to be abundant and although no state company was privatized and several government factories were created, the bulk of the new manufacturing companies were private. Unfortunately, the government continued to believe that cartels and trade barriers had a positive effect on the sector. The large number of state enterprises, trade barriers, and cartels once again suggested that a variable different from the ones presented by mainstream economics was ultimately responsible for the fast growth. These competition-hindering practices were evidently policy errors and the only way in which such errors could have still allowed for a fast growth was if support for manufacturing was the fundamental variable explaining growth.[51]

In the 1980s, there were still strong national security concerns, but they were noticeably lower than formerly, largely as a result of China's partial renunciation of communism. The growing trade surpluses of the preceding years had also largely diminished the balance of payment concerns. The desire to outperform China economically and technologically was still strong, but since the island had done precisely that during the preceding three decades, the government felt it no longer needed to make as many efforts as formerly. The developmental gap with the mainland had grown considerably and as the leadership felt partially satisfied with its success, it concluded that support for manufacturing no longer needed to be so intensive.

As a result, therefore, of a decrease in the motivations driving Taipei to promote factories during the 1980s, fewer subsidies were supplied to the sector. This in its turn correlated with a slower rate of factory output that averaged about 11 percent annually and with a slower rate of economic growth that averaged about 8 percent.[52]

There was no privatization, but state factories expanded at a slower pace. The bulk of new enterprises (in manufacturing or in any other sector) were private. During the 1980s, there was some liberalization of foreign trade and some deregulation of the economy. The exchange rate became also more market-driven as in 1980 the fixed exchange rate was supplanted by a tightly managed float system.

The second oil shock took place in 1980 and Taipei responded similarly to the way Tokyo did with the first oil shock. The government thought that the oil problem would get solved if the island produced mostly non-oil-intensive goods, so policy makers concentrated on promoting factory fields such as semiconductors, computers, robotics, machine tools, telecommunications, and pharmaceuticals.[53]

However, notwithstanding the liberalization and deregulation of the economy and the shift to non-oil-intensive production, the economy decelerated significantly. Growth was still strong, but everybody would have benefited more if double-digit growth had continued. It is also worth noting that during the 1980s oil prices on average dropped considerably and inflation fell significantly. The rate of inflation during this decade (5 percent) was only half as fast as that of the preceding ten years. With all the favorable measures the government took and all the favorable circumstances of this decade, growth should have accelerated or at least remained at the same pace as in the 1970s. The fact is that there was a noticeable deceleration, and that coincided with a noticeable reduction in the factory promotion efforts of the state.[54]

The 1990s

In the 1990s, national security concerns became lower as tensions with China diminished. Peking's increasing market liberalization of the economy in the 1980s and its friendly approach to capitalist countries did much to reduce tensions. That effort was further pursued in the 1990s and there was even a massive flow of Taiwanese FDI to the mainland.

Notwithstanding the rapidly growing economic relations, Peking continued to threaten the island with invasion if it declared independence. During the first presidential elections in Taiwan's history in 1996, China carried out numerous intimidating military maneuvers and insinuated that an invasion could take place if the island pursued its democratization efforts. Military threats were weaker than before, but they were still significant. Armament production was therefore still great, although less than formerly. The policy of producing domestically as much as possible so that if a war or a commercial blockade took place the island would not experience an economic paralysis was pursued, although not as enthusiastically as formerly.

By the late 1980s, Taiwan had the second largest foreign exchange reserves, and constant large trade surpluses with the rest of the world had for long become the norm. As a result, balance-of-payments considerations no longer drove Taipei to promote factories as much.

With the rest of the world it had surpluses, but with Japan, which was its main trading partner, it had constant deficits. This situation irritated the government because of the belief that the island should have bilateral surpluses with every trading partner. However, the colonial legacy also played a role. During Japan's half-century rule over the island, the Japanese always viewed it as inferior in every respect and incapable of ever matching Japan. Taipei saw the trade surpluses as confirmation of Japan's colonial vision. Since the authorities were convinced that such a vision was false, they took numerous measures to prove it false and supplied abundant incentives to the factory fields that accounted for most Japanese imports. The subsidies that this motivation generated flowed mostly into fields such as hard disc drives, cathode-ray tubes, and computers.[55]

By the 1990s, Taiwanese policy makers were also intended on demonstrating that the whole idea of Japanese superiority was false. Since the matter of superiority was more evidently expressed in technology, and technology materialized mainly in manufacturing, support was channeled mostly for this sector. This motivation had its origins in the 1950s, but always remained a minor one. As the developmental gap with Japan rapidly shrank after the 1970s, the motivation also lost weight.

There was also the continuation of the desire to outperform China economically and technologically. However, Taiwan had outperformed China again during the 1980s, and felt more satisfied with its achievements, so it concluded that it no longer needed to make as many efforts.

During the 1990s, there was strong government support for manufacturing, but since the national security concerns, the balance-of-payments worries, the desire to outperform China, and the desire to catch up with Japan were lower than before, the support was lower. A lower level of subsidies coincided with a slower rate of factory output, which averaged about 8.0 percent annually. This was reflected in a slower rate of economic growth that averaged about 6.4 percent.

During the 1990s the Industrial Development Bureau under the Ministry of Economic Affairs concentrated on promoting technology-intensive fields such as advanced consumer electronics, advanced industrial materials, aerospace goods, communication equipment, medical equipment, pollution control instruments, precision machinery, automation equipment, semiconductors, specialty chemicals, and pharmaceuticals.[56]

Companies in these fields received tax exemptions and tax reductions, large government grants, subsidized loans from government banks, subsidized technology from government laboratories, guaranteed state purchases at favorable prices, and numerous other nonfinancial incentives. Large financial infusions were also made by the state by means of large state share holdings in these companies. During 1990 to 1997, the majority of bank assets were still in state hands and these banks were ordered by Taipei to allocate the majority of their funds to manufacturing.[57]

During these years, trade barriers were lowered significantly and the economy was further liberalized and deregulated. Practically no new state companies were created and the bulk of the support went to the private sector. The private sector's share of manufacturing output and of the economy as a whole was by far the largest since the 1950s.[58]

Also worth noting is that investment in infrastructure was much greater than ever before. As a share of GDP it was much larger than before, in particular since early 1993 when a gigantic infrastructure program was launched which in U.S. dollar terms was one of the largest in the world. There was a further drop in oil prices, and a significant appreciation of the Taiwan dollar, which reduced even more the price of importing oil and other primary commodities.[59]

In the early 1980s, outflows of Taiwanese FDI began as wages had risen rapidly in the preceding decades and had outpriced several labor-intensive fields of production. Since wages continued to rise rapidly, the outflows became even larger during the 1990s. Most policy makers and economists on the island saw this event positively and even encouraged it as a way to reduce production costs for Taiwanese companies.[60]

However, in spite of the liberalizing and deregulating of the economy, the large investments in infrastructure, the drop in oil prices, the drop in commodity prices, and several other favorable circumstances, the economy experienced a noticeable deceleration. With so many useful measures that Taipei applied and so many favorable circumstances that took place, common sense demanded that an acceleration of the GDP figures would occur. The fact, however, is that the economy moved in exactly the opposite direction. Such a paradox becomes only understandable when it is departed from the premise that manufacturing is the bottom line of economic growth and if it is also assumed that the rate of this sector depends on the level of government support.

On the other hand, it is also worth considering that during the 1990s there was practically no privatization and that state companies accounted for about one-sixth of GDP. It was only in 1994 that the first privatization since the departure of the Japanese took place. Only two companies were privatized and the change of ownership was not even complete. A private majority share holding was allowed, but the government retained large stakes.[61]

In spite of the trade liberalization, tariff and nontariff barriers continued to be very high by OECD standards. Notwithstanding the deregulation, there was still more intervention in Taiwan's economy than in OECD countries. With the numerous market distortions that were obviously doing no good to the economy, it becomes extremely hard to understand how the island attained a rate of growth thrice as fast as that achieved by OECD countries. Only when it is assumed that support for manufacturing is the bottom line of growth does the empirical data become coherent.[62]

The only logical reason manufacturing would be the bottom line of growth would be that it is fundamentally responsible for the creation of technology. If that is true, then a correlation should exist between levels of state support for the sector and the fluctuations of technology. The correlation is clearly observed in the case of Taiwan.

For centuries, technology barely made any progress, until the early twentieth century when it began to grow at a much faster pace. This coincided with practically no government support for the sector during all those centuries and with a relatively large factory promotion effort of the Japanese during their rule over the island. They took over the island in 1895 and promoted factory production until the early 1940s. During the 1940s, technology contracted and so did manufacturing production.

Then in the 1950s, the KMT government subsidized the sector much stronger than the Japanese and technology developed at an unprecedented rate. In the 1960s, factory output grew faster and so did technology. In the 1970s, the factory promotion efforts of Taipei were sustained and technology grew at about the same pace as in the 1960s. In the 1980s, the rate of manufacturing production decelerated some and the same thing occurred with technology; and in the 1990s, technology decelerated further, as did the rate of factory output.

It is worth noting that the capacity of the island to generate its own technology grew extensively during the second half of the twentieth century. During the 1950s, Taiwan imported practically all of the technology it consumed, but by the 1990s it was producing a large share of it and it was also able to match and even outperform the United States and Japan in several fields. While in the 1950s R&D expenditures as a share of GDP were close to zero, by the 1990s they had grown to about 2 percent of GDP. In 1980,

Taipei founded Hsinchu Science City in an effort to foster technology-intensive production and to lure back nonresident Taiwanese scientists. In this science park the government injected capital in every firm by taking a 49 percent equity stake in each venture and also by supplying other incentives.[63]

By the mid-1990s, about 13,000 researchers labored in this park, almost as many as in Japan's Tsukuba Science City. Taiwan's R&D efforts had grown rapidly and by the 1990s they were comparable to those of OECD countries. However, after the 1980s those rising efforts correlated with a decelerating pace of economic growth. This phenomenon was also observed in Japan and in most OECD countries. A rising share of R&D expenditures from GDP and growing technical sophistication coincided with a decelerating pace of technology and a slower rate of economic growth.

Interestingly, in the 1990s, about half of the companies operating in Taiwan's Hsinchu Science Park were run by Taiwanese scientists returning from the United States. They had studied at the best American universities and gained experience in the laboratories of companies such as General Electric, AT&T, and those in Silicon Valley. The scientists were brilliant and more plentiful than ever before, but the economy nonetheless decelerated.[64]

In addition, education had improved remarkably over the years. The government invested extensively in this area during the second half of the twentieth century and by the end of this period the Taiwanese were among the best educated in the world. GDP figures were precisely at the lowest level at a time when the population had attained the highest levels of education. The majority of the Taiwanese who studied abroad in the 1950s and 1960s (in particular in the United States) tended to stay there after their university graduation, thus depriving the island of the best qualified workforce. By the 1990s, however, the large majority were returning. From every perspective, the island was better endowed by the end of the century to attain faster growth, but the fact is that growth was slower.[65]

Such a paradox can be explained only if it is assumed that manufacturing is the essential creator of technology. Analysts have tended to believe that technology is fundamentally embodied in the level of education of a population or in the scientific level of a population, but this does not add up with the facts. However, if it is assumed that technology is embodied in factory goods and that technology is created or reproduced while it is being manufactured, then it becomes understandable why in the 1990s technology expanded at a slower pace and why the economy did likewise.

Technology, Investment, Savings, Exports, Authoritarianism, and Small Government

Since the creation or even just the reproduction of technology is extremely difficult to achieve, the effort is inevitably predisposed to be highly investment-intensive. Precisely because of that, government support becomes indispensable. Without a significant reduction in the huge production costs of the sector by means of incentives, investment automatically flows to the other sectors where the costs are much lower and the risks are much smaller.

Seen from that perspective, it becomes understandable why a lower level of government support in Taiwan during the 1990s coincided with a deceleration in the rate of technological growth. What was important was not how much of the total technology consumed was domestically created, or if Taiwan could technologically outperform the most advanced countries in the world. Neither was it essential to have a highly educated labor force, or to possess top-level scientists. Apparently, what ultimately mattered was how much support there was for manufacturing. During the 1990s, for example, South Korea and Singapore attained a faster growth in technology as well as a faster rate of economic growth, and that correlated with a stronger support for the sector and a faster rate of factory output.

If manufacturing is the fundamental creator of technology, and since technology requires huge investments, a correlation should exist between the levels of state support for the sector and the differing levels of investment. The historical evidence reveals such a correlation.

Until the late nineteenth century, for example, investment as a share of GDP was close to zero, which coincided with an almost complete absence of support for the sector. During the whole nineteenth century, Peking was deliberately attempting to suppress manufacturing. Then, during the first four decades of the twentieth century, the Japanese occupation authorities promoted factories in a relatively decisive way and all of a sudden investment rose to unprecedented levels, averaging about 8 percent of GDP. In the 1940s, support for the sector largely vanished and investment contracted. After 1950, the Nationalists decided to promote factories far more decisively than the Japanese had done, and immediately investment rose to higher levels. During the 1950 to 1997 period, investment as a share of GDP averaged about 25 percent.[66]

A similar phenomenon occurred in the other NICs. Government support for manufacturing was much stronger during the second half of the twentieth century than during the previous fifty years, which correlated with a much higher level of investment. Investment as a share of GDP averaged about 9 percent for these three economies in the first half of the century while during the second half the figure was of about 27 percent. Rates of factory output were about three times faster during the second half.

If manufacturing is the prime creator of technology and therefore wealth, a correlation should also exist between levels of government support for the sector and levels of savings, because the speed with which wealth is created largely determines the levels of savings. Until the late nineteenth century in Taiwan, savings as a share of GDP systematically remained close to zero, which coalesced with an almost complete absence of subsidies for the sector. Then in the first four decades of the twentieth century, the factory promotion efforts of the colonial authorities were far superior to anything previously experienced and all of a sudden savings as a share of GDP rose considerably, averaging about 6 percent. In the 1940s, factory output contracted and the level of savings dropped considerably.[67]

Since 1950, Taipei began to promote factories in a much more decisive way than before and savings began immediately to rise. Strong support for the sector was sustained during the second half of the century and savings as a share of GDP averaged about 23 percent. A similar phenomenon occurred in the other NICs. Support for manufacturing was much stronger during the second half of the century than in the previous fifty years and the share of savings from GDP was much higher during the 1950 to 1997 period. While the average share of savings for Hong Kong, Singapore, and South Korea was about 10 percent during the first half of the century, the figure jumped to about 30 percent during the second half.[68]

Many economists have concluded that economic growth is largely dependent on the level of savings. A larger pool of savings supposedly allows for a larger level of investment, and there is indeed a long-term correlation between savings and growth. However, in the short term, the correlation has at times broken down. Savings in Taiwan during the 1950s, for example, were of only 6 percent of GDP even though economic growth averaged about 8 percent. Between 1990 and 1997, on the other hand, the share of savings was of about 28 percent even though economic growth was only 6 percent. The absence of a constant correlation suggests that there is no causality linkage between savings and growth, and that savings are actually an effect of growth and not the other way around. The historical evidence also leads to the conclusion that savings are a by-product of manufacturing.

The long-term development of exports leads also to the same conclusion. Taiwan attained an impressive export growth during the second half of the twentieth century and many analysts concluded that exports acted as a propeller for the economy. The average export growth of the period was very strong and fast enough to have pulled the economy, but the figures were not consistent. During the 1950s, for example, exports grew more slowly than

GDP (6 percent–8 percent respectively). Then, during the following two decades, they grew on average by about 28 percent per year and the economy grew only by 10 percent annually. First, it was too weak to pull the economy and then it grew disproportionately too fast relative to the GDP figures.[69]

Exports seem also to have been the result of growth and not the other way around. During the 1950s, for example, because of extreme national security concerns Taipei gave strong support to manufacturing in order to produce an abundance of weapons and civilian goods. Since an import-substitution policy was being applied, strong support for factories delivered fast growth even though relatively little was exported. In the next two decades, the government radically changed its policy and enthusiastically endorsed export promotion, which was accompanied by a slightly stronger support for factories. Stronger support delivered a slightly faster rate of GDP together with an explosion of exports.

Also worth noting is that the island's exports were increasingly put under import restraints in North America and Western Europe from the 1960s. Exports nonetheless continued to grow at an impressive pace. During the second half of the twentieth century, many developing countries complained that growing nontariff barriers in OECD countries hampered their possibilities for growth. The case of Taiwan, however, demonstrated that it was possible to export at a double-digit pace even under rising protection. The other NICs experienced a similar situation. Their goods were increasingly blocked, but they nevertheless managed to export to OECD countries in large amounts. By 1990 the four NICs accounted for about two-thirds of all the manufacturing exports that the total of developing countries made.[70]

The case of the NICs demonstrated that the bottom line for a successful export effort was government measures that concentrated on promoting factory production. Without strong support for the sector, the exports were simply not possible. The fact that during the second half of the twentieth century the immense majority of exports were factory goods makes it very evident where the base of the export success resided.

Numerous analysts have also wrongly taken other aspects of the Taiwanese economy as being essential for growth. During the 1950 to 1997 period, the island was characterized by the predominance of small companies. By the 1990s, about 95 percent of output came from firms that by international standards were classified as small or medium size. This was a very particular trait of Taiwan.

Large companies for centuries have been regarded with suspicion in most countries of the world because of their capacity to influence policy making for the sake of promoting their own interests. Many analysts therefore concluded that the predominance of small firms was largely responsible for the fast growth. However, if that were true, then Japan and South Korea during that same half-century period would not have attained fast growth because these two economies were largely characterized by their abundance of extremely large companies. Hong Kong and Singapore, which had a combination of large and small companies, also attained impressive rates of growth. The absence of a correlation clearly indicates that a causality linkage between small firms and growth does not exist.[71]

Taiwan experienced a very authoritarian form of government during most of this period and many concluded that such a political regime contributed significantly to rapid growth because it was not manipulated by special interest groups, and resources were therefore not allocated incorrectly. If that were valid, then the island should not have experienced fast growth between 1990 and 1997, because by then a very Western-like democratic system had taken root. With the death of Chiang Kai-shek in the mid 1970s, authoritarianism was considerably reduced. Economic growth, however, did not decelerate during the second half of this decade. It was actually faster than during the first half. In the late 1980s Chiang's son died and Western-style democracy began rapidly to take root. By the 1990s, interest groups had become highly vocal and demanding, but the economy continued to grow quickly.[72]

The other NICs also experienced differing forms of authoritarianism during the second half of the twentieth century, which at first glance seems to reinforce the authoritarian-growth thesis. However, once the data is thoroughly analyzed it becomes evident that such a thesis is weak. South Korea, for example, experienced events similar to those in Taiwan. Authoritarian government largely crumbled in the late 1980s and during the 1990s a Western-style democracy was in place. The new democratic institutions were so strong that even the ex-military rulers were put on trial for their misdeeds and sent to prison. That was largely the result of the lobbying power of interest groups. Something so drastic as the jailing of ex-presidents did not occur or come close to occurring in Taiwan. However, South Korea attained an even faster rate of GDP growth between 1990 and 1997 than Taiwan did.

Authoritarianism was actually stronger in the four NICs during the first half of the twentieth century, and economic growth was nonetheless much slower than during the second half. During the 1850 to 1999 period, authoritarian rule was even more encompassing and in spite of that the economy expanded at an even slower pace.

By the mid-1990s, about sixteen studies attempting to measure the link between forms of government and growth had been completed. Three of these cross-country efforts found a correlation between authoritarianism and growth. Three other studies, however, found an opposite correlation in which democratic institutions had paralleled fast growth while the rest of the studies found no conclusive results. Ten studies were unable to determine if an authoritarian form of government was positive or negative for the economy.[73]

The historical evidence clearly indicates that there is no causality linkage between forms of government and economic growth. A variant of this argument nonetheless asserted that the advent of democracy in Taiwan in the late twentieth century was responsible for the deceleration of the economy as interest groups had increased their demands. In 1995, for example, the island inaugurated a national health insurance program. Social welfare expenditures indeed increased considerably during the 1990s and it was evident that these expenditures were subtracting the possibilities for larger investments, but much suggests that this was not the cause of the deceleration. During the first half of the twentieth century, for example, social welfare expenditures as a share of GDP were only a tiny fraction of those of the 1990s, yet economic growth was only one-sixth as fast. In the second half of the nineteenth century, social welfare expenditures were completely nonexistent, yet the economy remained totally stagnant. The absence of a correlation clearly revealed an absence of a causality linkage.[74]

Largely as a result of the increased social welfare expenditures, government spending as a share of GDP became larger than ever during the 1990s and came to average about 18 percent. Many therefore argued that this was the cause of the deceleration. A long-term analysis, however, reveals that such an explanation does not match the facts. The share of government expenditure during the second half of the nineteenth century, for example, was only a tiny fraction of that of a century later in the four NICs and growth was nonetheless very slow.

Despite the large increase in the size of the state, during the latter half of the twentieth century, state expenditure was one of the smallest in the world, averaging about 13 percent of GDP. With just that small share, Taiwan attained the second fastest rate of economic growth. The three other NICs also spent as small a share and economic growth was equally impressive. Many therefore came to believe that small government was largely responsible for the fast growth of the NICs.[75]

During the 1850 to 1999 period, however, the size of the state was just about 1 percent of GDP and the economy of all the NICs was almost completely stagnant. The small size of the state was evidently not even an important variable affecting growth. The empirical data does suggest that with only a small government expenditure it is possible to attain impressive economic growth when the bulk of that expenditure is utilized to promote manufacturing. That is precisely what the NICs did between 1950 and 1997.

During the second half of the twentieth century, Taiwan experienced an unprecedented rate of population growth. From just about six million in 1949, the total number of people in the island rose to about 22 million in 1997. Population growth averaged about 2.3 percent annually. This was a faster rate than anything previously experienced and many analysts concluded that fast population growth was largely responsible for the acceleration of the economy.

The economy, however, grew almost four times faster. It is therefore highly unlikely that such a demographic growth could have pulled the economy to an 8.4 percent rate. It is even highly unlikely that it was an important contributory factor. In the 1950s, for example, population grew by about 3.5 percent annually and the economy grew by about 8 percent. In the 1970s, population growth slowed down considerably to just 2.1 percent per year. Had it played at least an important contributory role, the economy should have decelerated considerably or grown at about the same pace. The fact is that it grew faster (10 percent).[76]

Taiwan transformed its economy considerably during the second half of the twentieth century. In 1949, agriculture accounted for almost two-fifths of GDP and by 1997 the figure had shrunk to just 2 percent. The enormous contraction of agriculture had the effect of completely eliminating the idea among policy makers that agriculture was the motor of growth. By the late twentieth century it had become so small that it was not even possible to think of it even as a complementary causality factor of growth. The fact that the economy continued to grow quickly strongly suggested that this domain never had an important effect on growth. Many, however, interpreted the situation from another perspective. Many concluded that since agriculture had become so weak and the service sector had grown so much, it was services that had taken the role of propeller of the economy, or it had become one of the engines of growth.[77]

In the 1990s, services became the largest economic sector of the island. The fact that by then Taiwan's level of development was similar to that of the OECD countries with the lowest level of development drove policy makers to the conclusion that the island was in broad general terms a developed nation. Since for some time it had become a common belief among economists and policy makers in OECD countries that services were determinant for growth, Taipei began to believe the same. In the 1990s, therefore, measures were increasingly taken to promote services. The government, for example, took particular interest in finance and transportation for it wished to see the island become a regional center in those fields. The authorities also wanted to turn the island into a regional headquarters for multinationals.[78]

Independent of how trendy was the idea that services are determinant for growth, the fact is that the empirical data was not compatible with such an idea. In OECD countries, the data revealed an inverse correlation. The larger the share of services, the slower the GDP rates. On top of that, service sector productivity systematically attained the worst rates compared with the other sectors. In Taiwan the same phenomenon occurred. During the 1980s, the share of services from GDP increased significantly and economic growth decelerated noticeably. In the 1990s, the share of services grew even more and economic growth decelerated considerably more. On the island also, productivity in services was systematically the lowest. All other sectors on average attained faster productivity rates. In strictly logical terms, therefore, it becomes practically impossible to see how services could have acted as an engine of growth or even as an important growth factor.

The historical data seems rather to suggest that services are a by-product of manufacturing. It was manufacturing that in Taiwan, the other NICs, and the OECD countries uniformly attained the fastest rates of productivity. The economy constantly grew the fastest when this sector grew the fastest and the creation and embodiment of technology was bonded to this sector. Being the generator of technology, it was only inevitable that it would also attain the fastest productivity rates.

The historical evidence also reveals that the growth of services was largely dependent on the development of manufacturing. Until the late nineteenth century, for example, support for manufacturing in Taiwan was nonexistent and services were totally undeveloped. Their rate of growth was almost zero. Then, during the first half of the twentieth century, there were relatively strong factory-promotion efforts, and services grew for the first time in the island's history. During the second half of the century-factory output grew at an impressive pace, and services also grew at an unprecedented pace.

The dependency of services on factory goods was clearly appreciated by the fact that so many of those services would not even have come to exist without factory goods. The island, for example, had become by the late twentieth century a major supplier of transportation services, which would have been impossible without the gigantic fleet of sea vessels that Taiwanese companies acquired. Health services improved greatly and by 1997 the island had one of the lowest infant mortality rates in the world and one of the highest levels of life expectancy. This too was impossible without the huge amount of pharmaceuticals and medical equipment that became available.

Living conditions improved dramatically in Taiwan during the second half of the twentieth century. Up to the 1940s, rice alone constituted about 70 percent of what the islanders ate, and hunger and malnutrition were endemic. By the 1990s, however, the Taiwanese were among the populations in the world with the best nutrition, and morbidity and mortality had dropped remarkably. Human rights, civil rights, and women's rights had made gigantic progress. Democracy had made progresses beyond what most experts had thought possible and even environmental protection improved in an unprecedented way. It was not just that real incomes rose at a dazzling pace, but that even income distribution improved noticeably. And all that correlated with a massive government effort to promote factories. If manufacturing is the prime creator of wealth, as history strongly suggests, then it is clear why all of those positive events coincided with an impressive growth of factory output.[79]

However, much evidence suggests that living conditions could have improved still more during that period in Taiwan. The case of Hong Kong clearly indicates that faster growth was possible, and the fact that very large investments in nonmanufacturing sectors were made during the whole period clearly suggests that much larger resources could have been transferred to the factories. At first, Taipei was still strongly convinced that agriculture was important for growth and therefore allocated a large share of resources to this domain. However, even by the 1990s, when agriculture had become insignificant in size, the state continued to subsidize it on the grounds that the possibility of war or a blockade by China required that the island possess a ready source of food.

Had all of these resources been allocated to manufacturing, factory output would have surely expanded much faster, and GDP rates would have done the same thing. History suggests that agriculture would have also grown much faster. Until the nineteenth century, agriculture expanded at a very slow pace, even though the bulk of the island's resources were allocated to this domain. During the first half of the twentieth century, a much larger share of the island's resources was allocated to manufacturing and for the first time in history farm output grew at a relatively fast pace. In the 1950 to 1997 period, a still larger share of the nation's total resources was invested in factories, and agriculture grew at an unprecedented pace.

Had the large share of resources invested in primary activities, services, and construction been invested in manufacturing, much indicates that a much faster growth of these sectors would have taken place. Living conditions in an across-the-board manner would have most likely also improved much more. It is not yet clear how much faster that growth would have been, but much suggests that it could have been twice or perhaps even thrice as fast. Under those circumstances, democracy, the rule of law and the protection of the environment would have most likely also made proportional advances.

4
The Newly Industrialized Countries
The Case of South Korea and Singapore

Korea's Civil War and Its Aftermath

During the second half of the twentieth century, South Korea attained one of the fastest rates of economic growth in the world. Only the three other NICs managed to grow faster. This was the more outstanding because prior to the mid-century, growth had never been fast. GDP averaged a rate more than five times faster than during the first half of the century. Such an impressive acceleration clearly indicated that a variable such as culture could not possibly account for the fast growth because over time culture changes little.

By the late twentieth century, it had become a common belief among numerous analysts that South Korea owed its fast growth largely to its workaholic culture.

South Koreans were indeed very driven and hard-working, but they also possessed those same assets during the first half of the century, and during that period that same drive failed to deliver the same results. GDP averaged only about 1.5 percent annually during the first period while in the 1950 to 1997 years, the average was about 7.4 percent (see tables in appendix). During the first half century, the average number of hours worked per year was actually much greater than during the second half. The fact that the same culture was also present during the second half of the nineteenth century and the economy nonetheless performed even worse than during the 1900 to 1949 years substantiates even further the belief that there is no causality linkage between culture and growth.[1]

It is evident that the main determinant of the acceleration was a variable that could be manipulated artificially and at short notice. It had to be a variable that was almost totally absent in the second half of the nineteenth century, that was only moderately present in the following fifty years, and that was abundantly present between 1950 and 1997. The only variable that exhibited such a pattern of behavior during that period is manufacturing. During the second half of the nineteenth century, government support for the sector was practically nonexistent. In the following fifty years the state's promotion efforts were modest and factory output averaged about 2.1 percent annually. During the 1950 to 1997 period, they became very strong and manufacturing averaged about 11.1 percent per year.[2]

The strong level of support that Seoul gave to the sector during the second half of the twentieth century was the result of radical changes in the political climate of the peninsula and the region, which created very pressing national security concerns.

In 1945, the peninsula became divided along the 38th parallel between a communist north and a capitalist south. The leader of the North (Kim Il Sung) nourished strong desires to unify the peninsula under his rule and systematically prepared for an eventual invasion. The South instead gave investment priority to agriculture and other primary activities. That coincided with a stagnant economy between 1945 and 1949 in the South.

In 1950, the North launched a surprise invasion and the South was caught totally unprepared. Even though the population of the South was about twice as large, the North managed to conquer rapidly most of the South. Only because of the intervention of UN troops, composed primarily of American soldiers, were the Communists pushed back and in July 1953 an armistice was signed, leaving the borders at exactly the same place they had been before the invasion.[3]

The war shocked the political class in South Korea and brought it to the conclusion that there was a need to make major policy changes. On the one hand, it was evident that the country needed to produce many more weapons. Superior weapons were the main reason why the North had so easily defeated the South's forces. It was also clear that giving investment priority to agriculture could not uplift the country economically so as to diminish the internal communist threat. Such a policy had been practiced during all the preceding history and it had never delivered fast growth. The endemic poverty and unemployment of the time, plus the popularity of socialism among a large share of the population, pressured policy makers to search for an alternative policy that could deliver rapid growth.

Promoting manufacturing seemed the only alternative because nobody believed that services or construction could have the capacity to propel the economy. In addition, the country's foreign exchange reserves were very low and exporting factory goods seemed the only possibility for increasing exports rapidly. Large exports were needed in order to pay for the vast amount of machinery and equipment required for the production of weapons and related goods. It also became clear to most of the South's leaders that there was a need to produce a much larger share of the goods the nation regularly consumed, so that if war occurred again, the economy would not be paralyzed because imports were blockaded. Since by then the majority of goods the South consumed were manufactures, and were mostly imported, it became evident that a large share of these imports had to be supplied by domestic production.

Because of the colonization of the peninsula until 1945, Japan had a strong role-model effect on Korea. Since Japan had attained development by strongly promoting manufacturing, Seoul was also induced to believe that it should follow the example of Tokyo.[4]

Pressured therefore by national security concerns but also by balance-of-payments difficulties and the legacy of Japanese colonial rule, the government of Syngman Rhee began to support manufacturing in a more decisive way. From 1954 to 1961, his government allocated a much larger share of the nation's resources to the promotion of military and civilian manufacturing, which coincided with a much faster rate of factory output than at any other moment in the peninsula's history. That coalesced with unprecedented rates of economic growth.

From 1950 to 1953, the economy had contracted, which correlated with a contraction of manufacturing. Many weapons, as well as numerous other factory goods, were utilized to fight the Communists, but practically all of them were fabricated in the United States and other allied countries. Support for the sector was actually regressive and that coincided with a contraction of manufacturing. Then, during the rest of the decade, the factory promotion efforts of the state were relatively strong and manufacturing averaged almost 6 percent annually and GDP almost 4 percent.[5]

Many argued that this relatively fast growth was the result of the war reconstruction efforts that were substantially financed by Washington. The country had indeed been razed by the war and there was much to reconstruct, and it is also a fact that American aid was considerable. In the 1950s, aid accounted for the entire government's budget. However, Taiwan received about the same amount of aid as a share of GDP, but it attained a much faster growth. Growth in Taiwan between 1954 and 1959 was more than twice as fast as in South Korea, and Taiwan had absolutely nothing to reconstruct. Japan did have to reconstruct during the early 1950s, but by 1954 it had already surpassed by a large margin the best levels of production attained in the early 1940s. The aid that it received as a share of GDP was also much lower than what was given to Seoul. In spite of that, Japan grew about two times faster than South Korea during 1954 to 1959.[6]

It is evident that the reconstruction-aid argument could not explain the growth that took place during those years. It is most likely not serendipity that Taipei and Tokyo supplied a much stronger level of support to manufacturing during those years and that factory output grew at a rate which was about twice as fast as that of South Korea. It is also a fact that the authoritarian regime of Rhee was never very enthusiastic about promoting factories. The idea that agriculture was essential for growth remained very strong, and Rhee was also convinced that there was not a need to invest as much in weapons and civilian factory goods because the country could always count on the Americans for protection. As the war came to an end in 1953, Washington decided to leave a large contingent of soldiers and weapons in South Korea so as to guarantee that the North would abstain from further invasion efforts. Rhee therefore believed that if war broke out again, the Americans would supply plentiful weapons and other needed goods. His support for military and civilian manufacturing was therefore halfhearted. It was much stronger than ever before, but considerably weaker than in Japan and Taiwan.

American aid was by all means helpful and contributed to growth, but much suggests that only the share of it that was used to promote factories had an effect on growth. Unfortunately, much of the aid was not allocated to manufacturing.

On the other hand, the evidence also indicates that the aid was not essential for growth, as the case of Hong Kong demonstrated. During the 1950s and the whole second half of the twentieth century, the colony attained the fastest rates of GDP in the world even though it benefited from practically no aid. The impressive growth of Singapore during that same period also demonstrated that aid was not even a factor of secondary importance. The island received no aid and it nonetheless managed to attain the third-fastest growth in the world. Britain, on the other hand, which received a considerable amount of American aid in the aftermath of World War II, attained only a modest rate of growth.

As the case of Hong Kong and Singapore suggested, it was more useful to endorse a decisive factory promotion policy and mobilize as many resources as possible into manufacturing than to receive an abundance of aid which was not accompanied by a strong policy of support for the sector.

Since South Korea's economy grew much more slowly than its capitalist neighbors and national security concerns continued to be very pressing, many in South Korea and in particular among the military grew increasingly unsatisfied with the Rhee regime. In 1961, a group of top military under the leadership of Park Chung Hee launched a putsch and overthrew the government. These men were convinced that the country needed to produce many more civilian and military factory goods in order better confront the threats of possible aggression from the North and an internal communist subversion.[7]

The generals shared with the Rhee regime the same national security concerns, the balance-of-payments preoccupation, and the desire to imitate and catch up with Japan. There was, however, a difference on the degree of these motivations. Since they were more strongly motivated in early 1962 they started to implement their new program of development and resources were allocated to manufacturing on a much larger scale. To finance the numerous new factories that they desired, they considerably enlarged the budgetary allocations to this sector. They also nationalized banks and ordered them to lend the majority of their assets to this sector at subsidized interest rates and with longer than normal periods of maturity. The government also began to borrow very large amounts from international financial organizations and from foreign commercial banks, and allocated the large majority of these loans to the sector.[8]

The government gave manufacturers an abundance of grants, very low taxation, and numerous nonfinancial incentives. Most of these incentives were free or subsidized land for the factory, ready-made factory installations at very subsidized lease or sale prices, and subsidized utility rates. There were also guaranteed purchases at prices that ensured a high profit. Seoul concentrated the bulk of its support on promoting the private sector, but it also decided to create numerous state factories. It also raised trade barriers considerably to protect nascent companies, allowed for competition-hindering practices in the belief that investment would be increased, and manipulated the exchange rate to promote exports. Factory output began immediately to grow at a much faster pace and the economy did likewise. All in all, and despite the last two years of the Rhee regime, manufacturing output averaged about 15 percent annually during the 1960s and GDP averaged about 8.4 percent.[9]

Seoul made a major policy error as it created numerous state companies, as it nationalized banks, as it raised trade barriers, as it allowed for the existence of cartels, as it distorted the exchange rate, and as it guaranteed purchases. Hong Kong very clearly demonstrated that none of this was necessary for the attainment of speedy rates of factory output. The colony showed that the essential element resided in allocating as large a share of overall resources to this sector in the most market-oriented way.

In spite of the much larger share of resources that were allocated to this sector in South Korea during the 1960s, a very large share was allocated to agriculture. The idea that agriculture was important for growth still had some weight with the generals, and above all it was seen as very important for the national security of the country. In case of war, they wanted to have a ready supply of farm goods.

History and cross-country analysis strongly suggest that what South Korea needed in order to promote manufacturing was not state factories, trade barriers, cartels, managed exchange rates, and guaranteed purchases, but more fiscal, financial, and nonfinancial incentives. The empirical evidence also suggests that these market distortions hampered quality and efficiency and therefore had a negative effect on the economy. In South Korea and Taiwan, numerous state companies attained levels of quality and efficiency that were as good as those of the best private companies in the world, but there were also numerous others that underperformed. In Hong Kong, however, there were no enterprises that could not compete. South Korea and Taiwan could have attained a much higher return on investment had they abstained from these distorting practices. They could have also attained a larger agricultural output had they invested more in factories.[10]

When the generals published their first development plan in 1961, most economists abroad, particularly in international organizations, concluded that the plan was unrealizable. They predicted that the half peninsula would never attain a very fast economic growth or a very fast export growth. They also predicted that it would never succeed in becoming a major producer of internationally competitive factory goods. Their predictions were based on the fact that throughout all of the country's preceding history, none of this had ever been achieved. Because of the strong national security concerns, the generals disregarded the advice of the experts and went ahead with their plans. In no time, the country began to produce internationally competitive factory goods in large amounts, exports began to grow at an impressive pace, and the economy boomed.[11]

However, the generals did not completely disregard the advice of economists. The experts had argued that the country was well endowed for producing primary goods and that it should concentrate on this sector. The generals partially accepted this vision of things, which is why they allocated a relatively large share of resources to agriculture and related domains. The experts also said that if the country had a small chance of becoming a major producer of manufactures, it was only if it would limit itself to producing the goods that up until then had already been produced in small quantities. Seoul followed this advice, which is why during the 1960s it concentrated almost exclusively on promoting textiles, apparel, processed wood, and metal products, the goods the Japanese had mostly produced during the years of colonial rule.

The 1970s and the 1980s

In 1971, the United States reduced its ground forces in South Korea by a third and the generals saw it as the beginning of a full withdrawal in the near future. The idea that they could soon be without the protection of the Americans drove them to the conclusion that they had to prepare for that eventuality, and that the country therefore had to produce more weapons and more civilian factory goods. Until then, the Americans had supplied up practically all of the heavy weapons, and the generals decided that the country needed to produce at least a large share of these armaments. Since manufacturing these armaments required components such as metals, machine tools, electronics, and chemicals, a decision was also taken to produce them. Since the country was still importing most of the factory goods it consumed (in particular heavy manufactured goods), it was also decided to produce most of these goods so that the nation could be largely self-sufficient in case war disrupted trade.[12]

There were also balance-of-payments difficulties that pressured the government to promote new factory fields. The country had experienced trade deficits during the 1960s and the heavy dependence on imported capital goods was fundamentally responsible for the deficits. It was concluded that if these capital goods were produced domestically, the trade imbalances would disappear.

By 1971, the government and the country were already noticeably in debt and a much larger debt had to be taken in order to finance the numerous new factories that were to be founded. In order to be able to repay those debts, the generals ordered that all new fields orient the bulk of their output toward the export market so as to earn foreign exchange. Trying to export as much as possible instilled a very high level of competitive pressure on the new fields as it forced them to sell in the most competitive markets of the world.

Labor costs had rapidly risen during the 1960s and by the early 1970s labor-intensive fields such as textiles and apparel were rapidly losing competitiveness. Policy makers concluded that those fields would soon become totally outpriced from world markets and that there was a need to venture into new fields that were more capital-intensive.

Therefore, primarily as a result of national security concerns but also because balance-of-payments difficulties, and worries over competitiveness, Seoul launched a effort to develop heavy manufacturing and a few other new fields. Priority was given to iron, steel, nonferrous metals, shipbuilding, machinery, chemicals, electronics, and petrochemicals.[13]

This new effort translated into a stronger level of government support for manufacturing, which coincided with a faster rate of both factory output and GDP. Manufacturing grew during the 1970s at an average annual pace of about 18 percent and GDP grew by more than 10 percent. As the generals launched this new effort, most foreign analysts but also many within South Korea predicted that the effort would fail and the money would be wasted.

They substantiated their arguments on the fact that the country had never produced those goods, and concluded that such a state of affairs would remain a constant in the future. It was believed that the country was not naturally and culturally endowed for producing those goods. Such a vision of things was once again proven totally wrong as in no time vast amounts of those goods began to be produced competitively, and exports grew in an unprecedented fashion.[14]

To finance the new fields, the government ordered the banks (which were all in state hands) to allocate a larger share of their funds to manufacturing, giving preference to the new fields. They were also ordered to subsidize these loans even more than before. Loans were supplied at negative real interest rates and periods of maturity were stretched even more. During the 1970s, about three-fourths of bank lending was utilized to finance factory production, which was larger than the share of the preceding decade. The government also enlarged its direct budgetary allocations to the sector and supplied more grants. To finance the larger budgetary expenditures and the larger loan portfolio of the state banks, foreign borrowing increased noticeably, as did domestic borrowing. Foreign debt and the debt of the government rose considerably. By the end of the decade, foreign debt as a share of GDP was almost 50 percent, which was among the highest in the world. Government borrowing increased so much that budget deficits became much larger.[15]

Tax incentives were particularly abundant for the new fields, as were nonfinancial incentives. There was also a continuation of trade protection, cartels, the creation of more state factories, and a manipulative exchange rate policy. Hong Kong, which grew during the 1970s almost as fast as South Korea, very clearly demonstrated that in order to attain fast growth there was not a need to distort market forces in any of these ways. Much suggests that other measures with negative effects were budget deficits, a large government debt, negative interest rates, and guaranteed purchases. The case of Hong Kong clearly made it evident that none of this was necessary for the promotion of factories, and it also showed that these measures ultimately hampered the cause of manufacturing by lowering competition and reducing efficiency.

These measures were clearly policy errors, and in spite of so many of them, fast economic growth was attained. This situation very explicitly indicated that the variable that was fundamentally responsible for growth was not any of those that mainstream economists and developmental theorists had enumerated. The fact that these policy errors were more pronounced during the 1970s and that growth was nonetheless faster further substantiated this idea.

During this decade, there was even less support for agriculture and a much smaller share of the nation's resources was allocated to this domain. This situation also invalidated further the thesis of developmental theorists who asserted that the country's best possibility for growth resided in promoting primary production as much as possible. The abundance of policy errors and the application of measures that were incompatible with what most specialists asserted actually acted as a strong piece of evidence substantiating the manufacturing thesis. Only if it is assumed that manufacturing is the prime creator of wealth because of its intrinsic technology-generating powers does the empirical data begin to make sense.

Orthodox interpretations also fail to explain why in spite of the increase in inflation during the 1970s the economy accelerated in speed. The 1960s had already presented a paradox because inflation had risen quickly (about 15 percent annually) and economic growth had nonetheless been very rapid. In the 1970s, inflation averaged about 18 percent per year and the economy grew even faster. Most economists believe that low inflation is a precondition for growth, and much suggests that the pursuit of low prices is a rational goal. However, the case of South Korea demonstrated that if there was a causality linkage between low inflation and growth, then it was only a very weak one.[16]

Robert Barro, a Harvard economist, analyzed one hundred economies over the period 1960 to 1990 in an effort to measure the relationship between inflation and growth. He found no causality linkage between the two. Much suggests that governments should try to maintain very low levels of inflation on a constant basis, but the empirical data strongly indicates that low prices are not a major growth factor, and probably not a minor factor affecting growth. While inflation did not correlate with growth, manufacturing did. Seoul supplied to this sector during the 1960s abundant subsidies, which was paralleled with very fast growth. Support was even stronger during the 1970s and GDP also attained a faster rate.[17]

From observing the inflation and growth figures, some analysts came to the inverse conclusion and actually came to believe that fast inflation allowed for faster growth. However, this interpretation was also inconsistent with the facts, and the case of South Korea itself demonstrated its invalidity. During the 1980s, prices fell considerably and inflation averaged only 5 percent annually. If there had existed a causality linkage, then growth should have considerably decelerated, but the fact is that the fast GDP rates were maintained.

The case of Hong Kong, Singapore, and Taiwan even demonstrated that it was possible to attain impressive rates of growth while simultaneously experiencing very low levels of inflation. In Hong Kong and Taiwan growth averaged about 10 percent annually during the 1960s with an inflation rate of just 2 percent, and in Singapore GDP averaged about 9 percent during this same decade while inflation was just 1 percent per year.[18]

Mainstream economists argued that the high inflation between 1960 and 1979 was the result of fiscal imbalances, the policy of lending at negative interest rates, and inefficiencies resulting from monopolies and cartels, trade barriers, and state companies. On this matter, they were largely correct because in the 1980s budget deficits were considerably reduced, several banks were privatized, interest rates were gradually liberalized, the number of cartels was reduced, and trade barriers were diminished. Inflation immediately fell by a large margin. Much suggests that if the government had gone further in that direction, inflation would have been even lower.

There was also never a logical reason for these competition-hindering measures. The generals thought, for example, that because commercial banks had always proven unwilling to lend to manufacturing in large amounts, the only way of channeling the majority of bank assets into this sector was by nationalizing them. The case of Japan, however, clearly demonstrated that private banks could be easily induced to lend the majority of their assets to manufacturing by means of regulation and incentives. Japan also showed that there was no need to order interest rates to be negative. All that was needed was that banks be regulated into lending to manufacturing at lower rates than to the other sectors, but while interest rates remain positive.

The government not only created numerous state factories during the 1960s and 1970s, of which the steelmaker Posco became the most well known, but it also ordered private companies into particular fields. In the 1960s, for example, the Daewoo Company produced mostly textiles and in the following decade it was ordered into fields such as automobiles, shipbuilding, and machine tools. The case of Hong Kong demonstrated that fast factory output and fast economic growth was not dependent on state firms, and the case of Japan showed as well that it was not necessary to order firms into new fields. The efforts of Tokyo proved that what was needed in order to eliminate the natural fears of the private sector of venturing into high-risk fields, was to supply as many fiscal, financial, and nonfinancial incentives as possible. The more the incentives, the lower the risks became. There is simply a point where the incentives outweigh the risks and at that moment the private sector becomes highly enthusiastic about venturing into new fields, independent of its complete lack of experience in that domain.[19]

The government could have mobilized more resources during the 1970s into manufacturing by attracting foreign direct investment in large amounts. However, Japanese colonial rule in the peninsula drove most policy makers to the conclusion that FDI tended to be exploitative and abusive. National security concerns drove them also to see it as unreliable in times of trouble. Even without FDI, it was still possible to channel much larger resources into factories, but a number of ideological interpretations of reality misled Seoul into believing that the best possible results would be attained by allocating a large share of resources to the primary sector, to construction, and services.

There was an increase in the share of overall resources allocated to manufacturing during the 1970s, but the increase was not very large. The new heavy fields were mostly financed with the resources that in the 1960s were used to finance light manufacturing. That is why light factory production increased much more slowly. Part of the reason the fields of the 1960s were promoted less enthusiastically during the 1970s was that it was believed their decline was inevitable.

The case of Hong Kong showed that there was no such thing as an inevitable disappearance of labor-intensive fields. By the 1990s, textiles and apparel (in spite of having seen their birth a century earlier) continued to account for a very large share of the colony's factory output and they continued to grow relatively fast. That coincided with the fact that by then fiscal, financial, and nonfinancial incentives continued to be abundant for these fields. However, by then they were no longer labor-intensive; they had progressively become capital-intensive as ever-growing inputs of technology were instilled into them. There is much suggesting that light manufacturing could have continued to grow very fast during the 1970s in South Korea had it continued to receive as much support as before, and heavy manufacturing could have simultaneously grown as fast as it did. Both domains could have grown very fast if the overall level of support for manufacturing had been higher than it was.[20]

During the 1980s, national security concerns continued to be almost as high as during the 1970s. North Korea continued to invest a very large share of GDP in its armed forces, with Kim Il Sung continuing in power, and continuing to nourish the idea of a reunification of the peninsula under his command. Since the Americans changed sides during the 1970s with respect to their diplomatic relations with the two Chinas, the possibility that they could do the same with respect to the two Koreas became a very threatening scenario for Seoul. There was a small reduction of tensions in the peninsula, but the idea that Washington could disassociate from South Korea as a result of some strategic trade-off with the North's patron (the USSR) or with China acted as a strong threat. This drove Seoul to be almost as obsessed during the 1980s with producing a vast amount of weapons and related goods as in the preceding decade. This situation also pressured policy makers to try to produce as much as possible of what the country regularly consumed so as to be relatively self-sufficient in case of war.

Most of the same balance-of-payments concerns were also present during this decade and the desire to catch up with the ex-colonial master was just as strong. Seoul was originally driven to imitate Japan in its factory promotion efforts because the archipelago was the most developed economy in the region and Tokyo had abundantly subsidized that sector during its many decades of fast growth. Later, as South Korea overcame poverty, there was a small change in that motivation. Seoul, which was still hurting from the humiliation the Japanese inflicted on the peninsula during the years of colonial rule, supplied support to the sector so as to catch up technologically with the archipelago and so demonstrate that Koreans were as technically talented as the Japanese. Even though this nor any of the other motivations was deliberately intending to promote manufacturing, they ended up doing largely that.

Since the motivations were almost as strong as during the 1970s, the level of support for the sector ended up being almost as strong. This coincided with a rate of factory output and a rate of GDP that was also almost as fast. Factory output averaged about 14 percent annually and GDP a little less than 10 percent.

Several new fields were promoted, including semiconductors, computers, and telecommunication equipment. Once again, numerous analysts in the West, in Japan, in developing countries, and even in South Korea predicted that the effort would fail and that it was highly unlikely that the country would ever succeed in producing these goods. Once again, however, this vision of things proved not to be compatible with reality. In no time, the half peninsula became a major producer and exporter of these goods.[21]

The idea that certain countries are capable of producing manufactured goods while others are not was actually born in the sixteenth century, and since then it continuously proved to be false. During the second half of the twentieth century, this became even more evident, as the cases of South Korea and other East Asian economies demonstrated that from one moment to the next, countries could become major producers of factory goods. By the late twentieth century, it had become increasingly evident that high-quality manufacturing and even state-of-the-art manufacturing could be produced anywhere in the world.

The empirical data suggests that such a feat does not depend on the level of a nation's development, nor on geography, culture, or any other variable of the sort. Everything indicates that it is fundamentally dependent on the level of government support for the sector. Since technology is the determining variable that makes the production of factory goods possible, and technology finds itself practically entirely embodied in machinery and equipment, all a country needs to do is to import it. Once that is done, goods can immediately be produced with levels of quality almost identical to those in the countries with the highest levels of development.[22]

Even though economic growth on average was impressive during the

1980s, there was nonetheless a year in which South Korea experienced a contraction. That recessive year was 1980 and it coincided with the only year during the 1960 to 1997 period in which support for manufacturing was regressive. In late 1979, Park Chung Hee was assassinated and a new group of generals, under the leadership of Chun Doo Wan, took power. The next year was a period in which the dispute of power among the two factions in the military paralyzed policy making almost completely and manufacturing was not promoted.

By 1981, the new group of generals had consolidated power and they vigorously reinitiated the strong factory promotion efforts of the past. The economy began immediately to grow at a double-digit pace. The new generals were as convinced as the preceding ones that strong support for the sector was needed, but they strongly disagreed over how this support should be supplied. The new group was particularly worried about the high inflation of the 1970s and was convinced that if inflation was not brought down, it could eventually hamper economic growth. They concluded that to lower inflation the economy needed some liberalization, privatization, deregulation, and fiscal rectitude. They privatized some banks, allowed new private banks to open up, partially liberalized interest rates, and fostered competition among banks as well as in most other domains. Trade barriers were lowered, some FDI was allowed in the country, and budget deficits were lowered.[23]

Inflation did come down noticeably but economic growth did not improve. What's more, it slightly decelerated. After having averaged about 10.3 percent annually during the 1970s, it averaged 9.7 percent during the 1980s. The lower rate of inflation was evidently the result of the liberalization measures, but growth was obviously not. It was evident that the determinant growth variable was not among the ones that were applied, but the liberalizing measures did have a positive effect on the economy. In the 1970s, a rate of almost 18 percent of factory output was needed to get a GDP rate of about 10 percent, while in the next decade only like 14 percent of factory output was needed to attain almost the same GDP rate. Productivity and efficiency was significantly increased and everything indicates that it was the direct result of the liberalizing measures.

The whole situation was once again highly suggestive that the bottom line of growth resided in manufacturing. It is also worth noting that despite the liberalization measures, trade barriers remained high, competition-hindering practices were abundant, and the bulk of state companies were not privatized. Banks continued to be ordered to lend the majority of their assets to the fields the state selected; interest rates remained partially regulated; FDI was allowed only in small amounts; the government continued to run budget deficits; and regulations on business continued to be abundant.

For example, the average Korean firm needed at that time about 120 documents in order to export, and that even though Seoul was highly interested in promoting exports. Regulations of all sorts and endless bureaucratic procedures were needed for most economic activities.[24]

With so many policy errors, it becomes hard to understand how South Korea could have attained the fastest rate of economic growth in the world during the 1980s. It was far faster than any OECD country, any developing country, and even the other NICs. Hong Kong and Singapore, which did not commit these policy errors, attained slower GDP figures.

This paradox only becomes understandable once it is taken into consideration that government support for manufacturing was stronger than in the NICs and stronger than in practically any other country of the world. South Korea's rates of factory output were about the fastest in the world during this decade. Only China attained a slightly faster factory output, but because it distorted market forces so abusively it extracted from it a lower GDP rate than South Korea. The small economic deceleration the half peninsula experienced relative to the 1970s also becomes understandable when it is taken into account that Seoul's factory promotion efforts in the 1980s were slightly weaker.[25]

The 1990s in South Korea

In the 1990s, more liberalizing measures were undertaken and better levels of efficiency were observed, but the economy experienced a noticeable deceleration. During 1990 to 1997, GDP averaged about 7.4 percent per year. Once again, such a situation presented a paradox and once again the manufacturing thesis was the only one that could supply a satisfactory explanation. Factory output during these years averaged about 8.3 percent annually, and Seoul's support of the sector was noticeably less strong.[26]

By the 1990s, it became common for most economists and social scientists to argue that South Korea largely owed its fast growth to its emphasis on education. The investment efforts on this matter were indeed more decisive than in practically any other nation. So high were they that by the mid-1990s South Korea possessed the largest amount of Ph.D.s per capita in the world, even though it was still considerably less developed than the United States, Japan, Germany, most other OECD countries, and even the other NICs.

However, if education was pivotal for growth, why is it, then, that the economy decelerated? It was in the 1990s when the population attained the highest educational levels in the country's history, and it was then that the GDP figures were at the lowest levels since the 1960s. If education was determinant for growth, it should have been in the 1990 to 1997 years when growth was the fastest, especially because by then the economy was also more market-driven.[27]

In Taiwan, Hong Kong, and Singapore, by the 1990s the population was also better educated than before, but the GDP figures were far from being at the highest levels. Japan's educational system was supposed to be the best in the world by the 1990s, but the archipelago attained its slowest growth figures in all of the twentieth century. The absence of a consistent correlation clearly indicated that a causality linkage between education and growth did not exist. Much suggested that education was a by-product of growth and that growth was the result of some other variable.

Many argued that this other variable was infrastructure. Seoul invested significantly in this domain from the early 1960s and impressive progress was made on that front, but here again such a thesis failed to add up with the facts. At a time (1990s) when infrastructure was at its highest level of development, economic growth expanded at the slowest rate since the 1960s. In the other NICs and Japan, the same paradox was observed. Infrastructure attained unprecedented levels of development by the 1990s, but the GDP figures were far from being even as high as in other less infrastructure-developed decades.

In spite of the impressive infrastructure development of South Korea, by the 1990s it was still far from meeting demand. Since economic growth systematically grew faster than infrastructure, congestion was still endemic. Roads were continuously congested, ships had to wait days to dock, and delays at airports were routine. With so much congestion, it was hard to understand how South Korea could have attained one of the fastest rates of growth in the world. The case of China made it even more evident that a causality relationship was absent. By the 1990s, China had an infrastructure that was terribly primitive in comparison to that of South Korea, but in spite of that it got a faster rate of growth.[28]

Interestingly, Switzerland, which had perhaps the most advanced infrastructure in the world and no congestion, attained one of the slowest GDP rates in the world. Economic growth averaged about 7.4 percent annually in South Korea during 1990 to 1997, 8.0 percent in China, and 0.5 percent in Switzerland. By then, Switzerland's population was also better educated than ever before and better educated than the vast majority of nations of the world. In addition, the economy had been noticeably deregulated. None of this, however, managed to save this small European country from its worst economic crisis since the 1930s. The absence of a correlation between infrastructure and growth clearly indicated the impossibility of a causality linkage.[29]

While even an amalgam of the market liberalization, education, and infrastructure arguments fails to coincide with the facts, the theory of manufacturing succeeds very well. Support for this sector was still very strong in South Korea during the 1990s, in China it was even stronger, and in Switzerland it was terribly weak. This coalesced with very fast rates of factory output in the half peninsula, with even faster rates in China, and with sclerotic rates in Switzerland.[30]

During the 1990s, the ideological motivations that in the past had driven the government to enthusiastically promote factories were still present. The north of the peninsula continued to be ruled by Kim Il Sung and his son, who still nourished some desires to forcefully unify the peninsula. There were still considerable balance-of-payments concerns, and the desire to catch up developmentally with Japan was still strong. That translated into the need to produce armaments in large amounts, to produce most civilian goods so as to be self-sufficient in case of war, to export as much as possible, and to produce goods as technologically advanced as Japan did. The bulk of those goods were factory goods.

However, the rapidly growing economic and technological superiority of the South over the North and the rapid improvement in the balance-of-payments lowered the pressure to produce weapons and civilian goods. Tensions with North Korea were also significantly lower than before and the developmental gap with Japan had narrowed. Policy makers concluded, therefore, that the country no longer needed to allocate so many resources to manufacturing. Strong support for the sector coincided, therefore with fast economic growth, and the weaker support compared with the preceding decade coincided with a slower rate of GDP.

The desire to catch up with Japan was one of the strongest motivations by the 1990s. During the years of colonial rule, the Japanese had classified the Koreans as inferior. The differing Korean governments tried since then to disprove such a vision of things by attempting to match the Japanese on the areas where superiority was more clearly expressed. Since technology expressed superiority in the most evident way and technology was constantly tied to factory goods, the efforts to catch up ended up by basically promoting manufacturing. In 1993, for example, the Ministry of Science and Technology unveiled a massive program intended to channel state funds into high-tech fields such as automobiles, bio-engineered medicines, computers, electronics, and telecommunication equipment.[31]

From the 1960s, the Japanese systematically predicted that the Koreans would never catch up with them, but by the 1990s they had achieved parity in fields such as iron, steel, other metals, ships, automobiles, semiconductors, electronics, chemicals, and computers. By 1960, for example, South Korea produced less than 1 percent of the world output of ships. By 1993, it won nearly 40 percent of world orders, surpassing even Japan and becoming the largest shipbuilder in the world.[32]

During the 1990s, there was a further liberalization of foreign trade, FDI was allowed in larger amounts, banks were more deregulated, competition-hindering practices were reduced, a privatization program was initiated, regulations on business were diminished, and the exchange rate became more market determined. It became very hard to understand, therefore, why in spite of these numerous positive policy measures, the economy decelerated. If these measures were determinant for growth, then the economy should have gained speed.

On the other hand, it is worth noting that in spite of the liberalization, the economy remained largely distorted. Trade barriers, in particular nontariff barriers, remained very high, a discriminatory treatment of foreign companies was still very evident, and banks were still ordered to lend the majority of their funds to the fields Seoul chose. On top of that, only a few firms were privatized, and regulations on business continued to be excessive.[33]

For example, it was not until 1994 that a plan to sell some of the approximately 133 state firms got under way. Seoul was not even interested in selling them all, was not in a hurry, and in many cases was only willing to sell a small share of a company to the private sector. By then, several of these state enterprises still operated as monopolies (Posco, Kepco, Korea Telecom).

Regulations on business were still excessive by 1995. Building a new plant, for example, still required about a thousand government approvals. It was therefore logical to wonder how it was possible that such a market-distorted economy managed to attain one of the fastest rates of growth in the world. This situation once again suggested that free markets were not the determinant aspect for the attainment of growth.[34]

The empirical evidence suggested that even though free markets were helpful for the attainment of growth, the bottom line resided in manufacturing. During 1990 to 1997, an abundance of fiscal, financial, and nonfinancial incentives were offered to this sector. There were ample tax exemptions and tax reductions, grants, subsidized loans, free land for the factory, subsidized utilities, and guaranteed government purchases at high prices. Private banks were regulated into lending the majority of their assets to manufacturers, at below-market rates, and with longer periods of maturity than those determined by the market. About 55 percent of bank lending during these years went to finance manufacturing. In addition, when manufacturing companies ran into financial difficulties the government continued to guarantee a rescue. Although abundant, incentives were noticeably less numerous than during the 1980s when, for example, about two-thirds of bank loans were used to finance factory production.

Technology, Productivity, Investment, Savings, and Globalization

The way in which technology developed during the second half of the twentieth century in South Korea gives further credence to the idea that manufacturing is the bottom line of growth. Only if this sector is fundamentally responsible for the generation of technology could it be credited with being at the core of growth, because technology is the bottom line for the creation of wealth. For that to be true, a correlation should exist between the sector and technology, and the historical evidence shows such a parallelism.

Up to the early years of the twentieth century, for example, government support for manufacturing had systematically been almost nonexistent, which was paralleled by a terribly slow development of technology. Then in 1910, the Japanese conquered the peninsula and began to promote factory production, and suddenly technology began to grow at an unprecedented pace. All of the technology was imported, but it was imported at a much faster pace than ever before. From the early 1940s until 1953, factory output contracted and technology did likewise. Living conditions worsened as the goods that embodied technology became scarce.

During the remainder of the 1950s, however, technology grew at an unprecedented pace, and that correlated with the most decisive factory-promotion efforts up to that date. In the 1960s, Seoul's promotion efforts were even stronger and that was reflected in much faster technological growth. In the 1970s, the rate of technical progress accelerated some more, which coalesced with a small increase in the rate of factory output. During the 1980s, factory output slowed down a little and the pace of technology remained largely unchanged. In the 1990s, support for the sector diminished noticeably and the pace of technology did likewise.

During the second half of the twentieth century, South Korea made impressive progress not just in the pace at which technology was imported, but also in the capacity to create its own technology. From having imported practically all of the technology it consumed during the 1950s, it went to become a major exporter of it by the 1990s. By the end of the century, the half peninsula had become highly innovative. The Koreans were convinced that being able to produce their own technology was very important for the sustainability of growth. However, at the time when the country was producing the largest amount of technology, economic growth was at its lowest point since the 1960s. During the 1960s, for example R&D as a share of GDP accounted only for about 0.4 percent while in the 1990s the share had grown to about 2.1 percent.[35]

By the 1990s, South Korea was not only producing a very large share of the technology it consumed. Its inventions had become so sophisticated that they were even capable of matching in several fields the very best of the most advanced OECD countries. If being inventive was important for growth, then the GDP figures should have been faster than before. The fact that growth rates were the slowest since the 1960s suggests that it was not important. As in the case of Japan and the other NICs, economic growth failed to add up consistently with the capacity to self-produce technology in large amounts. However, the case of Japan and the NICs did add up with the manufacturing thesis. Technology systematically grew in proportion to the level of government support for this sector, independent of whether all of the technology was imported or produced domestically.[36]

Productivity behaved in a similar way. Independent of whether technology was all imported or produced domestically, productivity systematically reflected the differing rates of factory output. Until the early twentieth century, for example, it systematically endured stagnation, correlating with a systemic absence of government support for the sector. Then from 1910 to the early 1940s, there was an unprecedented support for the sector and productivity increased at an unprecedented pace. In the following years, it stagnated but between 1954 and 1959, productivity grew at a fast pace and that coincided with a relatively strong factory growth. In the 1960s, the factory promotion efforts of Seoul increased considerably and so did the rate of productivity. During 1970 to 1989, manufacturing grew still faster, as did productivity. In 1990 to 1997, the rate of productivity decelerated, which correlated with a noticeable decrease in the level of subsidies for the sector.

Since productivity is fundamentally the result of technology and technology is fundamentally generated by manufacturing, it is inevitable that productivity will be correlated with the fluctuations of this sector. That would explain, for example, why productivity in South Korea was higher in the 1970s (7 percent) than in the 1990s (5 percent), even though in the last period the economy had been considerably liberalized and deregulated. The decade in which market forces and competition were most repressed was the 1970s. It is evident that liberalization had a positive effect on productivity, but if it had been essential, it should have been in the 1990s when the figures attained the fastest rates. The fact that support for manufacturing was stronger in the 1970s suggests that the essential element determining productivity resided in this variable.

The other NICs also showed a similar correlation. By far, the fastest rates of productivity in all of their history occurred during the second half of the twentieth century, running in parallel with the strongest levels of government support for the sector. The manufacturing thesis would also explain why Taiwan, which by the 1990s had considerably liberalized its economy, attained slower productivity rates (5 percent) than in the 1960s (9 percent), when competition was much lower. Taipei's support was much stronger in the 1960s and factory output was almost twice as fast as during the 1990s. It would also explain why Hong Kong, which by the 1990s was the most market-driven economy in the world, attained slower rates of productivity (4 percent) than South Korea (5 percent), which was still strongly regulated.

Most likely it was not by chance that Seoul's factory promotion efforts were larger and that factory output grew faster than in the colony.[37]

Substantiating further the manufacturing thesis is the way in which investment and savings behaved. If this sector is the prime creator of technology, then investment should have systematically reflected the fluctuations of factory output because the creation (or even just the reproduction) of technology is an extremely investment-intensive effort. The historical evidence reveals precisely such a correlation. Whenever the state increased its factory promotion efforts, investment rose, and whenever support decreased, it fell.

Investment as a share of GDP, for example, systematically remained close to zero up to the early twentieth century in the peninsula, and government support for the sector was almost totally absent during all that time. Then, during the decades of colonial rule, investment rose by a large margin and so did the factory promotion efforts of the Japanese. From the early 1940s to 1953, manufacturing production contracted and investment as a share of GDP was negative. In the remaining years of the 1950s, support for the sector was higher that ever before and investment rose to unprecedented levels.

Then, during the 1960s factory output grew much faster than ever before and investment was also unprecedented, averaging about 23 percent of GDP. In the 1970s, support for manufacturing was even stronger and investment accounted for about 32 percent of GDP. In the 1980s, Seoul's promotion efforts were slightly lower and investment accounted for about 31 percent. In the 1990s, factory output grew noticeably slower and investment was of about 29 percent.[38]

The speed with which wealth gets created, is what basically determines the level of savings. Therefore, if manufacturing is the prime creator of technology and therefore wealth, a correlation between this sector and savings should exist. Such a correlation exists and savings behaved throughout the peninsula's history in an almost identical way as investment, always reflecting the differing levels of government support for the sector. By the late twentieth century, much was said about the Korean's cultural propensity to save a lot. However, the fact is that throughout all of their history up to the early years of this century savings remained perennially close to zero and it was only since the 1960s that they rose to high levels.[39]

Much suggested that investment and savings were a by-product of the level of government support for manufacturing and that the development of the other sectors was also a side effect of this support.

For centuries, it was argued that agriculture was the fulcrum for growth, but the fact is that the only time when this domain grew rapidly was when the state's factory promotion efforts were strong. Up to the early twentieth century, farm output grew at a snail's pace even though the bulk of investment was concentrated in this domain. Only until the decades of colonial rule did agriculture grow at a pace that was no longer quasi-stagnant, and that coalesced with a relatively large promotion of factories. However, even though the Japanese tried to use Korea as a supplier of raw materials, agriculture and the other primary activities never managed to grow as fast as during the last four decades of the twentieth century. It was during the 1960 to 1997 period that, for the first time, the government did not allocate investment priority to the primary sector. It was precisely then when agriculture grew the fastest. Farm output grew so fast during the 1960s that by the end of the decade the United States phased out its food aid, for it was no longer needed. The aid had its origins in the starvation that followed the end of the war in 1953.[40]

Not only did farm output grow at an impressive pace, but the rapidly growing foreign exchange reserves resulting from the impressive export growth of manufactures allowed for a massive import of food. By the 1990s, South Korea was the sixth largest net agricultural importer in the world. Agriculture grew on average by about 0.9 percent per year during the first half of the twentieth century while manufacturing expanded by about 2.1 percent. During the second half of the century, factory output averaged about 11.1 percent annually and farm output grew by about 2.9 percent.[41]

This is understandable once it is considered that the fundamental mechanism by which agriculture makes progress is by increasing its technological input, and technology has systematically found its materialization in factory goods such as farm machinery, fertilizers, pesticides, and irrigation equipment. South Korea produced these goods to an unprecedented extent during the second half of the century and Korean farmers increased the utilization of these goods at a very fast pace. The same occurred with the other primary activities and the other sectors. There was an unprecedented production of fishing vessels, mining equipment, and forestry instruments, as well as construction materials and construction equipment. Service factory goods such as transportation vehicles, telecommunication equipment, and education utensils also grew at an impressive pace.

The long-term development of each one of these domains and sectors correlated with the differing levels of government support for manufacturing. It was during the second half of the twentieth century that the factory promotion efforts of the state were the strongest and it was then that fish catches, mineral extraction, wood cutting, construction, and services grew at the fastest pace.

The fact that productivity in manufacturing was systematically faster than in all the other sectors substantiates further the idea that this sector is the prime creator of technology. Since the other sectors are mere recipients of technology, it seems only inevitable that they attained slower rates of productivity. During the 1960 to 1997 period, for example, manufacturing productivity grew annually by almost 9 percent while productivity in agriculture grew by only 3 percent per year. As for services, performance there was even slower. However, it is worth noting that primary, services, and construction productivity during this period was much faster than ever before, coinciding with an unprecedented utilization of factory goods in those sectors.[42]

The historical evidence suggested that the development of the primary sector, construction, and services directly depended on the support manufacturing received. The empirical data suggested the same about exports. For centuries, exports were practically nonexistent and support for manufacturing was also almost non-existent. Then since 1910 there was a massive increase in factory promotion efforts, and exports suddenly grew at an impressive pace relative to the past. Factory output contracted from the early 1940s to 1953 and exports did likewise. In the remainder of the 1950s, factory output grew at an unprecedented pace and exports followed. From the 1960s onward, the level of factory subsidization was far superior than ever before and exports also grew at a much faster pace. During the 1960 to 1997 period, exports grew at an annual average rate of about 20 percent.

After having accounted for just 2 percent of GDP in 1960, exports grew to account for about 36 percent by the mid-1990s. This impressive growth of exports was almost exclusively the result of factory goods. In 1960, primary goods accounted for about 85 percent of exports but by the mid-1990s they accounted for less than 3 percent. Many economists and policy makers became convinced that the impressive export growth of the 1960 to 1997 period had been the main propeller of the economy. If exports had been the main determinant of growth, it would be normal to find a tight correlation between the two. However, the empirical data does not reveal such a close parallelism. During the 1960s, for example, exports grew at the phenomenal rate of 34 percent per year and GDP averaged only 8.4 percent. During 1990 to 1997, on the other hand, exports grew by just 7.4 percent annually, but the economy grew almost as fast as in the 1960s (7.4 percent). If exports were determinant, then the growth figures should have been much faster in the 1960s and slower in the 1990s. The fact that they were not, clearly indicates that exports were not determinant for growth and that they were actually an effect and not a cause.

Because of the impressive growth of exports during the second half of the twentieth century, South Korean goods became increasingly blockaded in OECD countries. By the 1990s, about one-fifth of exports to developed countries were subjected to nontariff measures such as quotas, antidumping actions, voluntary export restraints, strict import inspection, countervailing measures, complex certification procedures, and delays for customs clearance. Many argued that these barriers hampered growth and were the cause of the deceleration of the 1990s. However, in the 1970s, trade barriers on Korean exports were higher than in the 1960s, and in spite of that, economic growth was faster. GDP averaged 8.4 percent during the 1960s and 10.3 percent in the 1970s. If the trade barriers were really a major hindrance, then the 1970s should have experienced a deceleration. The absence of a correlation clearly indicated that there was not a causality linkage. Surely the OECD trade barriers had an effect on the Korean economy, but much seems to indicate that it was only a very marginal one.[43]

The deceleration of the 1990s was also blamed on the rapid growth of social welfare expenditures that took place in those years. The pension system got started in 1988, a few years later a national health insurance system was inaugurated, and in 1995 unemployment insurance was introduced. It is evident that those expenditures were not invested and therefore reduced the possibilities for larger investments, but the historical evidence seems to indicate that if they had not been made, the overall level of investment would have been the same. During the first half of the twentieth century, for example, social welfare expenditures as a share of GDP were almost nonexistent, and in spite of that, investment was much lower than in the 1990s. Growth was also much slower. More to the point is the situation of the second half of the nineteenth century when social welfare expenditures were totally nonexistent, yet investment and growth were even lower than during the 1900 to 1949 period.[44]

Much suggests that the nonexistent or very low levels of investment that prevailed up to the mid-twentieth century were because government support for manufacturing was nonexistent or very low. In the 1990s, it was not social welfare expenditures that were hindering a higher level of investment, but a number of ideological motivations that drove policy makers to believe that there was no longer a need for allocating as large a share of overall resources to the manufacturing sector as before.

Others thought that the deceleration was fundamentally the result of rapidly rising labor costs. Since military rule ended in the late 1980s, wages rose quickly and grew faster than productivity. A fast growth of real wages could not have possibly been the cause because in the 1960s labor costs rose at an unprecedented pace, yet GDP figures were much faster than before. During the 1970s, wages grew even faster and the economy actually accelerated. In the 1980s, wages grew at a still faster pace, and in spite of that growth remained impressive.

That wages during the 1990s grew faster than productivity is also not a reason for believing that it was the cause of the deceleration. In OECD countries during these years, productivity grew faster than wages, and in spite of that, the economy decelerated. The economy not only decelerated relative to the 1980s, but GDP rates in OECD nations were only one-fourth the rate that South Korea attained. It is evident that such a situation had practically no incidence on growth. It would have surely been better if wages in the half peninsula had grown less fast than productivity, but the absence of a consistent correlation suggests that it was not determinant for the attainment of growth.[45]

Largely because of the fast growth of wages but also because of the globalization phenomenon, there was a massive emigration of Korean companies to the rest of the world during the 1990s. Labor-intensive production emigrated in an all-out way to places like China, Southeast Asia, and Latin America. It was therefore argued that such a large outflow of the production capacity of the country was the main cause of the deceleration. Here again, a consistent correlation was impossible to observe. It is worth noting that notwithstanding the large emigration of companies in the 1990s, growth was still one of the fastest in the world. Latin America, which during these same years received large inflows of FDI and experienced almost no outflows, attained only modest GDP figures. Latin America, which attracted many of the emigrating South Korean companies attained a growth of only 3.5 percent annually while the half peninsula had more than 7 percent.[46]

Much suggests that what was fundamentally responsible for the deceleration was not the emigration, but the decrease in government support for manufacturing. Support for this sector in Latin America, for example, was modest relative to South Korea, and the case of Singapore demonstrated that a deceleration of factory output was not inevitable. In the 1980s, factory output in Singapore grew at an annual average rate of about 8.5 percent. This was a considerable deceleration from the preceding decade and was accompanied by a large emigration of companies to neighboring Malaysia and Indonesia. Concerned with this situation and fixated with the idea that the share of manufacturing should not fall below one-quarter of GDP, the Singaporean government noticeably increased the level of subsidies for the sector in the 1990s. The rise in support coincided with a faster rate of factory output and a faster rate of economic growth.[47]

During the second half of the twentieth century, living conditions in South Korea improved at an impressive pace and far faster than ever before. Starvation, hunger and malnutrition had for centuries ravaged the population. By the end of the twentieth century, all of that had disappeared. Diseases, epidemics, illiteracy, and ignorance had for centuries devastated the population and all of a sudden all of those problems were superseded. Human rights, labor rights, civil rights, and women's rights were almost nonexistent during practically all of the nation's history, but by the end of the century legislation on all of those matters was extensive and it was seriously enforced.[48]

Throughout the history of the country, neither the government nor the population had ever bothered to protect the environment. Flora and fauna were exploited and nobody even came to the idea that these living organisms had rights unto themselves. By the 1990s, a large share of the population had accepted this idea and Seoul increasingly took measures to protect the environment.

Real incomes grew rapidly during this period and lifted the whole population out of poverty, but in addition, income distribution also improved considerably. By the 1990s, income distribution in South Korea was similar to that in OECD countries, the group of nations that had it the least unequal. The wealth of the nation increased so much that after having been one of the most in-debt economies in the world, it went on to become a major lender. In 1995, for example, South Korea changed its status with the World Bank from being a borrower to being a donor.[49]

And all of these achievements coincided with one of the most enthusiastic factory promotion efforts the world has ever seen. Much suggests that this was the cause, and that the achievements could have been larger had the government supported this sector even more. Had it not allocated so many resources to primary activities, services, and construction, the economy could have easily grown by 10 percent during the whole half century. Had the abundant market distortions been eliminated, a faster rate of GDP would have been attained even if the rate of manufacturing had not increased.

Economic History of Singapore

Singapore attained also one of the fastest rates of economic growth in the world during the second half of the twentieth century, and as a result, living conditions in every aspect improved massively. By the end of the century, Singapore was probably the most developed country in the world. Economic growth during this period averaged about 7.6 percent annually. This spectacular growth coincided with a very strong government support of manufacturing and with a very fast rate of factory output which averaged about 9.5 percent annually. (See tables in appendix.)

During this period, Singapore practiced economic policies that were among the most liberal in the world. Market distortions were very few, although they were a little higher than in Hong Kong. Numerous studies classified Singapore as the second-least distorted economy in the world, where government intervention was among the lowest. The island practiced free trade totally during the whole period; government expenditure as a share of GDP was very small; the budget was almost constantly in surplus; and public-sector debt was almost nonexistent. There were also no restrictions on FDI; cartels and monopolies were only rarely allowed; distortions on capital, labor, and prices were small; and exchange rate distortions were few.[50]

As in the case of Hong Kong, Singapore managed to extract from its manufacturing investments a very high economic growth and everything suggests that such high levels of efficiency were the result of its very liberal economic policies. The island submitted its manufacturing sector to a much higher level of competitive pressure than South Korea and Taiwan, and because of that it was capable of maximizing its investments more. For every unit of factory output, it extracted more economic growth than South Korea, Taiwan, and the large majority of the other nations in the world.

However, the fact that during the first half of the century economic policies were even more liberal and the GDP figures were nonetheless considerably slower clearly indicates that the bottom line of economic growth did not reside in these policies. The fact that economic policies were even more liberal during the second half of the nineteenth century and GDP figures were much slower than in the 1900 to 1949 period corroborates this idea even more. The historical evidence strongly suggests that the bottom line resided in manufacturing. The British colonial authorities, for example, supplied only a low level of support for this sector during the second half of the nineteenth century. In the 1900 to 1949 period, subsidies increased considerably and the rate of factory output did likewise. During the second half of the twentieth century, the factory-promotion efforts of the state increased by a large margin and the rate of output of this sector accelerated much more.

Logically analyzed, the empirical data leads to the conclusion that liberal policies are a very positive undertaking so long as they are accompanied by government support for manufacturing. Without support the sector does not grow, and if it does not grow, the economy remains stagnant no matter how highly competitive the environment. That is precisely what took place during the first half of the nineteenth century on the island, and the economic situation of the time was complete stagnation.

The strong support supplied to the sector during the 1950 to 1997 period was the result of a number of ideological motivations that were common among the island's policy makers. Never was there a clear understanding of the causality linkages between this sector and technology. Never was there a scientific appreciation of the wealth-creating powers of factories.

However, because of a number of ideological motivations, subsidies were supplied in large amounts—motivations such as the scarcity of natural resources, the British legacy of using the island as a regional production processing center, the conviction of the leadership that there was something about this sector that was important for development, and balance-of-payments concerns.

Singapore remained a British colony until 1959 when it gained self-rule. Until 1965, however, it was linked to Malaysia in a sort of federation that gave the island almost complete autonomy in economic policy-making matters. During the 1950s, therefore, economic policy was mostly decided by the British authorities, who continued to apply a number of measures intended to make the island the manufacturing center of the region. By then, the British still believed that exporting abundantly from Britain was conducive to a positive economic performance and that the colonies were mostly to be used as providers of raw materials. However, taking the raw material from East Asia to Britain and then processing it so that it would later be exported to that same region carried with it heavy transport costs. On top of that, labor costs in Britain were much higher than in its colonies, and competing with the Japanese in Asia, who had much lower transport and labor costs, was very hard.

To cut transport and production costs, the authorities offered fiscal, financial, and nonfinancial incentives to foreign and domestic manufacturers that could process the raw materials produced in the neighboring colonies, as well as those who could fabricate import substitutes for the regional needs.[51]

Power was not transferred all at once in the late 1950s. There was a progressive and peaceful transfer of policy-making responsibilities to the locals. The largest and strongest political movement in the island was the People's Action Party (PAP), which was led by the charismatic and strong-willed Lee Kuan Yew. Lee and the PAP's leadership were convinced that the island had no future in trying to produce primary goods because of the scarcity of natural resources. They were also convinced that there was some sort of linkage between manufacturing and development. In addition, foreign exchange reserves were low and the only sector that could procure an abundance of exports was manufacturing. Large imports of foodstuffs and other commodities could only be guaranteed by generating an abundance of foreign exchange, which was seen as a national security necessity. Therefore, as policy making fell increasingly in the hands of the PAP during the 1950s, support for manufacturing increased. This coincided with an unprecedented growth of factory output and GDP. Factory output averaged about 6.1 percent annually during this decade and GDP about 4.3 percent.

The government began to create industrial states in which ready-made, multistory factory installations were supplied to foreign and domestic manufacturers at subsidized prices. The government began to absorb also the direct training costs of manufacturers by creating industrial training facilities.

In 1957, the Singapore Industrial Promotion Board (IPB) was set and began to offer tax incentives and some grants to would-be factory investors, as well as free land for the construction of the production installations that did not fall within the category of the ready-made facilities. Once the British transferred all policy-making powers to the new state, PAP policy makers increased the level of support even more.[52]

In the 1960s, the IPB was transformed into the Economic Development Board (EDB) and the new government body was supplied with an abundance of resources so as to finance a much larger output of manufacturing. The EDB created many more ready-made factories and leased them or sold them at even more subsidized prices. It also procured abundant grants so that the occupiers of these subsidized factories could pay for a very large share of the costs of the machinery and equipment. Larger tracts of land were put at the disposal of the EDB so that they could be supplied free of cost to the manufacturers who were not interested in the multistory factory facilities. Utilities were offered at subsidized prices and the state absorbed the infrastructure costs linking the city with new factories in separated zones. The EDB absorbed an even larger share of the training costs of new factories. Tax exemptions and tax reductions were enlarged. Banks were also regulated so that they would lend a large share of their assets to manufacturing.

Not all of the support, however, was supplied indirectly. Contrary to Hong Kong, which limited its promotion efforts to fiscal, financial, and nonfinancial incentives, Singapore decided to make direct investments in the sector. Part of the EDB funds were utilized to create state companies in fields where foreigners and Singaporeans were not interested in investing, or in national security-related fields. Dozens of state-owned factories were created during the 1960s. The much stronger level of support for the sector during this decade was paralleled with a much faster rate of factory output, which averaged about 12.0 percent. This, in its turn, was followed by a much faster rate of economic growth, which averaged about 8.8 percent.[53]

In the 1970s, the PAP leadership continued to be as strongly motivated toward the sector as during the preceding decade. The political breakup with Malaysia in 1965 reinforced the ideas that up until then had been driving the political class to procure strong support to the sector. Because Malaysia is a resource-rich nation, the PAP leaders thought that the linkage procured a sort of primary-sector security. Once the political linkages were dissolved, it seemed as if Singapore could no longer count on the primary goods of its neighbor. So the PAP leadership concluded that the island had to concentrate even more on manufacturing, for it was the only means to earn large amounts of foreign exchange to pay for foodstuffs and other primary goods.

Without the political breakup and because of the erroneous conclusion that primary goods were important for growth, it is highly likely that the

Singaporean government would not have promoted factories as much as it did during the 1970s. The political separation with Malaysia also drove the government to pursue a more export-oriented policy. Until the breakup, many factories had directed most of their exports toward the Malaysian market, as had been the case during colonial rule. After the separation, Kuala Lumpur raised trade barriers for Singaporean goods and the island was forced to look for alternative markets. The only alternative for the export of large amounts of goods was OECD countries, which were very far away and demanded very high levels of quality. The new orientation of exports therefore increased competitive pressure as it forced production to become more efficient in order to reduce transport costs and in order to increase quality.

The very strong support of manufacturing during the 1970s correlated once again with a very fast growth of factory output (11.8 percent annually), and the higher levels of competitive pressure coincided with a larger extraction of growth from every unit of factory output. GDP averaged 9.2 percent annually. With an actually slightly lower factory output than in the 1960s, it was possible to attain a slightly faster rate of GDP.[54]

The political separation also drove the government to venture into new manufacturing fields because it thought that the traditional fields would not procure enough export possibilities, and because it was also convinced that sustained growth was only possible with a more diversified manufacturing base; both ideas were inconsistent. Hong Kong clearly demonstrated that it was possible to export massively and grow very rapidly on a sustained basis with just a few manufacturing fields. What ultimately mattered was the overall rate of factory output. However, this diversification policy attracted deep criticism from foreign and domestic analysts. They argued that because the island lacked experience producing these goods, the chances were very high that the effort would fail completely.

Beginning in the 1950s, but more particularly in the 1960s when the government began to try to produce factory goods in much larger amounts, numerous foreign and domestic analysts asserted that the effort was doomed to fail because the island had no experience producing large amounts of manufactured goods. The goods that were produced in these decades had already been produced during colonial rule. The difference was just a matter of quantity, and in spite of that, many still argued that it would never become reality. These people were proven totally wrong, as almost from one day to the next the island began to produce large amounts of factory goods that were internationally competitive. In the 1970s, the critics were even more convinced that the effort would fail because the goods had never been produced, and once again they were proven totally wrong. In no time, the island was producing vast amounts of the new goods and exporting them to all corners of the world.

The government provided a huge amount of fiscal, financial, and nonfinancial incentives to foreign and domestic manufacturers in the new fields it targeted. It also made direct investments and created its own factories in these fields. Because the state did not wish to generate a large public-sector debt and was afraid of the risks involved in the new fields, it aggressively promoted FDI. Foreign companies offered the advantage of being experts in the fields that were targeted and of having better export possibilities in OECD and other countries. During the 1970s, FDI accounted for about 70 percent of investment in manufacturing.[55]

After having produced mostly textile, garments, petrochemicals, ships, smelted tin, and milled rubber during the 1960s, the list became enlarged in the 1970s with electricals, electronics, TV sets, and several others. In the 1980s, more new fields such as computers, VCRs, and computer parts were promoted, and in no time the island became a major international producer of these goods. In the 1990s, more new fields, such as semiconductors, telecommunication equipment, and pharmaceuticals, were targeted by the state and almost from day one the island began to export them in large quantities. Again and again, numerous analysts asserted that the efforts would fail because the country had no previous experience producing those goods, and again and again those predictions were proven wrong.[56]

The case of Singapore and the other NICs very explicitly demonstrated that there is no such thing as a natural proclivity of a country to produce this or that. Countries that for centuries produced only primary goods can suddenly become major producers of factory goods. Countries that for long produced only certain types of manufactures can in no time begin to produce vast amounts of numerous other factory goods very competitively. A historical cross-country analysis suggests that what is determinant for deciding what a country ultimately produces fundamentally depends on the level of government support for manufacturing. The particular level of support of the different factory fields will also determine the development of those fields.

In the 1980s, the government continued to be strongly motivated toward promoting manufacturing. The scarcity of natural resources made factory goods the main possibility for exports and for generating large amounts of foreign exchange that were needed to import large amounts of foodstuffs and other commodities. The PAP leadership also continued to believe that there was some sort of causality linkage between prosperity and manufacturing. By then, the leadership was also aspiring to turn the island into a regional technological center. Since technology found its materialization almost exclusively in manufacturing, the high-tech ambitions of policy makers were transformed into support for the sector.

However, the impressive achievements of the preceding decades also drove the government to the conclusion that there was no longer a need to allocate as many resources to the sector. On the one hand, foreign exchange reserves had grown abundantly, and on the other, the country was rapidly moving toward a developed status. Since it had become a standard belief among most economists and policy makers in developed countries that the most important sector for an advanced economy was services, many in the PAP began to embrace similar beliefs.[57]

The end result was a level of support for manufacturing that was less strong than in the 1970s. While in the 1970s factory output averaged about 11.8 percent, in the following decade it averaged only 8.5 percent. Such a deceleration coincided with an economic slowdown. GDP averaged only 7.2 percent in the 1980s after having grown by 9.2 percent in the preceding decade.

It is worth noting that in the 1970s oil and most other commodities experienced a very rapid rise in prices and that Singapore imported almost 100 percent of its primary sector needs. Many economies throughout the world experienced a significant economic deceleration and it was argued that the higher cost of commodity imports was fundamentally responsible for the slow down. However, the majority of these countries were considerably less dependent on primary sector imports than Singapore. If that had been the cause of the slowdown, then Singapore should have endured the steepest deceleration, if not a contraction.[58]

The fact is that the economy accelerated its pace and GDP grew much faster than the bulk of nations. For a nation that lacked natural resources so totally that it even had to import a large share of its water needs from Malaysia, the events of the 1970s become impossible to explain under orthodox assumptions. Only if it is assumed that manufacturing is the fundamental source of wealth creation does it make sense. Under this premise, it becomes understandable why, in spite of the large commodity price increase, economic growth remained very strong and even accelerated a little.[59]

In the 1980s, oil and most other commodity prices fell significantly, but Singapore's economy did not profit from the much lower prices it paid for these imports. Orthodox interpretations asserted that such a situation reduced production costs and therefore acted as a stimulant to growth. The fact that GDP decelerated considerably once again made it evident how inconsistent these interpretations were. However, if it is assumed that manufacturing is the main creator of wealth, then it becomes understandable why in spite of the favorable commodity prices, the economy declined. Since the factory-promotion efforts of state during the 1980s were considerably weaker, the rate of growth of this sector was inevitably driven to be slower than in the 1970s, which was directly reflected in the GDP figures.

In the 1980s, policy makers continued to offer an abundance of fiscal, financial, and nonfinancial incentives to foreign and domestic manufacturers. The aversion to foreign borrowing and a large public-sector debt continued, and consequently emphasis was again put on attracting large amounts of FDI. During this decade, FDI accounted for more than two-thirds of factory output. FDI was also sought for its strong capacity to export. In these years, foreign companies accounted for about 80 percent of exports. The state also continued to make direct investments in the sector. Numerous state factories were founded during this decade. By 1989, the state-owned Temasek Holding Co. managed about 450 companies and employed almost 58,000 people.[60]

Social welfare expenditures increased considerably during the 1980s. The most noteworthy of these programs was public housing. While in 1960 the share of the population that lived in public housing was about 9 percent, the extensive efforts of the following decades and in particular the 1980s drove the figure to about 85 percent in 1989. Many foreign and domestic analysts argued that increased social welfare expenditures had taken away resources for investment and was therefore largely responsible for the deceleration.[61]

The expenditures in social welfare were evidently reducing the possibilities for investment, but if they had not been made it is highly unlikely that the funds would have been utilized to promote manufacturing. The decision to invest less in the sector was not the result of social pressures that forced the government to transfer resources to welfare. It was the result of an inadequate understanding of the causes of economic growth. The government thought that since it had already developed a large, diversified, and competitive manufacturing sector, it no longer needed to support it as much. It thought the sector would be able to take a larger share of the responsibility for financing its expansion. It thought that it was mature enough and that it could grow by itself.

If the government had understood that this sector is fundamentally responsible for the generation of technology, it perhaps would have also understood that without support the sector does not grow. Since the creation or even just the reproduction of technology is an extremely investment-intensive process, the risks involved in this sector are extremely high. That is why only when the state supplies incentives do the risks diminish. Therefore, the greater the incentives, the lower the risks. Without support, the private sector instinctively distances itself from this sector because its possibilities for making a profit are much lower than in the other sectors, which need much lower investments. Since technology is so hard to develop, rates of return in manufacturing also take much longer to materialize than in the other sectors. Without the support of the state, it cannot be expected that the private sector alone will absorb the massive risks involved; and that is precisely what took place in Singapore during the 1980s.[62]

Less government support translated into more risks for the private sector, and capitalists therefore opted for the other sectors. The government was also wrong in believing that large investments in social welfare were the best way to quickly improve living conditions. The bottom line for improving living conditions lies in economic growth. The faster the growth, the more the improvements, and much evidence suggests that growth would have been much faster if the social welfare expenditures had been utilized to promote factories. Even after making the expenditures in social welfare that it undertook, the state could still have promoted factories more enthusiastically without even having to increase public sector debt or run budget deficits, because during the 1980s the government ran very large budget surpluses.[63]

The 1990s in Singapore

In the 1990s, balance-of-payments concerns became less pressing because foreign exchange reserves had grown to a huge size. However, the natural-resource poverty of the island continued to make the government conclude that it was in constant need of exporting abundantly in order to pay for its total dependence on imported primary goods. Since manufacturing continued to be seen as the sector with the largest capacity for generating foreign exchange, support continued to be strong. The PAP leadership continued also to aspire to turn the island into one of the leading technology centers of the world. Since technology is so intimately bonded to factory goods, this motivation ended up channeling large subsidies for the sector.

The motivations of natural-resource scarcity and technology development remained largely unchanged during the 1990s compared with the preceding decade, but the idea of some sort of causality linkage between manufacturing and growth experienced a change. The deceleration of the 1980s reinforced the belief of many policy makers that large investments in this sector were necessary for the maintenance of fast growth. It was therefore decided that everything would be done to keep the share of manufacturing in GDP from falling below 25 percent. To achieve that goal, it was decided to increase the level of support for the sector, and more incentives were offered to foreign and domestic manufacturers. More fiscal, financial, and nonfinancial incentives were supplied to the private sector and more direct state investments were made.[64]

The stronger level of subsidies coincided with a faster rate of factory output that averaged about 9.1 percent annually. This in turn was reflected by a faster rate of economic growth that averaged about 8.3 percent. This acceleration actually constituted an impressive piece of evidence in favor of the manufacturing thesis. By the 1990s, Singapore had attained a level of development which was slightly above the OECD average. Measured in purchasing power parity terms, the island's GDP per capita was slightly above the OECD average. The OECD average for the 1990 to 1997 years was about US$19,000 and that of the island was about US$20,000. During these years, labor costs and average production costs were similar to the average in OECD countries.[65]

By then, it was increasingly argued in OECD countries that high labor costs and high social-welfare expenditures were hampering growth. If high production costs were behind the slow growth in OECD countries, they should have also hampered the growth possibilities of Singapore. The fact, however, is that the island attained one of the fastest rates of growth in the world. While OECD nations averaged a miserly 2 percent annual growth during the 1990s, the island had 8 percent. By then, it was also frequently held that an important factor explaining the slow growth of OECD countries was that they were already developed economies, so demand could not grow quickly because most of their needs were already largely satisfied. If that had been true, Singapore also should have grown slowly because by the 1990s the needs of the Singaporean population were actually more satisfied than were those of the OECD populations.

It was evident that these arguments could not explain reality. The fact that Singapore attained one of the fastest rates of growth in the world clearly demonstrated that fast and even spectacular growth was not limited to developing countries with a large unsatisfied demand. Also, the fact that government support for manufacturing in OECD countries was weak during these years clearly suggested that what these nations needed in order to overcome their high unemployment and underemployment levels was to supply a much stronger dose of support to the manufacturing sector. The case of Singapore further demonstrated that high labor costs and high production costs were not even a minor deterrent for the attainment of very fast growth.[66]

Until the 1980s, Singapore greatly benefited from FDI. The phenomenon of globalization drove nations that had for decades been totally closed or largely closed, to open up their economies to foreign capital. As this phenomenon began in the early 1990s, many economists and policy makers in Singapore and abroad predicted that the economy of the island would suffer because the world stock of capital was limited and numerous other nations would grab a share of it. Since practically all of these nations had much lower labor costs as well as overall lower production costs, and several of them even had very high educational levels with top-quality scientists, it was thought that foreign companies would abandon Singapore. The fact is that inflows of FDI grew actually at a faster pace than during the 1980s and the FDI that went to ex-socialist and developing countries was much smaller in per capita terms than what the island received. Such a phenomenon coincided with an increase in the level of incentives the island offered to foreign companies in the 1990s and with a relatively weak incentive policy in ex-socialist and developing countries.[67]

In OECD countries, many blamed the economic deceleration of the 1990s on globalization, arguing that the accelerated emigration of domestic companies to lower-cost nations was hollowing out the production base of these nations. This phenomenon was credited with being responsible for the increasing number of unemployed and the low-paying menial jobs that proliferated. If that had been the cause of the growing problems of OECD nations, then Singapore should have experienced a similar phenomenon, because from the 1980s it began to experience a considerable emigration of companies looking for cheaper production sites. In the 1980s, thousands of companies closed their operations on the island and moved to nearby Malaysia and Indonesia where labor costs were just a fraction. In the 1990s, the pace of emigration of labor-intensive production accelerated even more and companies began to decamp also to more distant countries such as China, Vietnam, and India.

However, Singapore's economy did not decelerate. It actually accelerated its pace, and its rate of growth was among the fastest in the world. In addition, unemployment and underemployment were practically nonexistent. There was even overemployment and the island increasingly had to import nonskilled and skilled labor. It is worth noting that in the 1950s, at a time when there was no emigration of Singaporean companies abroad and when labor costs were cheap relative to OECD countries, unemployment was about 14 percent and underemployment was very high. This all coincided with a much lower level of support for manufacturing than in the 1990s.[68]

What the empirical data suggests is that the globalization phenomenon cannot even be held responsible for a fraction of the problems OECD countries endured between 1990 and 1997. The case of Singapore demonstrated that not even a mass emigration of companies, plus very high labor costs, correlated with slow growth. It was evident that this was not the cause of the rising unemployment problems in developed nations. The case of Singapore suggested that as long as there was strong support for this sector, nothing could practically impede the attainment of fast growth.

Many analysts argued that the reason Singapore outperformed OECD countries so much during the second half of the twentieth century was the particularly harmonious relationship between capital and labor, which translated into very few days lost to strikes. During that period, the economy was indeed disturbed very little by labor unrest and since the 1970s days lost to strikes were actually zero. This argument, however, forgets that during the 1960s the relationship between capital and labor was very confrontational. During those years, strikes were of regular occurance. By then, numerous labor unions were influenced or controlled by communist agitators. However, notwithstanding those disturbances, economic growth was strong and averaged almost 9 percent annually.[69]

Also worth noting is that even though capital-labor relations in most developed countries were not characterized by harmony, they were not confrontational. Also, there were several OECD countries that by the 1990s had very harmonious relations. Japan, Switzerland, and the Scandinavian nations had a relatively long tradition of cordial capital-labor relations. What's more, they actually had a much longer tradition than Singapore, and in spite of that, economic growth was very weak. Governments should try to foster harmony between capital and labor, but this was clearly not the cause of Singapore's success. It is probably not chance that accounts for the fact that government support for manufacturing in Japan, Switzerland, and the Scandinavian countries during the 1990s was weak and that rates of factory output were also slow.

Because of their growing economic problems, OECD countries increasingly resorted to protectionism in order to slow down the emigration of their companies and to diminish the imports from non-OECD nations. However, from the 1970s, GDP figures decelerated more and more. It was evident that raising nontariff barriers could not impede the deceleration, and less still deliver fast growth. The case of Singapore demonstrated that fast and even impressive growth in an already developed economy was perfectly possible without the slightest tariff or nontariff barrier.[70]

The case of Singapore also demonstrated that small government and healthy public finances were compatible with fast growth. The Maastricht Treaty criteria for monetary unification and its demand for healthy public finances drove many in Western Europe during the 1990s to the conclusion that it was hampering economic growth. It was argued that not allowing budget deficits to exceed 3 percent of GDP was inhibiting governments from spending more to reactivate the economy. By then, government expenditure as a share of GDP in Western Europe was about 44 percent. In Singapore, on the other hand, it was only about 17 percent and it ran a budget surplus of about 10 percent of GDP.[71]

Because government expenditure in Western Europe was almost three times larger than in Singapore, it was evident that policy makers had a higher capacity to influence the economy, and in spite of that GDP figures were about four times faster in Singapore. However, if it were true that the restrictions on running large budget deficits were hindering faster economic growth, then Singapore should have performed worse than Western Europe because it was even running large surpluses. It was probably not serendipity that West European governments spent the majority of their revenues on nonmanufacturing programs such as social welfare, housing, health, education, and infrastructure. Only a very small share was used to promote factories. In Singapore, on the other hand, the majority of government revenue was utilized to promote manufacturing.

The case of Singapore clearly demonstrated that a large government expenditure for a developed economy was not inevitable, nor was it needed for the welfare of society. While West European governments were increasingly incapable of providing housing, health, and most other basic needs to their populations, the Singaporeans were increasingly attaining more and more of those things.

The only way to provide a population with housing, health, education, and the other things that it needs for a pleasant existence is by attaining economic growth. The faster the growth, the faster the improvement of living conditions, and as the case of Singapore suggests, governments can better attain growth by channeling the majority of their expenditure into manufacturing and not into other activities.

By the mid-1990s, the Singaporean government continued even to create its own factories, to supply ample grants to private sector manufacturers, and to regulate banks so that they would lend a large share of their assets to the sector. In the early years of this decade, for example, it initiated a multibillion U.S. dollar project intended on turning five diminutive islands to the south of the main island into a united piece of land in order to accommodate a vast expansion of petrochemical output. With ten-year tax holidays, subsidized rents, ample grants, and other incentives, numerous refineries, petrochemical plants, and factories making end-use products set operations in the artificial island.[72]

European monetary unification criteria also demanded low inflation levels (not more than 3 percent), and many analysts argued that the tight monetary policy that was needed to achieve that goal was hampering the growth possibilities of Western Europe. Once again, the case of Singapore showed that this argument was not consistent. During the whole second half of the twentieth century Singapore's government applied a coherent monetary policy, yet the economy grew at an impressive pace. The responsible monetary policy delivered a low inflation rate that averaged about 2.4 percent annually, while GDP averaged about 7.6 percent. In the 1990s, very low inflation was experienced, again accompanied by very fast growth. The absence of a correlation clearly indicated that a causality linkage did not exist between a tight monetary policy and slow growth.[73]

Misunderstanding Causality

The manufacturing thesis was also reinforced by the way technology developed. A strong correlation was systematically observed, and not just between 1950 and 1997. The growth rate of technology during the second half of the nineteenth century was very slow, and that coincided with weak government support for manufacturing. Then, during the first half of the twentieth century, support for the sector increased considerably, factory output grew much faster, and technology did likewise. During both periods, all of the technology was imported, but in the second period (1900-1949), the rate of technological development of the island was much faster. In the second half of the twentieth century most of the technology continued to be imported, but the pace of its growth accelerated by a large margin, and that once again correlated with a considerable increase in the factory-promotion efforts of the state. By the 1990s, Singapore was among the most technologically developed economies in the world and it was also among the half dozen economies with the highest per capita output of factory goods.[74]

The strong correlation between manufacturing and technology suggested a causality linkage, but there was more that substantiated such an idea. During the second half of the twentieth century Singapore developed technologically much faster than OECD nations even though these countries made the bulk of the inventions of this periods. OECD countries had practically all of the best scientists in the world as well as practically all of the funds, the technicians, and the equipment for advancing at the fastest pace. Singapore, which had practically nothing of this by the mid-twentieth century, grew faster, caught up, and even surpassed most of these nations.

Many argued that it was easy for the island to grow faster because it just imported the large stock of existing world technology, but the fact is that most developing countries that could have done the same thing did not. Practically all other developing nations attained a much slower pace of technological development than Singapore. Their slower technological growth, however, coincided with a much lower levels of government support for manufacturing. The slower growth of technology in OECD countries also correlated with much weaker factory promotion efforts compared with the island.

Since the creation or even just the reproduction of technology, by its very nature, absorbs a vast amount of resources, it is inevitable that if manufacturing is the fundamental creator of technology, then this sector and investment should correlate. The historical data gives credence to such an inference. Whenever the government promoted the sector, investment rose, and the stronger the support, the higher the level of investment.

During the first half of the nineteenth century, for example, investment was nonexistent. In the following fifty years, it rose considerably and that was paralleled by a total absence of support for the sector in the first half and by a considerable increase in the second half. During the years 1900 to 1949 the factory promotion efforts of the colonial government were much larger than before, which coalesced with much higher levels of investment. Investment as a share of GDP averaged about 6 percent in the 1850 to 1999 period and 13 percent in the following fifty years. During the second half of the twentieth century, the island's policy makers largely increased the share of overall resources that were allocated to manufacturing, and investment as a share of GDP rose to average about 30 percent.[75]

By the 1990s, Singapore had the highest level of savings in the world, accounting for about 46 percent of GDP. Many argued that a culturally determined propensity to save abundantly is what made the high levels of investment possible. If it was culturally determined, then levels of savings should have been largely constant through time. The fact, however, is that during most of the island's history, levels of savings were very low. Until the late nineteenth century, they were practically nonexistent and during the first half of the twentieth century they averaged only 4 percent of GDP. Even by the 1960s, they accounted for only about 12 percent.[76]

Also worth noting is that during the 1960s economic growth was actually faster than in the 1990s even though during this last period savings were almost four times higher than in the 1960s. There was a long-term correlation between savings, investment, and growth, but in the short term the correlation at times broke down. The fact that savings were not always high clearly indicates that they were not culturally determined and the absence of a constant correlation suggests that savings were an effect of growth and not the cause. Since growth seems to be the result of manufacturing and the levels of support for this sector varied so much through history, it becomes therefore understandable why levels of savings varied so much over time.[77]

By the 1990s, it had also become very popular to assert that Singapore's fast growth was the result of a culturally determined workaholic population. A long-term historical analysis, however, revealed that the empirical data could not substantiate such a view. The number of hours worked during the first half of the twentieth century, for example, was much higher than in the second half, and in spite of that, economic growth was considerably slower. The inconsistency of this argument becomes even more evident when the second half of the nineteenth century is compared with the period 1950 to 1997. In the first period, the island was impregnated with the same culture that prevailed in the second period, and the average number of hours worked per year was almost twice as many as during the second period. Economic growth nonetheless was very slow. It was precisely when the Singaporeans were working the least that they were attaining their fastest rates of growth. The argument of culture acting as a propeller of growth was also widely utilized to explain the fast growth of East Asia, and in every single economy it failed to correlate consistently with the historical evidence.[78]

There were others who argued that the island's fast growth was the result of the quasi-authoritarian regime that prevailed during the second half of the twentieth century. Even though since the 1960s parliamentary elections (free of fraud) were regularly conducted, the PAP ruled uninterruptedly in a rather authoritarian way. Many economists and policy makers in Singapore and abroad came to believe that the absence of a liberal democratic Western-style system was largely responsible for the fast growth. Interest groups, according to this argument, in their intensive lobbying efforts in Western nations inhibited the adoption of rational policies and therefore hampered growth. However, if that had been a major determinant of growth, the economy should have performed similarly during the preceding periods, because during British colonial rule authoritarian government was even more pronounced. Interest groups then had even less ability to interfere with policy making. The fact that growth was much slower during British rule suggests that this was not a causal factor.[79]

Many also believed that Singapore owed its success to its absence of natural resources. It was thought that the scarcity of resources drove the population to be more inventive, hard-driven, entrepreneurial, and resourceful. A study by two Harvard economists of nearly a hundred countries for the 1960 to 1990 period found that GDP growth was faster for countries with a low endowment of natural resources than for those with a higher endowment. There was indeed a correlation, but it was a weak one. By the late twentieth century, the United States possessed vast amounts of natural resources and it was also the wealthiest country in the world. This made it evident that lacking resources was not essential for growth. Substantiating further the idea that there is no causality linkage were the cases of Canada and Australia, which were among the wealthiest and possessed as well a huge amount of resources.[80]

However, something about this idea seems to have contributed to Singapore's fast growth. Because of the scarcity of natural resources, the PAP was driven to see manufacturing as the only viable alternative for exporting in large amounts. There were other motivations that drove the government to promote factories in a decisive way, but the absence of natural resources played certainly an important role. Because the promotion of factories was ultimately a politically motivated act that responded to particular circumstances, in several other resource-scarce economies growth was not fast. That is also why in numerous resource-rich nations economic growth was strong.[81]

The governments of Singapore during the second half of the twentieth century supplied the strongest support for manufacturing during the entire island's history. This level of support was also one of the highest in the world. However, independent of how strong this support was, there is strong evidence that this support could have been much stronger, and that factory output and GDP could have also grown much faster. Being resource-poor save the island from investing in the primary sector but it did not saved it from allocating vast amounts of resources into services.

During the whole period, the different governments were convinced also that services played an important role in growth and consequently they took numerous measures to promote this sector. By the 1990s, services accounted for almost two-thirds of GDP. As in practically all other countries, services constantly showed a strong dependence for their very existence on factory goods, and they also had a poor productivity performance. Much suggested that services were not a cause of growth but actually an effect of it, and therefore a by-product of manufacturing.

The historical evidence reinforced such a thesis, for it showed a strong parallelism between the two. During the second half of the nineteenth century, for example, support for manufacturing was weak and services grew slowly. In the following fifty years, the subsidization of factories rose considerably, which correlated with a much faster development of services. During the 1950 to 1997 period, support for the sector was much stronger than ever before and services grew at an unprecedented pace. Services grew on average by 7.2 percent annually while during the preceding fifty years they grew by only 3.8 percent.[82]

It is worth noting that during the first half of the century, services were more enthusiastically promoted than later, for Britain was mostly interested in utilizing the island as a trans-shipment, ship repair, and storage facility. Manufacturing was relegated to a small subsidiary role. In spite of that, services grew slowly. During the second half of the twentieth century, on the other hand, at a time when a much larger share of the island's overall resources was allocated to manufacturing, services grew at a much faster pace. This apparent contradiction becomes understandable only if it is assumed that manufacturing is the prime creator of technology. As such, it makes possible the existence of services and the other sectors, and also makes possible their expansion.

History suggests that allocating the largest possible share of overall resources to manufacturing is how the promotion of services, construction, and primary activities can be most rationally undertaken. The government of Singapore could have easily allocated more resources to this sector during the 1950 to 1997 period. Hong Kong did precisely that and it attained a faster rate of factory output and a faster rate of GDP. There is nothing, however, that indicates that the example of Hong Kong was the limit of possibilities.

Singapore also could have extracted more from its manufacturing investments if the government had not allowed for the existence of any form of competition-hindering practices. Although few, there were nonetheless some cartels and monopolies that could not be justified on logical grounds. Neither could the hundreds of state-owned companies be logically justified. It was not just that they should have been privatized, but also that they should not have been created in the first place. If the government was interested in targeting particular fields that were very investment-intensive, it should have largely increased the fiscal, financial, and nonfinancial incentives to the private sector. As long as the possibilities for profits are large, there is absolutely no manufacturing field that will not attract strong interest from capitalists and entrepreneurs, and capitalists and entrepreneurs are much better capable of maximizing resources.

Living conditions improved considerably in every aspect in Singapore during the second half of the twentieth century. Even income distribution improved significantly. Several development theorists have argued that fast economic growth has some negative side effects, such as the worsening of income distribution. The experience of Singapore, however, as well as the three other NICs that attained the fastest economic growth in the world during this period, shows precisely the opposite. Income distribution improved so much that it became among the least unequal in the world.[83]

By the 1990s, income distribution in Singapore was similar to that of Western Europe. The most interesting aspect of this phenomenon was that this success was achieved while taking very few measures to restructure the distribution of wealth. This situation suggested that some other variable spontaneously made the restructuring possible. It was surely not culture, for until the mid-twentieth century the same culture had coincided with a very distorted distribution of income. The fact that the governments of these four economies promoted factories more enthusiastically than any other government in the world leads to the suspicion that these measures caused the improvement of income distribution.[84]

If it is assumed that manufacturing is the prime creator of technology and therefore wealth, it becomes understandable why wages in this sector were systematically higher than in all the other sectors. In the four NICs, factory wages were the highest during the 1950 to 1997 period. The strong support of the sector translated inevitably into a much larger share of the population working in factories and therefore earning the highest possible wages. Under those circumstances, the share of income of the working class from the total income of the nation had to increase, and income distribution had no alternative but to improve. Ample evidence suggests that the impressive progress on income distribution, education, health, real incomes, law and order, democracy, and environmental protection could have been greater in

Singapore and the other NICs if support for manufacturing had been stronger. In spite of having promoted this sector more than practically all other nations of the world, it is evident that it could have been promoted at least twice as much. Had that occurred, factory output would have almost surely grown twice as fast and the economy would have shadowed.

By the 1990s, Singapore had become a highly environmentally conscious country, where the government and the general population took numerous measures to safeguard the living conditions of plants and animals, as well as to protect ecosystems. By then, Singapore was a highly developed nation that could afford to fund all of those measures. Such a high level of development, however, would have not been possible without the strong factory-promotion efforts of the state, and if the support had been stronger the protection of nature would have almost surely been even more thorough.

5
The United States of America During the Second Half of the Twentieth Century

Small Government and Large Government

The economic performance of the United States during the second half of the twentieth century was almost identical to that of the preceding fifty years. Economic growth averaged about 3.0 percent annually while during the 1900 to 1949 period the figure was about 3.1 percent. (See table in appendix.)[1]

During this time government expenditure grew considerably. From about 18 percent in 1950, it accounted for about 35 percent of GDP by 1997. This was the second fifty-year period in the history of the country in which a large increase in government expenditure had not delivered an improved economic performance. During the first half of the century, the participation of the state in the economy had grown even more compared with the previous half century. During these years, the share of government expenditures tripled in size and GDP figures actually experienced a pronounced deceleration relative to the 1850 to 1899 period.

Several times during the twentieth century, the majority of American policy makers were convinced that an increased state participation in the economy had positive effects. A long-term analysis, however, of the economic history of the United States and of numerous other nations very clearly demonstrates that such a belief was not solidly founded. A correlation was missing and therefore the possibility of a causality linkage. By the end of the cen-

actually become convinced that the causal linkage ran in the opposite direction and that government participation had negative effects on the economy.

A long-term analysis of the twentieth century seems indeed to indicate a causality relationship, but a further observation of the historical data makes it clear that such a relationship did not exist either. During the first half of the nineteenth century, for example, government expenditure as a share of GDP was extremely small, averaging about 4 percent. Economic growth, however, was much slower (2.3 percent annually) than during the second half of the twentieth century, even though the share of the state was many times larger.

Also worth noting is that during the second half of the nineteenth century, government expenditure grew considerably in size, doubling its share, and simultaneously there was a very large acceleration in the pace by which the economy expanded. During shorter periods, there were also moments in which an enlarged participation of the state correlated with an acceleration of economic activity. During the 1940 to 1969 period, for example, state intervention was greater than in the previous three decades and GDP figures were also faster.[2]

Enlargement of the state was evidently not conducive to an improvement of economic activity, but history demonstrated that small government was no guarantee either for the attainment of fast economic growth. It was evident that neither was determinant for growth. The fact that fast economic growth at times coincided with less government activity and at other times with more made it plain that a completely different variable was fundamentally responsible for fast growth. Even though not determinant, small government seems to have had a more favorable effect as it allowed market forces to operate in a freer environment and therefore made possible higher levels of efficiency.

While both schools of thought fail to concur with the facts, the manufacturing-growth thesis fits in very well on that front. State support for manufacturing and rates of factory output fluctuated in direct proportion to the GDP figures during the whole time. The combined manufacturing-promotion efforts of Washington and local governments were, for example, the weakest during the first half of the nineteenth century, which is why, notwithstanding the high levels of competition that prevailed, economic growth was slower than during all the following periods.

During the 1850 to 1999 period, there was a very large increase in the factory-promotion efforts of the state, which is why, notwithstanding the doubling in the size of government expenditure, economic activity accelerated by a very large margin. During this time, most of the increased expenditures were utilized to support manufacturing. In the following fifty years, the tripling in the share of state expenditures was almost entirely utilized to promote nonmanufacturing activities, and part of what had been formerly allocated to factories was transferred to other sectors. There was a subtraction of support for manufacturing, which correlated with a significant deceleration of the GDP figures.

During the second half of the twentieth century, the participation of the state in the economy grew once again by a large margin and once again practically all of the increase was utilized to finance nonmanufacturing activities. Support for the sector remained unaltered, factory output grew at almost exactly the same pace, as did GDP figures.

There is much to suggest that if the large increases in government expenditure during the two halves of the twentieth century had not taken place, the economy would have surely performed a little bit better as a lower distortion of market forces would have made higher levels of efficiency possible. Most of the increases in expenditures were used to finance social services because it was believed that the living conditions of the masses would be improved faster in this way. However, it was during the second half of the nineteenth century that living conditions improved at the fastest pace, for it was then that the economy expanded at the fastest rate. Economic growth is fundamentally responsible for the improvement of living conditions and much suggests that manufacturing is fundamentally responsible for growth.

A long-term analysis of the historical evidence gives strong reasons for believing that if the American governments of the twentieth century had maintained the same level of support for the sector as in the 1850 to 1999 period, economic growth would have been much faster. Much faster GDP figures could have been attained while government expenditure simultaneously would have remained much smaller; and much faster growth would have delivered a much faster improvement in living conditions.

The large increase in government expenditure during the 1950s, for example, was mainly used to finance social services. Expenditures for all forms of public welfare increased by almost 100 percent, notwithstanding the fact that the Republicans ruled during most of those years. Both parties were convinced that a large increase in social welfare was needed. However, despite the large increase in welfare expenditures, living conditions did not improve quicker than during the preceding decade. Economic growth was actually slower. After having averaged about 4.4 percent annually during the 1940s, GDP averaged 3.8 percent in the 1950s.[3]

Also worth noting is that the bulk of the increases in welfare programs occurred from the mid-1950s onward, but it was in the early 1950s when the economy grew the fastest. The same phenomenon had taken place in the preceding decade. Most of the expenditures on social matters occurred during the second half of the decade, but by far the fastest rates of growth happened in the first. In the early 1940s, at a time when welfare expenditures were practically nonexistent, the economy grew at a breathtaking pace and living conditions improved substantially. Support for manufacturing was very strong during the early years of these decades and the evidence suggests that this was the reason for fast growth. It was not because welfare expenditures were low.

The War Decades

In both cases, national security concerns forced Washington to allocate a large share of the nation's resources to the factories producing weapons and related goods, and in both cases the sector's rate of output grew very fast. The GDP figures, however, were not similar. Those of the early 1940s were considerably faster. GDP averaged about 15 percent annually in 1940 to 1943 while during 1950 to 1953, they averaged only 7 percent. It is most likely not an accident that World War II was a much greater conflict than the Korean War and therefore drove the government to promote war related factory goods in a much more aggressive way. The end result was that manufacturing grew by about 19 percent annually during the first period and by about 9 percent during the second.[4]

The abrupt fluctuations in GDP figures and the correlating abrupt changes in levels of support for the sector were very strong evidence indicating that the prime generator of growth was in manufacturing. In the immediate post–World War II years, the economy performed poorly, which coincided with a large reduction in support for the sector. In 1949, stagnation had again gained the upper hand and the government was rapidly running out of ideas as one reactivation effort after another failed. Then, in July 1950, North Korean forces crossed the 38th parallel and invaded the southern half of the peninsula.[5]

Washington immediately decided to endorse a large-scale military operation to liberate South Korea from the communist invasion and much larger expenditures were allocated for the production of weapons and related goods. Defense as a share of GDP went from about 4 percent in 1949 to 14 percent in 1952.[6]

Since the support for civilian manufacturing remained largely constant, this translated into a much faster pace of factory output, and as soon as this occurred, the economy began to grow quickly. In 1950 GDP expanded by 8 percent and the next year by 10 percent. In July 1953, the war finally ended and defense expenditures were cut by more than half. As in the aftermath of World War II, nobody in Washington seriously considered rechanneling those resources into civilian manufacturing, and the end result was a noticeable subtraction of support for the sector. This situation coincided with a recession in 1954.[7]

In the remainder of the 1950s, growth on average was modest and manufacturing also grew at a modest pace, coinciding with a much lower level of support for the sector than in the early years of the decade. During these years, military manufacturing once again received most of the support. The Korean War had heated up the cold war considerably and Washington became convinced that future direct confrontations with communism were inevitable. Defense expenditures therefore were not cut to the same low levels as after World War II. All in all, economic growth was on average strong during the 1950s (3.8 percent), which coincided with a fast rate of factory output (4.2 percent), as well as with a relatively strong promotion effort of the sector.

In the 1960s, the cold war became again heated as communist insurgencies in Indochina threatened to take power. The United States became increasingly involved in the conflict and defense expenditures once again rose considerably for the fabrication of a large amount of weapons. By some estimates the United States dropped more bombs over Vietnam than over Germany during World War II. On average, defense expenditures during this decade were about 8 percent of GDP, about the same as during the 1950s. Even though defense expenditures were almost identical as in the 1950s, manufacturing nonetheless grew a little faster during the 1960s. This faster rate coincided with Washington's stronger promotion of civilian manufacturing.[8]

This stronger support centered fundamentally on the field of space factory goods. In 1957, the Soviets had taken the lead in this field as they launched the first artificial satellite. This event humiliated Washington, as it had been repeatedly argued that the centrally planned system was incapable of being creative and inventive. New funds for the fabrication of rockets, satellites, and related goods were greatly and immediately increased. Only by catching up with and overtaking the Soviets could the pain of the humiliation be eliminated.[9]

In 1961, however, the Soviets again beat the Americans in the space race as they orbited the earth for the first time. This further humiliation made it very evident that too little had been invested in that field. New funding was immediately approved. President Kennedy became obsessed with beating the Soviets to the moon and he did not have much of a problem convincing Congress that only a massive government program like the atom bomb's Manhattan Project could deliver the desired results. With such strong support, space factory output began to grow immediately at a very fast pace. The new effort succeeded in delivering the desired results and in 1969 the Americans were the first to walk on the moon.[10]

There was hardly anybody among academic circles who saw the enormous expenditures in space goods as positive. Economically, most condemned it as badly allocated resources that would generate goods that were not the result of real private sector demand. It was seen as a wealth-subtracting activity that would inhibit useful and productive investments. For many, the money would have been better used if it had been deployed to finance social programs. For most economists, however, the government should have simply abstained from taking any action and should had left to the market to decide if spacecrafts, rockets, and satellites should be built. For the vast majority of academics and policy makers, it was hard to envision how those huge investments would ever be useful to society.[11]

Investments in space goods were seen in the same way as those spent on armaments. The efforts to produce these goods were seen as wasteful, as subtractions of wealth, and as obstacles to progress. In the vision of most intellectuals and politicians, the only way in which investment in these two types of goods could be justified was national security or national pride. However, notwithstanding the apparently large-scale waste of resources in weapons and space goods, the economy performed very well; and despite the increase in the share of "wasted" resources from GDP, economic growth even accelerated its pace. After having grown by about 3.8 percent in the 1950s, GDP expanded by about 4.2 percent annually during the 1960s.

The empirical evidence clearly demonstrates that there was something intrinsically wrong with the vision of the world of most intellectuals and policy makers of the time. If their premise was right, the economy should have performed poorly and below the rate of the 1950s, and the 1950s and the 1940s should have been characterized by stagnation or worse. Only by conceding that manufacturing is the principal agent that makes growth possible do the events of these decades become rationally explainable. Factory output corresponded very closely as it expanded by about 5.3 percent annually during the 1960s.[12]

For such a premise to be logically consistent, the historical evidence would have to demonstrate some form of strong causal linkage between this sector and technology. The empirical data would have to indicate that manufacturing is fundamentally responsible for the creation and diffusion of technology. Since technology is the prime generator of wealth, if such a historical correlation existed it would be understandable why, while large amounts of goods with no private sector demand were being produced, the wealth of the nation was rapidly increasing. The historical evidence reveals such a correlation.

The rate by which technology expanded was much faster during the 1960s, 1950s, and 1940s than during the 1930s, when practically no resources were being wasted in weapons or grandiose space projects. Also worth noting is that most of the patents during these decades were directly attached to manufactured goods such as aircraft, missiles, electric instruments, electronic machinery, communication equipment, satellites, tanks, chemicals, motor vehicles, and machine tools.

The technology that was created as weapons and space goods were produced soon found a civilian application, which is how society profited considerably from those investments. Satellites, for example, originated by largely enhancing knowledge about the enemy's forces and therefore improving the defense capabilities of the nation. Soon, however, they started transmitting a huge amount of information about the weather and climate, about geography and topography, and geology and cartography, and were used for radio broadcasts, television signals, electronic paging requests, newspaper layouts, and even credit card transactions.[13]

Armaments that were seen as an investment even less likely to deliver positive results ultimately ended up delivering a vast amount of very useful things that improved the lot of society. For example, the creation of the B-47 and B-52 intercontinental bomber aircraft in the post–World War II years soon led to the development of large jet-propelled passenger aircraft in the late 1950s. Suddenly, average Americans were capable of flying faster, farther, and with more comfort.[14]

Lasers are another illustrative example. Laser machines were first developed in the 1960s because the Pentagon was interested in developing precisely targeted death rays. As with the intercontinental bombers, the government paid for practically the entire research and development costs. Soon, however, laser technology began to be utilized for numerous civilian applications. In the 1960s, for example, a California company began to produce laser instruments named Geodolites, which could very precisely measure distances. By the 1990s, these instruments remained unsurpassed in their geographic and topographic-measuring abilities. Other civilian applications sprouted from laser technology, among the most noteworthy the machines that can perform eye surgery, heart surgery, and even cosmetic surgery.[15]

In the 1960s, the Pentagon's DARPA agency financed the creation of a computer-satellite system that would allow a far superior means of communication for the military and that could permit that the armed forces to remain networked in case a Soviet nuclear attack disrupted normal means of communications. Some years later, the worldwide Internet system of computer communication evolved from this effort.[16]

During the 1950s and 1960s, it was once again this sector that absorbed the bulk of R&D expenditures. Practically all public and private R&D expenditures were for the creation of new factory goods. Aircraft, missiles, electrical and communication equipment, chemicals, motor vehicles, and machinery alone absorbed about 80 percent of overall R&D expenditures.

When all types of weapons and space goods were added to the preceding list, plus medical goods such as pharmaceuticals and medical equipment, and all other fields of manufacturing, the figure came close to 100 percent.

The fields of manufacturing that received the most support from the state were the ones that innovated the most. The factory fields that grew the fastest absorbed the largest share of overall R&D expenditures. It was indeed armaments that were more subsidized, that were greatly promoted, which grew the fastest, which absorbed by far the largest share of R&D, and which innovated the most. In the 1960s, as in the two preceding decades, weapons continued to generate by far the most inventions. Space manufacturing generated the second largest amount of technological breakthroughs, and this was the field that, after weapons, was promoted the most.[17]

Precisely because manufacturing has such an impressive and perhaps unique capacity to create technology is it so investment-intensive. The creation of technology is the most resource-absorbing of human efforts, which inevitably predisposes this sector to be highly investment-intensive. As such, it is by far the most risky economic activity. Not only much more funding per unit of output is required than in the other sectors, but it also takes much longer for profits to materialize. The risk-averse nature of human beings therefore instinctively leads capitalists to distance themselves from this sector. Only when the government diminishes the risks by supplying incentives does the private sector see worthwhile profit possibilities, and therefore engages its resources in factories. For example, R&D expenditures represent a very large share of the overall costs of creating and producing new goods, and during the 1950s and 1960s Washington paid for about two-thirds of all R&D expenditures.[18]

This was not the only mechanism by which policy makers promoted the sector. There were also tax exemptions and tax reductions, subsidized financing, and guaranteed state purchases at prices that secured a profit. Weapons producers received the most incentives. The standard fixed-price contract for weapons producers provided for profit margins of about 8 percent. However, most of the time the government granted higher profits. In some cases, profits of up to 800 percent on investment were realized. Many of these exorbitant prices were the result of unlawful behavior on the part of the private sector. However, most were was the result of a deliberate government policy to increase the profit possibilities of producers so that private sector investment would be high and the procurement of weapons and related goods would be assured.[19]

For most people, spending about 8 percent of GDP in defense was already a massive waste of resources that was subtracting wealth from society, but when the press uncovered cases of vastly inflated defense contracts, the waste was seen by almost everyone as reaching monstrous proportions. Under those circumstances, it became even harder to explain why the economy performed so well during the 1960s and the preceding two decades. Many argued that the fast growth was the result of the fiscal stimulus of war.[20]

That is very unlikely to have been the cause because during the 1930s fiscal stimulation efforts failed to deliver the desired results. Also worth noting is that during the 1940s, 1950s, 1960s, and 1970s, Keynesian policies systematically proved to be an inadequate countercyclical stimulus. When the economy occasionally reacted to these efforts, it frequently happened with a considerable lag, suggesting that the reactivation was not really the result of it.

Others asserted that the fast growth of these three decades was the result of the tight collaborative efforts between the government, big business, and the military. During the 1950s and 1960s in particular, there was an increase in the number of businessmen employed in government and of high-ranking officers employed by big corporations (especially those in defense fields). It was thought that there was something positive about such a triangular collaboration. However, if this collaboration had really possessed growth-generating powers, it becomes very hard to explain why during the 1970s and thereafter, when those linkages were still in operation, the economy decelerated so noticeably. It is worth noticing that during the last three decades of the nineteenth century, GDP figures were much higher than in the 1940 to 1969 years, when that triangular collaborative structure did not exist. This was evidently not the cause.[21]

Deceleration and Causality

In the 1970s, the American economy decelerated considerably, averaging about 2.8 percent annually. Most economists and policy makers became convinced that the main variable responsible for the slower pace of growth resided in the abrupt and large increase in the price of oil around the middle of the decade, and in the price increase of most other primary-sector commodities. However, the fact that several other countries that were considerably more dependent on commodity imports managed to attain much faster rates of growth clearly indicated that this could not have been the fundamental cause. Japan grew by 5 percent and all four NICs grew by at least 9 percent. South Korea and Singapore even grew faster than in the 1960s.[22]

Other evidence indicating that such a thesis was not consistent was found in the price of raw materials during the Korean War. During these years, the price of primary commodities rose to very high levels in international markets, yet the American economy performed extremely well. Why would high raw material prices deliver fast growth at one moment and slow growth at another? Simply because there was no causality linkage.

Washington was nonetheless convinced that oil was essential for the well-being of the nation and that prices had to remain low for the economy to function well. It was so convinced that in the mid-1970s several government officials implied that the United States might even invade the most uncooperative oil producers in the Middle East if they persisted with their high-price policy.[23]

In the 1980s, oil and the price of most other primary goods fell significantly but in spite of their low prices, the American economy did not recover the dynamism of the 1960s. What's more, it could not even retain the rates of the 1970s and GDP figures actually decelerated further. In the 1990s, the price of oil and most other commodities fell again, and once again the economy took the road opposite to the one assumed by the oil-growth thesis. During 1990 to 1997, average GDP rates decelerated further. The evidence clearly indicated that the rise in commodity prices was not the determinant, nor even an important cause behind the significant deceleration of the 1970s.[24]

It was also argued that the rise in commodity prices plus a lax monetary policy delivered a higher level of inflation, and a faster pace of inflation ended up hampering economic activity. As a matter of fact, inflation did rise considerably from an average of about 3 percent in the 1960s to about 7 percent in the 1970s. There is much to suggest that the cause of such a rise was in the increase in the price of imports and in the lax monetary policy that allowed real interest rates to average zero. Capital was actually costless.

However, a causality linkage between inflation and growth is impossible to find. During the 1980s, tight monetary policies were applied and real interest rates became positive. From zero they went to average about 4 percent. Inflation did fall to an average of about 6 percent annually, but the economy did not accelerate and performed actually worse than in the 1970s.[25]

Many economists and policy makers nonetheless argued that even though inflation had decreased, it was still far above the rates of the 1960s, and for the economy to grow quickly, there was a need to bring it further down. In the 1990s, therefore, monetary policy was tightened even more and inflation actually fell to almost the same level of the 1960s. During 1990 to 1997, prices increased on average by a little more than 3 percent annually. The economy, however, performed at an even slower pace than in the 1980s.

The control of inflation is actually the only real success of the economics science. When governments have endorsed a responsible monetary policy and have balanced the budget, prices have always fallen rapidly. The twentieth century has provided numerous examples of the above. It is also beyond any doubt that low and stable prices is a desirable goal. Much suggests, therefore, that the efforts undertaken by the American government in the 1990s had positive effects for the U.S., but an even larger body of evidence indicates that inflation has practically no effect on growth. The fact that inflation in the United States was zero during the eighteenth century and that GDP managed to average only a little more than zero, further substantiates such a claim. The fact that inflation averaged almost 20 percent annually in South Korea during the 1970s and GDP expanded by more than 10 percent per year went a step further in that direction. Several cross-country studies have demonstrated that it is not possible to establish a correlation between inflation and growth.

After the end of World War II, the federal government ran budget surpluses for about a quarter of a century. Then, in the 1970s, budget deficits appeared and most economists saw in them a major cause behind the significant deceleration of the economy. From the early years of independence, the Americans inherited from the British the belief that healthy public finances were vital for an adequate management of the economy—that expenditures needed on a constant basis not to be higher than revenue. The immense deficits of World War II and the accompanying impressive growth figures clearly indicated that a causality linkage between fiscal imbalances and poor performance was nonexistent. However, most economists and policy makers viewed World War II as an exceptional situation in which economic laws were turned upside down and which could not be repeated.[26]

In the 1980s, budget deficits became larger. Most analysts during this decade became even more assertive in their claims about the negative effects of deficits as further deceleration of the economy seemed to confirm their view. So in the 1990s, Democrats and Republicans united their efforts to reduce the imbalances. From a deficit that averaged almost 4 percent of GDP during the 1980s, the figure dropped to just 2 percent during 1990 to 1997. The economy, however, did not respond positively and actually decelerated more. After having averaged about 2.4 percent annually during the 1980s, GDP slowed down to about 2.3 percent during the 1990s.[27]

There was nothing in the history of the U.S. or of any other country to suggest that budget deficits had a positive effect on the economy, but nor was there anything that indicated the opposite. Little suggested that healthy public finances were essential or even important for strong economic growth. The evidence was mixed, but it nonetheless showed that healthy finances were marginally positive for the economy. On several occasions in history, some nations ran deficits yet attained impressive growth, but by far the most impressive growth figures were reached while governments ran surpluses or balanced budgets. The United States, for example, attained its fastest GDP figures during the late nineteenth century when healthy public finances prevailed. The fastest rates ever experienced among OECD countries were attained by Japan during the 1960s, coinciding with budget surpluses by the central government which averaged about 1 percent of GDP.[28]

The fastest sustained rates of growth the world has ever seen over half a century were achieved by Hong Kong during the second half of the twentieth century, and during practically this whole period the colonial government ran surplus or balanced budgets.

There were others who argued that the pronounced deceleration of the 1970s was the result of strong labor activism which hampered productivity as it inhibited flexibility at the workplace and as it demanded high wages. The problem with that argument is that it fails to objectively analyze all of the available information. The fact is that labor activism was stronger during the 1950s and 1960s. For example, it was in 1955 that unionization as a share of the work force reached its historic peak level. In that year, it stood at about 33 percent of the total.[29]

From then onward it began to decline and even though the share was lower in the 1960s, it was nonetheless higher than in the following decades. If that argument pointing at labor were true, then it should have been during the 1950 to 1969 period when the economy performed the worst. That was not the case. GDP figures during the 1980s were even slower than in the 1970s, although the share of unionization of the American labor force had continued to decrease. By 1989, the share was just 15 percent. The Reagan administrations largely blamed the unions for undermining the economy and much was done to reduce their productivity-hindering activities, but that was totally incapable of restoring the dynamism of the 1960s and was not even able to sustain the rates of the 1970s. In the 1990s, labor activism dropped again and the American labor market was regularly cited by economists and policy makers throughout the world as an impressive case of flexibility and rational wage demands. However, during the 1990 to 1997 years, the economy continued to decelerate.[30]

There is much to suggest that labor flexibility has productivity-enhancing effects and that deregulation and liberalization measures are positive policy undertakings, but there is also evidence that they are not determinant for the attainment of fast economic growth. Had they been determinant it would have been in the 1990s when the best GDP rates were attained and in the 1950s when the economy grew at the slowest pace. The events in numerous other nations further substantiate this assertion for it was impossible to uncover a correlation between labor flexibility and economic growth. In the bulk of OECD countries during the second half of the twentieth century, the fastest rates of growth (1950s and 1960s) were obtained at a time when their labor markets were characterized by rigidity and immoderate wage demands.

At a time when they were the most flexible and wage demands were the most moderate (1990s), their GDP figures were at the lowest level.

Since the nineteenth century, the U.S. economy showed clear signs of being domestically driven. By the later part of this century, it had already become the largest economy in the world but its foreign trade as a share of GDP only accounted for about 12 percent. By the 1960s, the share had only barely increased and was almost identical to the late nineteenth century. Many therefore began to argue that the rapid internationalization of the world economy in the 1970s had significantly changed the relationship of the American economy to the rest of the world. Since the share of foreign trade only barely increased in the 1970s, it was argued that maintaining an insular position was incompatible with the new economic situation of the world.[31]

The rapid growth of Japan, Western Europe, the NICs, and numerous developing countries in the post–World War II decades had indeed radically changed the structure of the world economy, but there is much to suggest that such a change was not the cause of the significant deceleration of the 1970s. Several West European countries experienced an even steeper deceleration even though they were much more in tune with the changed world economy. Germany, for example, by then had a share of foreign trade almost three times as large as that of the United States, and in spite of that, the economy decelerated from 4.6 percent in the 1960s to 2.6 percent in the 1970s. In the United States, it just went from 4.2 percent to 2.8 percent. In Japan, the figures did not fall below those of the United States but fell by a larger percentage (11 percent to 5 percent), even though this was the country that best epitomized the new internationalized economy.

It is also worth noting that the American economy during the 1980s experienced considerable internationalization. Foreign trade grew to account for almost one-fifth of GDP and, in spite of that change, the economy made absolutely no progress. Quite to the contrary, rates of growth actually decelerated. In the 1990s, the economy became even more globalized and foreign trade came to account for about one-fourth of GDP, but once again the economy did not respond in the way most economists had predicted. GDP figures, instead of accelerating, became even slower than during the preceding decade.

Along with foreign trade, it was asserted that foreign direct investment (FDI) was also important for internationalizing the economy. Many therefore stated that the United States had not linked its economy with the rest of the world by means of FDI as much as many other nations had. Here again, such a position was not consistent with the facts. As with trade there were growing inflows and outflows (the emphasis was nonetheless on inflows) of FDI during the second half of the twentieth century, but the correlation went actually in the opposite direction. The larger the share of those flows, the more the economy decelerated. In the 1950s and 1960s, for example, inflows of FDI accounted for about 2 percent of GDP and the economy grew by about 4.0 percent. By the 1980s, inflows had grown to account for about 5 percent but growth had slowed down to just 2.4 percent.[32]

By the early 1990s, Japan had the lowest level of accumulated FDI per person among OECD countries, even though it had outperformed all of these nations since the end of World War II by a large margin. Due to its strong nationalism, Japan systematically blocked FDI and received much less than the other developed nations. However, it managed nonetheless to grow much faster. By the early 1990s, for example the United States had received in per capita terms almost ten times as much as Japan, but during the 1950 to 1990 period the archipelago grew more than twice as fast as the United States.[33]

Much evidence suggests that a larger share of imports and exports as well as larger flows of FDI help to increase the level of competition in the economy and therefore the level of efficiency. As firms fall under more competitive pressure, they are forced to cut costs, improve quality, and innovate. However, the empirical evidence clearly indicates that these positive effects are not essential for the generation of economic growth. If they were, the United States should have attained the best performance in its entire history during the late twentieth century and Japan should have grown much more slowly than the United States between 1950 and 1990. As for Britain, it should have actually grown the fastest among large OECD countries during the second half of the twentieth century, because flows of FDI were the largest and the share of foreign trade was among the largest. None of this happened and actually, the opposite took place. Britain was the worst performer among large OECD countries during this period.[34]

In the early 1970s, the American government put an end to the gold standard by suspending its linkage to the dollar. As the currency was allowed to float freely, it immediately lost much value. Many argued that the end of the gold standard and the devaluation that followed were largely responsible for the deceleration of the economy. A long-term analysis, however, reveals that it is highly unlikely that these events played even an important role in the slowdown. The United States, for example, adopted the gold standard in 1871, but during the preceding three decades it attained rates of growth that were faster than those attained in the 1970s. Then in 1933, the government stopped converting currency into gold. Although the economy did not perform well during 1933 to 1939, its performance was far above that of 1930 to 1932, when the dollar was linked to the precious metal. In 1946, the dollar was relinked to the metal but in the remaining of that decade the economy performed poorly.[35]

It was like the case of Britain in the eighteenth century, when the pound's linkage to gold at times coincided with fast growth and at other times coincided with stagnation or even worse. There was simply no correlation and therefore no possible causal linkage.

On the subject of devaluation, it was also impossible to establish a consistent correlation with the performance of the economy. Numerous countries at different moments were able to attain fast growth while they devalued their currency. For example, China in 1994 devalued its currency and in that year and those following it had a very fast rate of growth. In other cases, overvaluation coincided with fast growth, as the case of China during the 1980 to 1993 years demonstrated. In other cases still, the opposite took place. In the United States during the 1990s, the currency appreciated against most other currencies in the world and in spite of that the economy did even worse than in the preceding decade. The fact that the empirical evidence does not reveal a correlation of currency's value in one direction or another indicates that this variable cannot possibly be an important causal growth factor.[36]

There were many economists who argued during the 1970s that it was the concomitant effect of all the variables previously cited that was responsible for the deceleration. According to them, each variable individually could not significantly alter the economy, but the compounded effect of all of them could. However, not even when the problem was approached from this perspective was it possible for such a view to add up consistently with the facts. In the 1990s, for example, the inflation-adjusted price of oil was at its lowest point during the whole second half of the twentieth century while most other commodities traded at prices that were lower than those of the 1970s. Besides, inflation was almost at the same low level of the 1960s, monetary policy was handled very responsibly, the budget deficit had been significantly cut, and labor activism was at its lowest point since the end of World War II. Foreign trade accounted for the largest share in the history of the country, the share of flows of FDI was at the highest level in the twentieth century, and the currency had noticeably appreciated.

In spite of all this, GDP figures were slower than during any other decade of the twentieth century with the sole exception of the depression-ridden 1930s. It was very evident that none of these variables, individually or as a group, were responsible for the significant deceleration of the 1970s. The evidence also indicates that these variables, even when presented as a group, had only a marginal positive effect on the economy.

Armaments and Space Goods

While none of these variables managed to relate coherently to the facts, the thesis of manufacturing does. During the 1970s, government support to the manufacturing sector decreased as a result of a reduction of national security concerns. As tensions with communist countries decreased significantly, Washington decided to reduce the share the nation's overall resources that was allocated for the production of weapons and related goods. The Vietnam War came to an end in the early 1970s and both superpowers agreed that if they continued spending so much in armaments they would end up by wrecking their own economies. Both concluded that they needed a period of détente that would put a lid on the arms race. Weapons limitations treaties were signed and the United States lowered its defense expenditures from 8 percent in the 1960s to 6 percent in the 1970s as a share of GDP.[37]

The reduction of military and political tensions among the superpowers also affected the space race. In both countries, numerous economists and policy makers had abundantly criticized the government for making such vast investments in prestige projects while there were still so many unresolved problems on earth. Both governments therefore decided to cut their space budgets. The Americans in particular felt largely satisfied with their space achievements, as they had decisively beaten the Soviets to the moon in 1969. The Apollo program, which made that possible, ended in 1972 and from then on NASA lost a clear sense of purpose.

The decrease in Washington's promotion of weapons and related goods, as well as the production of space equipment inevitably translated into a very large drop in the rate of output of these goods. The resources that were no longer invested in these fields could have easily been transferred to civilian manufacturing, but practically nobody in Washington thought that the nation would obtain any benefit from such a transfer. Since these two fields accounted for a very large share of overall factory production, the decrease in support ended up delivering a significant deceleration in the overall rate of manufacturing output. The average annual rate of this sector went from about 5.3 percent in the 1960s down to 3.1 percent in the 1970s, which was reflected in GDP rates that averaged about 4.2 percent and 2.8 percent respectively.[38]

Just about everybody was convinced that since the weapons and space efforts were subtractions of wealth, the nation would be much better off without them, however, the economy slowed down considerably and wealth was created at a much slower pace.

Economists and policy makers were right to believe that the investments made in armaments were not the best way to maximize resources, but they were wrong to think that when those resources were transferred to civilian nonfactory activities, society was going to extract more from them. Had they been transferred to civilian manufacturing, much suggests that with the same rate of output as in the 1960s the economy would have performed better. Unfortunately, there was hardly anything within the ideas that molded the thoughts of intellectuals and policy makers that considered promoting the sector per se to be a wise policy undertaking.

When it came to space goods, they were even more incorrect in believing that investing in them was a waste of resources. Had they understood that manufacturing is the prime generator of technology, it would not have surprised them that if goods were fabricated at a slower pace, technology expanded also at a slower pace. This inevitably translated into a slower rate of creating wealth. A clear example of how the large investments in weapons and space goods created an impressive technology that significantly benefited the civilian economy was the military's creation in the 1970s of the Global Positioning System (GPS). This system consisted of a satellite network with the capacity to guide missiles, fighter planes, and warships to within 30 feet of targets.[39]

This superior technology soon began to be utilized by numerous civilian fields. Perhaps its most impressive use was in farming. By the 1990s, tractors began to be equipped with receivers that absorbed the information relayed from GPS satellites, displayed it in onboard computers with a mapping system, utilized laser crop monitors to measure field topography, and took soil samples to determine fertility, salinity, pH and water levels. With these high technology factory goods, farmers increased efficiency by being able to irrigate just the precise areas where water is not retained, to spray fertilizers on just the areas with the lowest levels of soil nutrients, and to spray pesticides on the exact areas where crops are infected.[40]

By the 1990s, some of the leading technologies of the world found their embodiment in computers and microchips. The place where these technologies saw the most progress was in Silicon Valley, California, reputed for its inventiveness and entrepreneurial drive. The production of these factory goods began in the post–World War II decades and, contrary to what most would like to believe, Silicon Valley was weaned on government defense contracts. It was the Pentagon that was largely responsible for the development of this economic zone, as it paid for much of the R&D costs and guaranteed purchases at prices that allowed for high profit margins. The technology that sprang from these manufacturing efforts (which was originally intended for military purposes) soon found endless applications in the civilian economy.[41]

The Soviet Union invaded Afghanistan in 1979 and the Carter administration considered that notwithstanding that event, defense expenditures would not increase. In 1981, however, the new Reagan administration decided that Soviet intervention could only be curtailed with force, and therefore defense budgets rose significantly. The much larger output of weapons immediately translated into a increased overall rate of factory production that coalesced with faster economic growth. However, in early 1985, the Communist Party chose a new Soviet leader, and as soon as Mikhail Gorbatchev began to dictate policy it became evident that he wanted more than just a détente like the one in the 1970s. His numerous conciliatory efforts soon ended by convincing even the Reagan administration that the military threat of the USSR had significantly diminished.

Wide-ranging arms reduction treaties were approved by both parties and defense expenditures in both countries began to decrease rapidly. These events coincided with an economic deceleration in both countries. It was though, in both countries, that the savings in defense would translate into more wealth for the civilian economy, but the end result was exactly the opposite. Despite the antagonism between the two systems, there was nothing in the theories supporting capitalism and communism that asserted that government promotion of manufacturing was essential for the attainment of economic growth.[42]

With such a view of the world, it was inevitable that the resources not spent on weapons would not be reallocated into civilian manufacturing, and the end result was a slower overall rate of factory output. The reduction in the second half of the decade canceled the defense increases in the first half and the expenditures of the American armed forces in the 1980s ended being almost the same as during the preceding decade. Defense once again absorbed about 6 percent of GDP.

As the cold war had reheated during the first half of the 1980s, the space race also had gained some steam. This time both superpowers aimed to create the first space station and in 1984 the space station project was unveiled by the Reagan administration. Economists, budget hawks, and even many space scientists immediately argued that the money could be more productively used elsewhere. With the arrival of Gorbachev and the decrease in tensions, the funding for the space station rapidly began to dry up. Had tensions remained high or had they even increased, the opposing arguments would have been put aside as during the 1960s; but, since the effort was almost exclusively driven by ideological motivations, the opponents of investing in space goods gained the upper hand.[43]

Support for space manufacturing was on average slightly lower during the 1980s than during the preceding decade. Since government support for non-defense-related civilian manufacturing remained largely unaltered and there was a small drop in the promotion efforts of weapons and space goods, it was inevitable that overall rates of factory output would decelerate. After having averaged about 3.1 percent annually, the sector went to average about 2.8 percent during the 1980s, and that was followed by a proportionate deceleration of the economy (2.8 percent and 2.4 percent respectively).[44]

Unemployment, Underemployment, and Income Inequality

The peace dividend, which almost everybody thought would arrive in the second half of the decade, never made an appearance. Unemployment and underemployment, which had been on the decrease during the 1940s, 1950s, and 1960s, began to rise in the 1970s and rose even more during the 1980s. After having averaged about 4 percent during the 1950 to 1969 period, unemployment rose to average about 6.0 percent in the 1970s and about 6.5 percent during the 1980s.[45]

Not just the official unemployment figures rose, but also the harder-to-measure underemployment. Low-paying jobs accounted for a larger share of the jobs created during 1970 to 1989. During 1950 to 1969, high-paying jobs had been on the rise, but during the following two decades their numbers decreased. Part-time jobs were practically the only ones that increased rapidly. In consequence, income distribution worsened. Worse still was that those jobs were not spiritually fulfilling. Most Americans considered that many of the jobs created in these years underutilized their abilities. People with a high school or even college education found no alternative but to take work that required an education below the completion of secondary school. A larger share of the work force was frustrated not just by the salaries, but also by the type of work.[46]

As unemployment and underemployment rose, income distribution worsened and average incomes grew slowly. For a relatively large share of the population, however, real incomes fell or stagnated and poverty began to rise. Cross-country studies show that developed countries need at least 3 percent GDP growth on a sustained basis in order to avoid a rise in unemployment and/or underemployment; such a rate was not attained during the 1970s or in the 1980s.

As in previous decades, the differing levels of support for manufacturing varied in direct proportion to the overall well-being of society. In the 1930s, for example, support was weak, and poverty rose in a way unprecedented in the country's history. In the 1940s, 1950s, and 1960s, the factory-promotion efforts of the state were relatively strong and poverty systematically declined. Then in the 1970s and 1980s, support for the sector declined noticeably, GDP fell below the 3 percent mark, and poverty began to rise.[47]

Many argued that the rise in poverty was the result of the rise in the share of GDP going to pay interest on the national debt. However, in other episodes of the country's history, a large national debt did not hinder fast growth. During the Civil War, for example, public-sector debt rose to unprecedented levels, yet fast growth was experienced. The high debt that remained as the conflict ended did not hinder the most spectacular growth in American history in the decades that followed. The even larger debt of World War II coincided also with impressive rates of economic growth, and the large debt that remained as the conflict ended did not hinder fast growth in the 1950s. The federal debt during the 1950s, for example, averaged about 53 percent of GDP while during the 1970s it was only 25 percent.[48]

It is true that in the 1980s the federal debt rose, but it was still way below that of the 1950s and still much lower than during World War II, when it grew to more than 100 percent of GDP. It was most likely not by chance that during the Civil War and the decades that followed, as well as during World War II and the 1950s, support for manufacturing was strong. There is much evidence that government debt, in particular a large one, is not positive for any economy. However, the historical data of the United States and numerous other countries clearly indicates that maintaining low or nonexistent levels of debt is not essential for the attainment of fast economic growth. A policy of strong support for manufacturing does not need budget deficits or public-sector debt. However, since the promotion of this sector is the main variable determining growth, it becomes understandable why, when factories were promoted with the help of a large government debt, the economy nevertheless performed very well.[49]

There were others who argued that the rise in poverty during the 1970s and 1980s resulted from increased drug consumption. This was indeed a new phenomenon, but it is very likely that the majority of the Americans who developed drug habits were frustrated with their inability to find a job. The deteriorating economy increased unemployment/underemployment and therefore poverty, and once in poverty, people turned to drugs in an effort to forget their depressing situation. In practically all countries of the world, drug consumption is more pronounced among the population with the lowest incomes.

Other analysts argued that the growth in poverty was the result of the increased female participation in the work force. Since the salaries that women received were on average below those of men and since their share of the workforce increased significantly, many therefore concluded that this was a major factor affecting poverty. However, the proportion of women in the work force during World War II reached levels similar to those of the 1970s and 1980s, yet economic growth was impressive and poverty actually fell at a very fast pace. More to the point is that by the 1980s women were considerably more qualified than in the World War II years.[50]

None of these arguments could consistently explain why the share of families below the poverty line had risen from 7 percent in the 1960s to 9 percent in the 1970s, and to 11 percent in the 1980s. Nor could these arguments explain why income distribution had worsened so much. By the 1980s, income distribution was at its worst in comparison to all previous decades of the century. It was only during the 1930s that income differentials were wider. The situation was considerably paradoxical when one considers that social welfare programs, explicitly created to fight poverty, had rapidly multiplied since the end of World War II. Perhaps most noteworthy of these efforts was President Johnson's War on Poverty, which debuted in 1964. It is true that during the Reagan administrations some of these programs experienced budget cuts but the vast majority remained intact.[51]

It therefore becomes very hard to explain why during the 1940s, for example, when only a fraction of the antipoverty programs that existed in the 1980s were in operation, poverty decreased so much. How is it possible to explain that at a time when practically no social welfare expenditures (early 1940s) were made, income distribution improved dramatically, and while a multitude of income redistribution programs were in operation (1980s), the opposite took place. It is worth noting that during the late nineteenth century up to the outbreak of World War I, there were practically no income redistribution programs and poverty decreased very rapidly, as did income inequality. This situation strongly suggests that some other variable was not only rapidly creating wealth, but was simultaneously distributing it in a more equal way.

The history of the United States and numerous other nations reveals a strong correlation between levels of government support for manufacturing and income distribution. Whenever the state undertook decisive factory-promotion efforts, the economy grew fast, poverty decreased rapidly, and income distribution improved. In the United States, that is what took place during the late nineteenth century, the early twentieth century, the 1920s, 1940s, 1950s, and the 1960s. However, whenever the efforts were weak as during the 1930s, 1970s, 1980s, and 1990s, the opposite happened and no social program managed to contain the growing disparities in income. The evidence seems to indicate that the power of manufacturing to create and redistribute wealth is so large that when this sector does not receive the necessary support, nothing can contain a rise in income inequalities.

Western Europe during the late twentieth century, had a much larger social welfare system than the U.S., and income distribution nonetheless deteriorated. This situation corroborates the above. The fact that during the 1980s wage inequalities among OECD countries were the widest in the United States and the lowest in Sweden and other West European nations indicates that income redistribution programs are effective to a certain degree. However, the fact that during this decade there was a widening of incomes in most of Western Europe clearly indicates that some other variable was at work. The fact that income inequalities were rapidly reduced in the 1950s, even though the social welfare programs of Western Europe were much smaller, further substantiates the thesis that a strong promotion of factories is the best policy for reducing wage disparities. While support for this sector was strong in that decade, by the 1980s it had become weak.[52]

The reason manufacturing possesses impressive powers for improving income distribution seems to be the same one that explains why this sector is fundamentally responsible for economic growth. As the prime generator of technology, it is predisposed to create the most wealth and in consequence to pay the highest salaries. At practically every moment in American history manufacturing paid on average the highest wages. During the 1980s, for example, despite much talk about the importance of services in a developed economy, it was manufacturing that paid the most. On average, its wages were more than 10 percent above those of services. In the 1990s, the gap was even wider.[53]

In most other countries, the same phenomenon has been observed. Since this sector pays the most, it is only inevitable that as it grows quickly and a larger share of the population works in it, a larger share of the work force therefore receives the highest possible salaries. Under those circumstances, the wages of the working class relative to the rest of society improve, and income distribution has no alternative but to improve. Under a policy of strong support for manufacturing, the poor increase their incomes at a faster pace than the wealthiest, but those with the highest incomes also experience a rapid growth in income. When this sector is not strongly promoted and average factory rates are such that they cannot deliver an average sustained economic growth of about 3 percent per year, this sector's share in the economy and in employment starts to shrink. As a larger share of the labor force is obliged to work in lower-paying sectors, income inequalities inevitably rise.

Causality Misinterpretations of Growth

Although such an interpretation of reality was not considered by most economists and policy makers during the 1980s, much was asserted about the causes of the economic deceleration and the worsening social conditions. Many asserted that it was the result of much lower levels of savings in that decade. Savings as a share of GDP averaged about 3 percent after having averaged about 8 percent between 1950 and 1969. Lower savings, according to this argument, translated into less investment and therefore slower growth. However, during the 1970s, savings had averaged about 7 percent of GDP, which was only a little less than in preceding decades. Had this been really the cause or a major factor affecting growth, the economy should not have decelerated so much in the 1970s, and in the 1980s the deceleration should have been much more pronounced.[54]

That savings were at a terribly low level by the late 1930s, and investment and growth nonetheless accelerated to impressive levels in the early 1940s reinforces further the idea that savings do not have a relationship of causality with growth. The evidence suggests that some other variable was behind the performance of the economy and that the level of savings was actually an effect of growth and therefore of that other variable.

In the 1980s, the United States experienced a negative trade account for the first time since the late nineteenth century. For about a century, trade surpluses had been constantly experienced and many came to see this reversal as the cause of the slow growth. The fact that protectionism was practiced by several of the nations that were responsible for the trade imbalances, drove many analysts and policy makers to the conclusion that it was an exogenous factor that was negatively affecting the economy. The American government therefore began to exert considerable pressure on several governments in East Asia so that they would liberalize their trade regimes.[55]

The pressure forced all of these countries to liberalize as least to some extent and Washington also removed some trade preferences to some of those nations. There was as well a rise of protectionism by the Americans, mostly in the form of nontariff barriers. None of this, however, was capable of generating fast economic growth or reversing the trade deficits. During the 1990s, more pressure was exerted on East Asian countries as well as most other countries of the world so they would liberalize their trade regimes. There was much liberalization, in part because of the pressure and in part because most governments became convinced that it was a positive undertaking. In addition, the American government again increased its nontariff barriers. All of that, however, once again did deliver positive results. The economy decelerated yet again during the years 1990 to 1997 and the trade deficits remained.

That during the 1930s the trade account was in surplus and the economy nonetheless performed miserably substantiates further the idea that there is no causality linkage between trade deficits and growth. The fact that numerous countries have experienced periods of fast growth while simultaneously enduring trade deficits points further in that direction. Negative trade figures are evidently not responsible for poor economic performance. They seem to arise when a nation's main trading partners attain faster rates of economic growth. This seems mostly to occur when a government supplies a weaker level of support to manufacturing than do their main trading partners. As their factory output grows faster, they are therefore increasing at a faster pace their capacity to export. That is true not just because manufactured goods are the most exportable of goods but also because these goods make possible the production of primary goods. Under those circumstances, primary sector exports increase also at a faster pace.

Washington was right to demand that its trading partners liberalize their trade regimes, for much suggests that these impediments were actually harming the economy of those countries; however, it was wrong to believe that such trade barriers were the cause of the deceleration of the United States. The fact that the United States attained much faster rates of GDP growth during the 1950s and 1960s, even though then– East Asian countries, developing countries, and Western Europe had much higher trade barriers, clearly indicates that the source of the problem did not come from abroad. Washington also committed an error in believing that by raising nontariff barriers the economy would improve. Much suggests that what they actually did was to harm it.[56]

In the 1980s, the United States experienced its worst financial crisis since the 1930s as hundreds of banks failed. Too much lending by mostly savings and loans banks to real state developers ended by delivering numerous nonperforming loans. During this decade, the Federal Deposit Insurance Corporation (FDIC) closed more than one thousand banks, which was more than during the preceding forty-six years since the founding of the FDIC. There was also the foreign debt crisis that exploded in 1982 when Mexico declared its inability to service its foreign debt. Many Latin American countries soon after did likewise. Since most of this debt had been contracted with American banks, the financial crisis in the United States became larger.[57]

This was indeed an exceptional situation that was evidently negative for the economy. However, the situation was rapidly brought under control and by the early 1990s American banks had very few nonperforming loans in their balance sheets. During 1990 to 1997, therefore, American banks were not only largely free of problem loans, but they also enjoyed among the highest levels of efficiency in the world. If the banking crisis of the 1980s had been really responsible for the deceleration of the economy, GDP figures should have grown faster during the 1990s. That was not the case, and the economy actually decelerated further.

Also worth noting is that nonperforming loans in the United States as a share of GDP averaged about 2 percent during this decade while in Japan during 1990 to 1997, they averaged about 8 percent. If such financial problems were a major factor affecting the economy, Japan should have then attained GDP figures much lower than those the United States got in the 1980s. That was not so; they were lower, but the difference was small. GDP averaged about 2.4 percent for the United States in the 1980s and about 2.1 percent for Japan in the 1990s. Still more indicative of the absence of a causality linkage between banking problems and economic performance was found in China.[58]

During the 1990s, conservative estimates calculated that nonperforming loans in China as a share of GDP were of about 20 percent. Some credit rating agencies like Standard & Poor, however, thought the figure was much higher. This was actually one of the highest figures in the world. If nonperforming loans had a negative effect on growth, then China should have perhaps attained the worst GDP figures in the world. The fact, however, is that it attained one of the fastest rates of economic growth in the world.[59]

The absence of a correlation left no doubt about the absence of a causal relationship. Financial crises were indeed a problem and much suggests that without them a better economic performance would be attained, but it was evident that their effects were only marginal on growth. Evidence also suggests that increasing competition in the banking sector as much as possible, increasing reserve requirements to high levels, and improving supervision on banks did much to prevent financial crises. However, even under those circumstances, which the United States had by the 1990s, it was evident that it was not enough to deliver fast growth.

The first computer was developed during World War II, but it was not until the early 1980s that computers began to be extensively utilized. The advent of the personal computer revolutionized the way business was conducted. Computers helped to rationalize production and many began to argue that the deceleration of this decade was to a large extent the result of the rapid increase in the utilization of technology by enterprises. It was argued that the rise in unemployment and the rapid loss of high-paying jobs were its consequence.[60]

However, there was practically nothing in the empirical data to suggest that American firms had added technology faster than in the past. On the contrary, much evidence actually pointed in the opposite direction. The rate of patent registration was much faster during 1950 to 1969, and in the 1970s it was at about the same speed as in the 1980s. Rationalization was faster in the 1950s and 1960s, yet unemployment fell and an abundance of high-paying jobs were created. This variable was evidently not the cause of the problem.

In the history of the United States and numerous other countries, technology has been frequently blamed for the shortcomings of the economy, although a rational appreciation of history actually reveals the opposite. It has been precisely during the moments when technology grew the fastest that employment possibilities expanded the most. It was when rationalization was the fastest that unemployment dropped the most.

The fact that technology grew very rapidly during the 1870 to 1913 period as well as during the years 1940 to 1969, while poverty and unemployment simultaneously declined rapidly, points clearly in that direction. Washington's factory promotion efforts were strong during those periods, suggesting that this variable was fundamentally responsible for the creation of technology and therefore for the creation of wealth.

Unfortunately, the different governments never came even close to understanding the causality linkages of this sector with technology. The support they supplied to the sector was almost exclusively the result of national security pressures and several other nonscientific motivations. Precisely because of that, when these pressures and motivations became less pronounced, support for the sector dropped, and along with it the rate of factory output, the rate of technology, and the rate of GDP.

That is precisely what took place during the 1980s. It was a deceleration in the rate of factory output and a deceleration in the rate of technology that delivered a slower rate of wealth creation and therefore lower employment possibilities. Establishing a causal link between this sector and technology has always been very difficult for intellectuals and policy makers. Since so much of the output of this sector in the 1980s and in the preceding decades was in the form of weapons, it became even harder to see clearly through the maze of information.

A clear example of how the sector creates technology and how this technology benefits society even when it sprouts out of a weapon-production effort is that of long-distance medicine. In the 1980s, the Pentagon financed a project to create the means for soldiers on the battlefield to be assisted from afar by a doctor. The effort created special cameras, video monitors, and computers that allowed for long-distance diagnosing and some treatment. By the 1990s, these factory goods were adapted to civilian uses and began to allow adequate medical diagnosing and even treatment to reach far-away rural areas where specialists are rare. The technology of these machines allowed a city specialist to discern, hundreds of miles away, almost anything that a physical exam would reveal.

With that same goal in mind, the Pentagon also financed a project that would permit long-distance robot-assisted surgery in the battlefield. By the mid-1990s, these same factory goods began to be utilized by civilian medicine, and people living in places where specialist surgeons were not available began to obtain the benefits of long-distance surgery.[61]

The 1990s

Communism worldwide, but more particularly among Soviet bloc countries, began to disintegrate rapidly in the late 1980s, and by 1991 there was not much left of it. The Soviet-led Warsaw Pact, as well as the CMEA organization, was dissolved and in that year even the USSR disintegrated into fifteen independent countries. The countries of Eastern Europe and those that emanated from the ex-USSR almost immediately embraced capitalism and democracy. From being the arch-enemy, these countries instantly became almost allies of the United

States. Under these circumstances, Washington saw no logic in spending as much in defense as during the 1980s and armament budgets began immediately to be significantly reduced. The end of the cold war brought a worldwide reduction of defense expenditures, which encouraged American policy makers to cut further in that direction.[62]

After having averaged about 6 percent in the 1980s, defense as a share of GDP dropped to about 4 percent during 1990 to 1997. Everybody began to talk about the peace dividend. Everybody was convinced that the large savings on defense would be put to better use and society would benefit abundantly. By reallocating the funds into education, infrastructure, and lowering the budget deficit, the economy was supposed to perform much better. The peace dividend, however, never materialized. Not only could the economy not accelerate its pace, it could not even maintain the pace of the preceding decade. After having averaged about 2.4 percent annually during the 1980s, the economy slowed down to about 2.3 percent during the 1990s.[63]

Expenditures in education rose, the budget deficit was considerably reduced, and there was even a little more spending on infrastructure, although that ultimately did not translate into faster growth. Common sense has long driven Americans and practically all other nations into believing that education is an essential element for the attainment of fast growth. However, a consistent correlation between this variable and growth has practically never been achieved in any country of the world. During the 1940s, for example, only one-fourth of adult Americans completed secondary school and only 3 percent received a college degree. By the 1990s, much progress had been made on that front and the figures had jumped respectively to about 80 percent and 23 percent.[64]

Despite very low educational levels of the population, economic growth during the 1940s averaged about 4.4 percent annually and notwithstanding the much higher standards of the 1990s, the figure was only 2.3 percent. Logic demands that if education were an important variable affecting growth, it should have been in the 1940s when the GDP figures were weak and in the 1990s when they were strong. During the late nineteenth century, the share of the population with secondary and higher education was only a fraction of that of the 1940s, yet economic growth was even stronger than during the World War II decade. This data further points to the lack of causal linkage between education and growth. It must also be added that the scientific value of the education of the 1990s was far above that of the 1940s and still further above that of the late nineteenth century.[65]

It was in the 1990s when the compounded expenditures on education by federal, state, and local governments, and by households, were the highest in American history. However, at this time the United States attained one of its lowest rates of growth in its history. About 7 percent of GDP was spent on education, while during the late nineteenth century not even a tenth of that was spent yet growth was about three times faster. From the historical data, one could actually conclude that spending a lot on education is harmful for the economy. This is evidently not the case, for education has numerous positive aspects, but it is also evident that by investing more in education nations have no guarantee of improving their economic performance.[66]

Nations should strive to educate their population as much as possible and there is much to suggest that the best way to do this is by reaching the fastest possible rates of economic growth. It is only when wealth is created that the possibilities for education expand, which is why it was always during periods of fast growth that education progressed the most, as during the 1870 to 1913 and 1940 to 1969 periods. Since the evidence indicates that manufacturing is fundamentally responsible for growth, it follows that the best policy for improving education is by promoting this sector as decisively as possible.

Expenditures in infrastructure during the 1990s only increased slightly, but the idea that this was a positive policy undertaking was widely accepted by economists and public officials. As with education, a correlation with growth was absent. In the 1930s, the share of overall resources allocated to this domain was very large, but the economy failed to perform well. In other places of the world, similar situations have occurred. In the 1990s, for example, Japan allocated the largest share of the nation's resources in its history to infrastructure, and the economy actually performed worse than in any other decade of the twentieth century. During the 1990s, Taiwan allocated the largest share of the island's resources to this domain since 1950, and the economy, instead of accelerating actually experienced slower rates of growth than in any other decade during the second half of the twentieth century.[67]

In the second half of the nineteenth century in the United States, it seemed as if there was a correlation between infrastructure and growth, because the large expenditures that were made on railroads were accompanied by fast growth. A similar phenomenon took place in Canada, Australia, New Zealand, and most of Europe, giving further credence to this idea. However, the fact that over the long term the correlation could not hold indicates clearly that it was some other variable that was fundamentally responsible for the growth; and the only variable that manages to be consistent in the long term is manufacturing. Factory output grew rapidly during the second half of the nineteenth century in the United States, and in numerous other countries, as the production of iron, rails, and locomotives was enthusiastically promoted. In the United States during the 1930s, government support for the sector was very weak and factory output grew very slowly. In the 1990s it grew slowly in Japan as Tokyo's promotion efforts were also weak, and in this same decade the factory promotion efforts of Taipei were weaker than in any other period of the second half of the twentieth century.

At times, large expenditures on infrastructure stimulated factory output, which is why, when the priority centered on railroads, it correlated with fast economic growth. However, on many other occasions, as during the 1930s in the United States and the 1990s in Japan and in Taiwan, increased budgetary allocations for infrastructure did not stimulate a faster factory rate of output, which is why it did not coincide with faster economic growth. On these occasions, the bulk of the infrastructure expenditures were utilized for building highways, sewage systems, ports, airports, and recreational areas, which are not manufacturing activities. During the second half of the nineteenth century, the subsidies flowed directly to the factories, while during the 1930s and the 1990s the subsidies flowed mostly into construction companies.[68]

Improving the infrastructure of a nation is a worthwhile goal, but history clearly reveals an absence of causality linkage with growth. Fast economic growth is what history unequivocally demonstrates as the best for reaching this goal, because whenever there was fast growth, infrastructure tended to expand rapidly. As with education, the evidence indicates that it is an effect of growth and not the cause.

Balancing the budget or even attaining a surplus is also worth pursuing, but the historical and cross-country evidence reveals that it is not a determining variable affecting growth. That would explain why, notwithstanding the considerable reduction of the budget deficit in the 1990s, it did not translate into faster economic growth for the United States. The only explanation for why more budgetary allocations for education, infrastructure, and a lower budget deficit did not deliver a faster GDP rate is the manufacturing thesis. Since the cuts in defense expenditures were not all transferred to civilian manufacturing, the sector's rate of output inevitably had to decelerate. After having averaged about 2.8 percent in the 1980s, the figure was about 2.6 percent during 1990 to 1997.[69]

During the 1990s, however, there was a noticeable change in the way the government managed the reduction of defense expenditures. A large share of the money previously set for fabricating weapons was utilized for civilian manufacturing. The complaints of labor unions, as well as numerous politicians and even some economists, that the country's industrial base was hollowing out drove Washington, and in particular the Clinton administration, to transfer a large amount of the defense money into civilian manufacturing. The Pentagon's DARPA program, for example, was transformed into ARPA so that it would concentrate on promoting civilian technologies. The government's Small Business Administration program even saw a small increase in the loans it offered. This program was not concentrating on manufacturing, but the share of loans it offered to this sector increased along with the other sectors.[70]

The level of government support for manufacturing in the United States during the 1990s was one of the lowest in the nation's twentieth-century history. However, its extent was very large when it is considered how, even among economists, it is frequently believed that subsidies to this sector are almost nonexistent. During these years, the federal government alone supplied about US$75 billion per year in grants and other forms of direct spending to private civilian manufacturers and about US$60 billion in tax breaks. If to that are added the abundant fiscal, financial, and nonfinancial incentives that state and local governments supplied, the figures almost double.[71]

This obviously did not include the huge expenditures that went into the production of weapons and NASA's space goods. The large subsidies that have flown into civilian manufacturing have been driven by several motivations. Federal subsidies have frequently been offered in order to promote exports, to catch up with competitors, or to maintain a technological edge. Since factory goods are the most exportable, since they are also the goods that generate the most envy, and since they embody technology at the highest levels, the subsidies have mostly been supplied to this sector.

Health issues have also pushed Washington to allocate extensive funding to the sector. Since the 1940s, there has been sustained public investment in the production of pharmaceuticals and medical equipment. In 1971, for example, Congress approved the National Cancer Act and by 1997 about US$30 billion had been spent in the war against cancer. In the early 1990s, about US$3 billion was budgeted by the federal government for the Human Genome Project in an effort to sequence the entire base of human DNA and increase the means to eventually cure the vast number of genetically determined diseases. Numerous other government-funded programs to combat particular diseases were also in place by the 1990s.

Most of the money flowed into laboratories of medical manufacturing companies in an effort to lower the R&D costs of these firms. The rest of the money went to university and government laboratories, which delivered the same effect of reducing the overall costs of discovering and producing new solutions to the nation's health problems. A manufactured good, such as pharmaceuticals, always ended up embodying a discovery that was made at a university or at a government laboratory. There were also large tax reductions to enterprises that worked toward cures for diseases that affected only a very small share of the population.[72]

In the early 1980s, AIDS was first identified in the United States, and its sexually contagious nature and therefore the possibility of an epidemic immediately drove the government to allocate considerable funding for research. Homosexuals particularly were hit by the disease and as a very organized group they successfully lobbied policy makers to increase funding for research. By 1996, relatively successful pharmaceutical goods made their appearance, and it was the federal government that granted the majority of the vast R&D funds that made these factory goods possible.[73]

At other times, the motivation that drove the state to promote civilian factory production was national security. By the 1990s, for example, the Jones Act, which was passed in 1920, continued to regularly supply subsidies to shipbuilders. On the grounds that it was a strategic necessity to possess a merchant navy that could back its military counterpart in times of war, ship producers received several fiscal, financial, and nonfinancial incentives.[74]

Microchip producers have also received much support from the government on the grounds that American "smart" weapons can only maintain a technological lead over their adversaries if they possess the most advanced microchip technology. Since the 1980s, semiconductor producers started receiving abundant subsidies as Japanese chip technology began to excel that of the United States. The Sematech project is the most noteworthy of these efforts.

National security also drove the government in the 1970s to create the Synthetic Fuels Program. The large increase in the price of oil convinced policy makers that for the economy and the armed forces to operate adequately, they needed a domestically generated and secure source of fuels in case foreign oil suppliers cut the flow. In the 1980s, the Reagan administration trimmed this program, but other sources of energy generation continued to receive subsidies. Factory goods that generated energy from wind, sunlight, and biomass were the largest recipients of funds from the Energy Department in the 1990s. The growing environmental concerns of Americans convinced the government that it needed to support this type of manufacturing.

On the other hand, the motivations driving state and local governments to supply fiscal, financial, and nonfinancial incentives to manufacturers were different having mostly to do with job creation or regional development. In their efforts to create high-paying jobs, for example, the subsidies ended up frequently in this sector, which tends to pay the highest wages. In their efforts to reduce unemployment or income disparities within a state or a smaller region, the authorities frequently supplied support for the sector. The subsidies were usually across-the-board to all sectors and some therefore fell to manufacturing. Desires to promote technology have also ended up significantly assisting factory production, and nonfederal governments have frequently shown as much interest as the central government on this matter. Technology parks are the most noteworthy of these efforts.

Since support for manufacturing from state and local governments has remained largely constant over time, it has been the actions of the federal government that have been basically responsible for the fluctuations in overall levels of support. The reduction in Washington's factory-promotion efforts during the 1990s coincided with a deceleration in the rate of factory output, which in turn correlated with a deceleration in the rate of GDP. Practically nobody managed to observe such a situation and less still to conclude that this was the cause of the slowdown.

Globalization and the Presumed Causes of the Problem

Many began to argue that the deceleration was the result of an increase in competition from newly opened ex-socialist countries and developing countries that had abandoned their introverted policies for export-oriented policies. It was asserted that cheap labor costs in these countries drove American companies to decamp from the United States and set up operations abroad. The low production costs of these countries also gave a large cost advantage to the domestic producers from these nations and American companies were frequently incapable of matching those prices, which drove them to further cut production.[75]

The increased competition was indeed there, but the case of other economies clearly demonstrated that this could not possibly have been the cause behind the deceleration. Singapore and Hong Kong, for example, by the 1990s had levels of development almost at parity with those of the United States. GDP per capita in terms of purchasing power parity was almost as high, as were labor and production costs. As a result, the lower production costs of numerous countries had almost an identical effect on these two economies. Emigration of domestic companies from these two city-states to low-cost developing countries was even more pronounced than the one the United States experienced, yet economic growth was much faster. While GDP averaged about 2.3 percent in the United States during 1990 to 1997, in Hong Kong the figure was about 5.0 percent and in Singapore it was about 8.2 percent.[76]

These economies demonstrated that independent of the very wide gap in production costs relative to less-developed countries, they could still attain not only very fast rates of growth but even among the fastest in the world. Singapore actually grew much faster than practically all developing countries, demonstrating that it was also false that for these nations to grow out of poverty there had to be a corresponding drop in the living conditions of developed countries. This situation made it evident that it was not a zero-sum game, as many analysts and policy makers in developed countries had come to believe. The fact that China, with its extremely low wages, and Singapore, with its very high wages, both managed to attain two of the fastest rates of growth in the world made it clear that poor and rich countries could simultaneously grow very rapidly. It is likely not by chance that support for manufacturing in China and in Singapore in the 1990s was among the strongest in the world.

There were others who argued that a technological revolution in these years rationalized production, eliminated high-paying jobs, and caused the deceleration. The rapid robotization of production as well as the increasing utilization of computers by enterprises drove many to this conclusion. Once again, the case of Singapore demonstrated that this thesis was not consistent with the empirical data. During the 1990s the utilization of robots and computers on the island increased faster than in the United States, yet fast growth was attained and high-paying jobs grew rapidly.[77]

Also worth noting is that by the 1990s Japan had about 60 percent of the world's industrial robots in use while the United States had only about 10 percent. That means that, in per capita terms, Japan had about twelve times more robots. If the rationalization argument were valid, then Japan should have grown at a rate far below that of the United States. However, it grew at an almost identical rate, and while unemployment in the United States averaged about 6 percent, in Japan it averaged only 3 percent. Japan should have endured very high levels of unemployment due to the numerous jobs that were eliminated by robots. Reality, however, turned out differently and it was actually the OECD country with the lowest level of unemployment. Also worth pointing out is that in the preceding decades, rationalization in Japan grew at a far faster pace than in the United States and not only were jobs created at a much faster pace than in America, but GDP figures were much faster.[78]

The fast rationalization of these decades and in particular the 1960s drove many to the conclusion that automation would create technological unemployment. On the contrary, what actually took place was a situation in which the faster the rationalization, the lower the unemployment. For Japan and most of OECD countries, 1950 to 1969 (but mostly the 1960s) was when production was more rationalized and when unemployment was at its lowest point during the second half of the twentieth century. In the United States, it was also in the 1960s when unemployment was the lowest and when high-paying jobs were created at the fastest pace.[79]

There is very little in the history of the United States, and most other countries to indicate that rationalization is harmful for the economy. What history actually suggests is that when rationalization is not fast enough, as during the late twentieth century, the creation of new jobs is not fast enough to exceed the ones that are eliminated. The empirical data also suggests that when support for manufacturing is strong, there is a rapid creation of employment. OECD countries promoted this sector in the most decisive way during the 1960s.

It was also asserted that the deceleration of the 1990s was the result of a reduction in capital available for lending. In 1988, the Basel Agreement on capital standards for banks became effective and OECD financial institutions (in particular) were required to raise their reserve requirements, which left them with less funds for lending. The Basel Agreement was indeed effective in forcing American banks to raise their reserve requirements, but they were just raised a few percentage points. The increase still left a large share of unused capital that could have been lent. The banks had become very cautious after the savings and loan fiasco, but there were several firms that bankers deemed creditworthy and that would have received all the loans that they would have demanded. However, those enterprises barely demanded loans for they were not interested in expanding operations. Many actually ended up downsizing their operations.

A shortage of capital never really took place during the 1990s, as evidenced by the unprecedented high levels of the stock market index. The large amounts of capital that developing countries and ex-socialist nations sucked up during these years never managed to significantly reduce the availability of capital for the United States or for other OECD countries. Capital proved to be constantly abundant but it just languished without being put to work. In addition, the reduction of the budget deficit made more funds available for capital markets, and the rising price of stocks was largely a reflection of that.[80]

Much information suggests that capital was not put to work because government support for manufacturing was weak. Throughout history, whenever the factory promotion efforts of the state were strong, capital was immediately utilized in large amounts. As long as policy makers supplied ample incentives and therefore guaranteed that investments in this sector would be profitable, capital always flowed to it in large amounts. When that took place, overall investment rose to high levels.

This would explain why in the 1990s reputable and creditworthy manufacturers were not interested in expanding operations. Since the incentives of the government were not large, the risks for them were too high and they therefore did not search capital for expansion. Between 1950 and 1969, on the other hand, firms showed much interest in expanding operations, which coincided with years when Washington supplied ample support to the sector.

Associated with the argument that the emigration of capital to developing and ex-socialist countries was the cause of the deceleration was the one that largely put the blame on the free trade agreement that first took place with Canada and then with Mexico. In 1989, free trade between the United States and Canada became a reality, and in 1994 Mexico was added to the association. It was argued that there was a large emigration of companies to Canada and Mexico and that the production capacity of the United States was reduced. Production costs were indeed lower in the neighboring countries and there was a considerable emigration of enterprises, but the case of Singapore and Hong Kong once again demonstrated that this could not have possibly been the cause of the slow growth.[81]

These two economies had actually a free trade relationship with the whole world and not just with a few neighbors, and they even experienced a much larger emigration of capital than of the United States did. In spite of that, they had much faster rates of growth than the United States. Singapore even accelerated its pace relative to the 1980s, even though the emigration of labor-intensive production in the 1990s was faster than in the preceding decade.[82]

History strongly suggests that promoting free trade is a positive policy undertaking and that the NAFTA agreement significantly contributed to improving efficiency among its three members. However, the empirical evidence also clearly indicates that free trade is not the fundamental nor even a very important variable responsible for economic growth. If it were, then GDP figures should have accelerated. For the United States and Canada, the economy decelerated. After having averaged 2.6 percent and 3.1 percent respectively in the 1980s, the figures dropped to about 2.3 percent and 2.1 percent in 1990 to 1997. Even Mexico, which was supposed to be the great beneficiary due to its very low production costs, only witnessed a small improvement in its growth figures. After having grown by about 1 percent annually in the 1980s, Mexico's rate averaged only 3 percent in the 1990s.[83]

This was insufficient to help Mexico reduce poverty and it was only a fraction of what China got, with no free trade agreement with any country and with a policy of high trade barriers. The much faster growth of foreign trade in the three NAFTA countries delivered no boost to the economy. It was evident that free trade was positive, but it was also evident that it was not essential for growth. What seemed to be essential was manufacturing. In none of the three NAFTA countries was there strong support for this sector. In Singapore and China, however, the factory promotion efforts of the state were very strong.[84]

Even though many questioned the value of NAFTA, most American economists and policy makers remained convinced that it was important for the well-being of the country. Most were particularly convinced that increasing exports was essential for accelerating GDP figures. The Bush administration made numerous efforts to promote exports, but they were actually small in comparison to those the Clinton administration made. In 1993, the new Democratic president launched an all-out coordinated and interagency effort to promote exports and all sorts of American business opportunities throughout the world. At no other time in American history had a government determined its foreign policy for the sake of serving the economy. Previously, national security had mostly determined foreign policy.[85]

All of these efforts contributed to accelerating the pace of exports. After having grown by about 3 percent annually during the 1980s, the rate jumped to about 6 percent in 1990 to 1997. The economy, however, did not follow and it actually ended up doing the opposite. Long before that date, the history of the United States and of numerous other countries had demonstrated that it is not possible to establish a consistent correlation between exports and growth. Frequently, they ran parallel but at times they did not, and that very clearly signaled an absence of a causal linkage. A consistent parallelism, however, was systematically observed between manufacturing and growth. Whenever this sector was promoted, GDP figures moved in direct proportion to the factory rates.

That would explain why during the late nineteenth century, when exports were not being promoted and when they accounted for a much smaller share of GDP than in the 1990s, economic growth was much faster. At that time, factory rates of output were about three times faster than in the 1990s and GDP figures were also about three times faster. That would also explain why between 1940 and 1969, when exports were not promoted either, the United States grew about twice as fast as during the 1990s. In those war decades, manufacturing expanded at about double the rate of the last decade of the century.[86]

There were many economists who held that the slow economic rates of the 1990s were the result of natural limits that levels of unemployment imposed on the economy. It was asserted that below a certain level of unemployment, tight labor markets would push wages up and along with them the rate of inflation. If inflation accelerated, then the economy would slow down. Most economists argued that the natural rate for the United States was about 6 percent, and 6 percent was on average the rate of unemployment experienced during 1990 to 1997.[87]

According to this argument, therefore, it was practically impossible to attain faster rates of economic growth without risking higher inflation, and once higher inflation appeared, GDP figures would inevitably decelerate. This argument was obviously forgetting that between 1950 and 1969, unemployment was much lower than in the 1990s, averaging just 4 percent, and even though wages rose rapidly, inflation remained below the level of the 1990s. Inflation averaged about 2.5 percent in the first period and about 3.3 percent in the second, and growth was almost twice as fast as in the 1990s. Interesting to note is that during the 1960s American economists asserted that the natural rate of unemployment was about 3 percent.[88]

Since an unemployment level of about 4 percent was not producing any negative effects, then the Non Accelerating Inflation Rate of Unemployment (NAIRU) figure was much lower. This accommodative reasoning reveals not only an incapacity to add up consistently with the facts but also a lack of respect for logic, and a failure to revise the existing information.

Even the 3 percent unemployment barrier was not compatible with the cases of several other nations. Germany, for example, during the 1960s experienced an unemployment rate of just 1 percent, yet it actually had a rate of inflation that was lower (2.7 percent) than the one experienced by the United States during the 1960s and 1990s. Wages rose rapidly as very tight labor markets pushed them up, but prices remained tame. On top of that, the economy grew also faster (4.6 percent) than in the United States in the 1960s and in the 1990s. In Japan, during the 1980s, labor markets were also very tight as unemployment averaged only 2 percent. Wages rose quickly but inflation was among the lowest in the world, averaging only 1.3 percent annually. And this situation was accompanied by a fast rate of economic growth that averaged about 4 percent.[89]

The case of Singapore also debunked the NAIRU argument. During the 1980s, unemployment was low, averaging just 4 percent, and wages rose quickly, but inflation was only 1.7 percent annually; and that was paralleled with very fast economic growth (7 percent). In 1990 to 1997 unemployment dropped to just 2 percent, wages rose once again very fast, and again inflation remained subdued, averaging about 2.3 percent per year. That was accompanied by an acceleration of economic growth (8 percent), that was one of the fastest in the world.

The empirical data strongly suggests that nations can attain fast and sustained economic growth, bring unemployment to almost nonexistent levels, and simultaneously experience very low levels of inflation. One thing does not inhibit the other.

The reason a rapid economic growth in the cases previously cited did not deliver high inflation seems to be the result of two factors. On the one hand, strong support for manufacturing delivered a rapid growth of technology that created the wealth that allowed for higher wages. The rapid growth of technology also increased productivity, which dampened inflation. On the other hand, monetary policy was conducted in a very responsible way.

Technology and Manufacturing

In all of these cases as well as in all others in history when governments promoted manufacturing decisively, technology grew fast and productivity did also. In the United States, for example, productivity averaged about 2.6 percent annually during 1950 to 1969, while in the following two decades the yearly growth of productivity dropped by about one-half, and in 1990 to 1997 it averaged just 1.2 percent.[90]

This situation is not surprising when it is assumed that manufacturing is the fundamental generator of technology. Since technology is the main variable responsible for the growth of productivity, it is inevitable that when this sector grows rapidly, productivity expands at a fast pace, and in turn it allows for rapid wage increases without posing an inflation threat. As long as wages grow more slowly than the rate of productivity, inflation cannot accelerate. This would explain all of those apparent paradoxes, in particular in East Asia during the second half of the twentieth century. In several of these economies, wages grew much faster than anything ever experienced in any other country, yet inflation remained very low.[91]

In all of those cases such an event was accompanied by the strongest factory promotion efforts and the fastest rates of productivity growth in history, and by a responsible monetary policy. In some cases, monetary policy was not conducted all that responsibly and in these cases inflation was not so low; however, this did not inhibit impressive rates of factory output, rapid rates of productivity, and fast GDP figures.

Another way this sector reflected its almost unique technology-generating capacities was how productivity behaved among all economic sectors. Manufacturing systematically attained the fastest rates. During the second half of the twentieth century, manufacturing productivity in the United States grew much faster than services, which was the sector frequently cited by economists as the most important for a developed economy.[92]

On average for the whole period factory productivity grew about three times as fast as service productivity (3.5 percent and 1.1 percent, respectively). Even by the 1990s, when economists' admiration for services reached its highest point, the performance of services was as poor as before, if not worse. While manufacturing productivity averaged about 3 percent for 1990 to 1997, that of services was close to zero. Several other countries showed a similar pattern of behavior. Productivity in factories in practically all other OECD countries during the second half of the twentieth century grew much faster than services. Since services are a mere recipient of the technology created by manufacturing, inevitably its productivity figures were much slower.[93]

The evidence linking manufacturing in a causal way with technology is quite compelling. Productivity is only one of many ways it is reflected. However, very few economists have come to consider some sort of linkage. Most do not even see much of a linkage between technological innovation and economic growth. Joseph Schumpeter was among the first to establish a strong causal linkage between innovation and economic growth, although he never managed to see that the fundamental source of innovation was in manufacturing. He concluded that it was the entrepreneurial drive of business people that drove innovation, and he was also wrong in believing that only innovation creates wealth.[94]

The history of the United States and numerous other countries reveals a multitude of episodes in which great wealth was created even though innovation was not observed. In these periods, technology grew rapidly but it was practically all imported. East Asia during the second half of the twentieth century is the clearest example. The fastest GDP rates in history were attained, wealth was created at an impressive pace, and living conditions improved at a faster pace than at any other moment in history anywhere in the world. However, very little innovation took place. Technology grew very rapidly, but most of it was simply imported. Still, in all of the fast-growing East Asian nations government support for manufacturing was more decisive than at any other moment in world history.

This would also explain why in the first century after independence the United States managed to grow, and at times very quickly, even though it imported most of its technology from Europe. Throughout history, economic growth has not always correlated with innovation but it has always coincided with technology, which has systematically moved in tandem with the rates of factory output.

Technology is almost exclusively responsible for improving living conditions. That is why even though nations frequently did not create any new technology, they nonetheless obtained great improvements in their living conditions, for importing technology had the same positive effects as creating it. Since technology was embodied in manufactured goods, nations succeeded in reproducing it and creating wealth by fabricating those goods domestically.

The ways in which technology improves living conditions and its linkages to manufacturing are perhaps more clearly expressed on matters referring to health. The public's health only barely improved during the seventeenth and eighteenth centuries in the future United States, which coincided with very weak government support of manufacturing and with very slow growth of technology. In the nineteenth century, however, the sector grew at an exponential pace compared with the past, technology grew at the same pace, and the health of Americans improved in a similar way. In spite of great progress, by the end of the century Oliver Wendell Holmes, a doctor and poet of the times, spoke of the medicines of his day in a very negative way. He said that if most were sunk to the bottom of the sea it would be all the better for humanity and all the worse for the fishes.[95]

Morbidity and mortality decreased remarkably, but the technological embodiment of medical goods such as medicines was still very weak. It was the spillover effects of the technology created in other manufacturing fields that were mostly responsible for the improvements in health. Technology (whether imported or created domestically) had increased very rapidly, but it had been almost all accrued to engineering goods such as metal, trains, and machine tools. It was not by chance that these factory goods had received the bulk of government subsidies. In the early twentieth century, the situation only slightly changed. Government support for medical manufacturing continued to be very weak and technological progress in this field continued to grow very slowly.

With the outbreak of World War II, however, vast government funds were spent to find a cure against bacterial infections, influenza, and several other diseases that were severely weakening the combat readiness of the troops. Since these diseases had ravaged humanity for millennia, very few thought a cure would ever be found. However, in just a few years, a cure was indeed found. On each occasion, the cure resulted from a manufacturing effort and became embodied in a factory good such as a vaccine or a drug. The results were so impressive that Washington decided to sustain public investment in biomedical research, and this was followed by a great increase in the appearance of pharmaceuticals with proven therapeutic efficacy.[96]

As soon as government support became strong for this field, medical technology began to grow at a much faster pace. There was not even a small lag of time between the time the state's support began and the appearance of impressive technological breakthroughs.

On the other hand, during most of the twentieth century, state support for trains was weak compared with the preceding century, and that coincided with a much slower pace of development of train technology. By the end of the nineteenth century, the United States possessed the most advanced train technology in the world, paralleled by the most decisive support in the world to this field during the preceding decades. By the late twentieth century, Western Europe and Japan possessed the most advanced train technology. Surely not by chance is it that during the second half of the century it was the governments of France, Germany, and Japan that promoted the production of trains most enthusiastically in the world.[97]

Technology in the United States and practically everywhere else systematically developed in the direction where there was support for manufacturing. This would explain why during 1940 to 1969, it was weapon technology that developed at the fastest pace. This would also explain why so many other apparently unrealizable goals of society were attained as soon as governments channeled abundant resources into particular fields of manufacturing.

When scientists and policy makers, for example, in the 1970s started demanding the elimination of chlorofluorocarbons due to their ozone-depleting nature, chemical companies said that it would be impossible to replace them by the end of the century (the date originally proposed). The R&D costs needed to create the substitutes were indeed enormous and chemical companies were convinced that they would run major losses if they paid for the R&D costs. As evidence mounted that the ozone layer was rapidly thinning, governments (OECD in particular) became more determined and budgeted large amounts of funds to search for a substitute. These funds considerably reduced the overall production costs of chemical firms and under those circumstances they no longer saw an obstacle to achieving the goal. By the early 1990s, well before the proposed date, the technology for a cost-effective substitute had been found.[98]

Another example is space technology. It grew slowly in the United States during the 1950s even though technology in general was growing rapidly, and that coincided with scant support for this field. In the following decade, however, Washington allocated extensive funding for it, space factory goods were produced at a very fast pace, and technology grew exponentially. In the 1970s, the rate by which space technology progressed decelerated considerably, which coalesced with a considerable decrease in the state's promotion efforts. In the 1980s, budgetary allocations for NASA remained at about the same level and technology grew at about the same speed. In the 1990s, the level of support for the field remained largely unaltered and space technology continued to grow at about the same pace.[99]

In the 1960s, an enormous goal such as sending the first person to the moon was completed in less than a decade. However, a much simpler agenda such as the creation of the second space station in history will take at least two decades. That's only inevitable when one considers that just in the first decade after the Reagan administration unveiled the space station project in 1984, Congress made about sixteen attempts to kill it. It did not succeed, but it did significantly reduce funding. The original plan called for it to be operational by 1994, but at the earliest only its first stage will be ready by about 2003. In the 1990s, NASA also failed to complete other projects. It attempted to build the next-generation space shuttle but since the government was not prepared to spend much on it, NASA proposed to private aerospace companies to ante up the bulk of the development costs. Boeing and other producers refused to risk so much capital and the new spacecraft was not created, nor was the technology that would have made it possible.[100]

When the first space shuttle was built, NASA paid for all of the development costs and under those circumstances private companies were more than happy to be part of the fabricating effort. The weight of governments in the creation of space technology and the speed by which that technology was created was almost as evident as in the case of weapons. Even by the mid-1990s, at a time when the share of the private sector in the financing of space goods had grown to unprecedented proportions throughout the world, it accounted for only about one-quarter of the worldwide costs of producing these goods.[101]

Innovation and Technology

Since the creation of technology is so resource-absorbing, investment-intensive, and mentally demanding, it is inevitable that much funding is required for it. With such a nature, it is also inevitable that the private sector instinctively refuses to commit its capital to that effort. Since manufacturing is the sector fundamentally responsible for the creation and reproduction of technology, the private sector therefore develops a natural tendency to abstain from it. The only thing that manages to reduce that risk-averse nature is for the government to supply incentives that significantly reduce the costs of production.

A look at the structure of R&D costs during the second half of the twentieth century significantly substantiates such an assertion. In 1950 to 1969, in its efforts to produce weapons, space goods, and several other goods at a fast pace, Washington decided to reduce the costs of production of private manufacturers by (among other things) decreasing their R&D expenditures. In these years, the government absorbed about 70 percent of the overall R&D costs of the nation. In the following decades, the government's share dropped considerably and by the 1990s it was only 40 percent.[102]

Most analysts and policy makers thought that overall levels of R&D would not decrease in the post-1960s decades, as the funds cut from defense would by means of market mechanisms find their way to the laboratories of civilian private companies. Many actually became convinced that R&D overall would experience an increase. While the private sector's R&D share did increase, overall levels did not. During the 1960s R&D as a share of GDP averaged about 2.8 percent, but by the 1990s the figures had dropped to about 2.3 percent.[103]

Even more important was that the rate of technological development decelerated considerably. This was only inevitable when seen from the perspective of the manufacturing thesis. The funds spent on weapons would have surely been more of a benefit to society had they been reallocated to civilian manufacturing, but since they were not, they ended up in other sectors, that do not have the capacity to create technology. As less was invested in the sector that is fundamentally responsible for the generation of technology, R&D expenditures had no other alternative but to decrease. Since support for civilian manufacturing was not strong the private sector was not going to incur the massive investments that the creation of any new civilian technology demanded.

Not all of the technology that was consumed in the United States during the second half of the twentieth century was produced domestically; much of it was imported. Here also the same phenomenon took place. When support for the sector was strong (1950s and 1960s), technology was imported at a fast pace, and when support decreased, technology was imported at a slower pace. That would explain why the overall pace of technological development was much faster during the first two decades than in the three that followed. Since manufacturing is the sector fundamentally responsible for the reproduction of technology, and even the reproduction is highly investment-intensive, it follows that this effort also intimidates the private sector; and without a reduction of the costs by means of subsidies, capitalists take the road of not investing.

Examples are abundant of how the creation and reproduction of technology are basically dependent on the extent to which government reduces the investment costs to manufacturers. Health matters are particularly illustrative. Viral infections, for example, ravaged the lives of Americans for centuries, and not until the 1940s, when Washington for the first time allocated abundant funds to fight against them, was some progress made. However, the progress was limited to preventive measures such as vaccines. In the 1980s and 1990s, with the rapid spread of AIDS, significant funds were budgeted for an effort to find a cure. Many medical specialists argued that it was a useless effort because viruses were extremely difficult to vanquish and the AIDS virus was perhaps the most difficult because of its rapidly mutating capacities. However, by the mid-1990s the first effective anti-viral drugs in world history made their appearance, making the disease in infected persons retreat.

Over and over again, policy makers have worried and complained about the vast costs that manufacturing projects carry, convinced that if they were not made, society would be better off. However, every time the investments were made, society always ended up with improved living conditions. As AIDS and numerous other diseases demonstrate, society was definitely better off by possessing a solution against them. It was obvious that those who had argued that a cure would never be found and that the funds would be wasted were completely wrong.

Others, however, argued that if the government had not intervened, the private sector would have allocated the funds on its own and a more efficient solution would have been found. Considering the precedents, this is highly unlikely. AIDS was indeed a new disease, but cancer and bacterial infections, polio, influenza, tuberculosis, and numerous others had been around for centuries. However, during all that time the private sector systematically refused to allocate funds to search for a cure, not in the United States, or anywhere else.

As long as the government would not reduce the gigantic R&D costs by means of subsidies, the private sector constantly refused to absorb those costs. It was only, for example, since the 1970s when noticeable progresses in the fight against cancer were made, coinciding with the precise moment when the government began to allocate large R&D funds for this field.

The United States might be a country with abundant entrepreneurial drive, but at practically all moments in its history, the private sector only rarely took the initiative to absorb the bulk of R&D costs when it came to opening new technological boundaries. Biotechnology, for example, was born in the United States in 1972 when the first gene was cloned in a laboratory, and almost a quarter century later only 1 percent of companies in this field were making money. During this time, however, huge investments were required for these enterprises. This situation was reflective of the inherent investment-intensive nature of manufacturing, and the massive risks that go with it.[104]

Throughout history, investments in manufacturing have constantly taken the longest to reach a profitable return. Not only are investments in this sector much greater than in all the other sectors, but it also takes much longer to recuperate the funds and make a profit. With those inherent characteristics, who could blame the private sector for not wishing to invest in it? Who could blame capitalists for only agreeing to invest in the sector once the government has given guarantees, by means of incentives, that their efforts will be remunerated.

The private sector is without any doubt best positioned to maximize the resources of a nation, but without government support it just does not venture into manufacturing, and if it is not invested in this sector, technology does not expand. If technology does not expand, wealth is not created. That is why even when the investments were made for the sake of fabricating weapons, society benefited abundantly. It was, for example, investments by the Pentagon that made possible the creation of artificial intelligence for use in smart weapons. Some time later, it began to be utilized by civilian manufacturers, and by the mid-1990s machines equipped with it were even capable of diagnosing diseases more precisely than the average doctor, notice sooner than the owner when a credit card had been stolen, and recognize hundreds of different voices.[105]

Society always benefits when it invests in manufacturing because the nature of this sector inevitably creates or reproduces technology, which is the fundamental means for improving living conditions. Since the 1950s, the huge investments that space manufacturing has demanded have been loudly criticized as a waste, incapable of ever generating real private sector demand. As time passed and society increasingly extracted very real benefits from the investments in space goods, the accusations of their being a wasteful allocation of resources decreased. However, even by the later part of the century, the majority of analysts and policy makers were still employing the term "waste." In their view, the benefits for society were only marginal and the private-sector demand was barely noticeable.

By the 1990s, one of the most dazzling technologies was embodied in portable personal telephones. This was perhaps the most sought-after technology of the decade. Annual demand grew at an exponential rate due to their obvious superiority over normal telephones. The most impressive of these phones were those that allowed for worldwide communication, but they would have been impossible to create without space factory goods such as satellites. By then much of the communication carried out by conventional telephones was also transmitted by satellite.

Obviously nobody was thinking of such telephones in the 1940s when the Nazis made enormous investments to create the technology for rockets; or when the Soviets in the 1950s made even larger investments to create the first spacecraft and the first satellite; or when the Americans in the 1960s mobilized gigantic funds for the creation of the space shuttle that took Neil Armstrong to the moon. However, that is the thing with manufacturing. When society invests in it, technology always gets created, and once it is created, its possibilities are almost endless. In the mid-nineteenth century, Alexander Graham Bell, for example, was laboring to create a machine that could translate words into images for his deaf wife, when out of that effort he created the technology that led to the telephone, his most noteworthy discovery.[106]

Another example of the spillover and unsuspected derivative applications that space technology delivered is fuel cells. Fuel cells were originally invented to power NASA's spacecrafts, and at first nobody thought they could ever be used to propel anything else. However, the growing environmental concerns in OECD countries about the pollution emissions and heat trapping gases that gasoline engines deliver drove society to search for alternatives to the combustion engine. Of the numerous options that have appeared, none comes even close to matching the impressive advantages of fuel cells. The majority of alternatives promise only to reduce the problem, and frequently they just shift the pollution from the cities to some other place. In addition, most are not capable of propelling vehicles as fast and for the same distances as gasoline motors. Fuel cells, however, by means of a chemical reaction between hydrogen and oxygen, by the late 1990s were already capable of powering an automobile in the same way as a gasoline engine and the only emission that came out of the exhaust was water vapor. This was the ultimate zero-emission engine. By then, Japanese, German, and American automobile makers were claiming that by 2004 they would have a cost-effective commercial line of fuel-cell cars ready for mass sale.

Not only was this marvelous environmentally-friendly possibility the direct offspring of space technology, but its further development and application to the automotive sector was also largely the result of government funding. Even by 1987, the price of each fuel cell was about US$20 million, rendering it totally out of reach for mass consumption. Since the private sector showed no interest in reconverting this technology, it was the Energy Department that funded much of the R&D to make it cost-effective so that the average consumer could buy it. It was also the Energy Department that funded the creation of the technology needed to miniaturize it, for it originally had spacecraft proportions. Practically all of the technology was created in private companies, but the state paid for the majority of the vast R&D costs.[107]

In 1997, the first world climate international convention took place as a result of growing scientific evidence strongly suggesting that world temperatures were rising and that the main cause were the heat-trapping gases emitted by automobiles and several other sources. The destruction of natural habitats, the loss of numerous species, an increase in droughts, an increase in hurricanes, and the disappearance of several kilometers of coastland due to rising sea levels were among the possible consequences of climate change. This situation was one of the most threatening that humanity had ever confronted, and it was fuel cell technology that most promised to prevent the worst effects of climate change. If this was not a major contribution to society, then it is hard to see what could possibly be. However, the technology would not have even existed by the 1990s had the government not invested extensively in NASA's space goods. Was it possible then to argue that the money spent in NASA was wasted or that it could have been much better used in some other activity?

The reason Congress has made so many attempts to kill the space station is its astronomical cost. Its estimated cost is about US$100 billion. The United States, however, is not absorbing all the costs. Western Europe, Japan, Russia, and Canada, among others, are contributing. Nevertheless, the individual cost for the United States is very high, but only when seen from the traditional perspective that fails to recognize the impressive technology-creating powers of manufacturing. Once the world is seen from the perspective of manufacturing, then the amount of money is not large and the funds cannot possibly be wasted. The only way in which funds are wasted is when they are invested in the nonfactory sectors.[108]

Since the money will be utilized for the creation of factory space goods, the project is destined to create technology that sometime later will create goods and services that a large share of the world population will want to consume. Studies indicate that the zero gravity of space could provide the means to fabricate far superior pharmaceuticals, microchips, and numerous other goods.

A space station could therefore create the technology for a definite cure against cancer, Alzheimer's, Parkinson's, and many other diseases.

When things are seen from the perspective of manufacturing, no expenditure, independent of its size, can be seen as too large or as a waste. Once there is a clear understanding of the causality linkage between this sector and wealth creation, then the only thing that comes to the mind is why much larger expenditures have not been made. Viewed from this angle, one can imagine the exponentially higher levels of wealth that the United States could have been enjoying by the 1990s, had the government endorsed a decisive factory promotion policy earlier. If, for example, Washington had allocated since the end of World War II as large a share of the nation's resources to civilian manufacturing as it did to military manufacturing during that war, wealth most likely would have grown exponentially faster than it did. If factory output had continued to grow at about 19 percent per year, as it did during 1940 to 1943, the 1946 to 1997 period would have most likely experienced a growth of technology about six times faster.

Many will argue that such an idea is ridiculous and wishful thinking, but these are likely to be the same people who have systematically failed to deliver positive results to society. These are the people who have regularly endorsed policies that have at best delivered modest results, and at times have even driven nations into economic catastrophes. These are the people who never manage to be consistent with the empirical evidence.

The manufacturing thesis, however, adds up very well with the facts. Productivity, for example systematically reflected a causality relationship between this sector and technology. Technology is fundamentally responsible for productivity, so inevitably whatever is responsible for the generation of technology is also responsible for productivity.

Cross-country comparisons demonstrate that productivity constantly varied in direct proportion to the differing levels of government support for manufacturing. During 1950 to 1969, for example, support among OECD countries was the strongest in Japan, less strong in Germany, and less strong still in the United States. Factory output grew in proportion to those levels of support and so did productivity. In Japan, productivity grew the fastest averaging about 7 percent annually; in Germany, it averaged about 4 percent per year, and in the United States about 3 percent.[109]

In the following two decades, productivity decelerated considerably in the three countries, coinciding with a significant decrease in the factory promotion efforts of the three governments. In spite of decreased support, Tokyo still promoted the sector much more strongly than Washington and Bonn, and that was paralleled by faster rates of productivity for Japan, averaging about 3 percent annually. The much slower rates of factory output of the other two, about half those of Japan, were followed by much slower rates of productivity, which averaged about 1.6 percent for Germany and about 1.3 percent for the United States.[110]

Finally in the 1990s, the subsidies for manufacturing were slightly lower than in the preceding decade and roughly the same among the three governments, and productivity rates were almost the same. Rates of factory output for the three countries were almost identical and so was productivity. In these three countries, it averaged a little over 1 percent per year. Most other OECD countries showed a similar pattern of behavior. Productivity grew by far the fastest during the 1950 to 1969 years, when state support for the sector was by far the strongest. By the 1990s, productivity rates were at their lowest level in the second half of the century, coinciding with the years in which factory output grew at the slowest pace.[111]

The behavior of exports also reflected the technology-generating capacity of manufacturing. During the 1950 to 1989 period among OECD countries, for example, it was Tokyo that promoted the sector the most; therefore, Japan attained the fastest rate of growth of high-tech exports. Germany which attained a slower rate of factory output, saw its high-tech exports grow more slowly, and the United States, which promoted the sector even less enthusiastically than Germany, saw its high-tech exports grow even slower.

While the United States's share of these types of goods accounted in 1950 for more than half of the total exports of the world, by 1990 it had shrunk to about 24 percent. That of Japan, however, grew from a minuscule 5 percent to 20 percent; Germany went from 9 percent to 14 percent. Among West European countries, Britain experienced the most significant drop in its share, coinciding with the weakest promotion efforts in the region. In the 1990s, however, Japan and the United States exported high-tech goods at about the same pace. This was paralleled by almost identical rates of factory output.[112]

Since the speed of manufacturing growth determines the rate at which technology grows, the degree to which factory output was promoted determined the export possibilities of high-tech goods.

Factories and the Nonmanufacturing Sectors

The dependence of the other sectors on manufacturing for their ulterior development was also very clear evidence of how this sector is the main generator of technology. The progress of the primary sector, construction, and services were made possible on a constant basis by the utilization of factory goods. These sectors managed fundamentally to improve productivity and to revolutionize their operations by means of factory goods.

Services, which by the late twentieth century were seen as the most dynamic and important sector of the American economy, were illustrative of the above. It was, for example, switching boards, mainframe computers, fiber optic cables, satellites, fax machines, telephones, telex machines, video conference machines, and portable phones that made possible the vast array of telecommunication services that existed by the end of the century. Without these factory goods, the services would have never been even conceived, for it was these goods that embodied technology.

Without delivery cars, trucks, trains, airplanes, ships and numerous other factory goods, transportation services were also impossible. Without pharmaceuticals and medical equipment, the high-quality medical services that were supplied in the United States during the second half of the twentieth century would have not been feasible. Antibiotics, analgesics, anti-arthritics, anti-hypertensives, tranquilizers, vasodilators, and protease inhibitors made the difference. It was X-ray machines, operating instruments, ultrasound machines, laser equipment, magnetic resonance imaging machines, and positron emission tomographic devices that made possible the great advances in medical services. These goods were the depositories of technology, which is why the possibilities of services depended on them.

Pens, paper, books, calculators, laboratory equipment, computers, and numerous other factory goods made educational services feasible and were largely responsible for the great progress of the period. The revolution in entertainment services is unthinkable without television sets, cinematographic equipment, video cameras, video cassette recorders, satellites, parabolic antennas, and digital video devices. Fast food services were also impossible to materialize without ovens, stoves, processed food, cooking utensils, milkshake machines, drink dispensers, and cash registers.

It is also worth noting that when a given manufactured good was utilized in services, it tended to deliver a much lower productivity performance than when it was used in a factory. Computers, for example, which probably embodied the most revolutionizing technology of the second half of the twentieth century, systematically generated higher levels of labor productivity when they were utilized in factories than when they were utilized by service sector firms.[113]

Construction in all its forms was no different than services. It was only possible as a result of steel, cement, cranes, processed wood, synthetic materials, earth-movers, glass, screws, nails, electric instruments, electronic devices, dredgers, bulldozers, and many others.

The same principle applied to primary-sector activities. By the late twentieth century, a revolution had taken place in American agriculture, the result of genetically altered factory-produced seeds, computer-guided milking machines, satellite-guided tractors, etc. Fishing had also become high-tech, due to improved sea vessels, spotter planes and helicopters, directional sonar machines, nets, satellites that permitted ships to lay their nets precisely near fish schools, improved refrigerators, and other devices. Forestry had also experienced great progress with the appearance of microchip-based electronic saws and numerous other factory goods.[114]

Major progress also took place in mining, as bucket-wheel reclaimers, automatic shovels, giant dump trucks, giant earth-movers, computer-assisted mineral-finding equipment, microchip-based boring machines, computer-guided drilling devices, giant platforms, and other manufactured goods widely expanded the possibilities of this domain. Oil extraction was perhaps the most illustrative example of how factory goods can expand the scope of primary sector activities in an almost unlimited way.

The oil shock of the 1970s and the existing world reserves of those times drove many analysts to conclude that a scarcity of this mineral was inevitable, and that in time prices would inexorably rise. By the mid-1990s, however, the price of oil (inflation-adjusted) was lower than before the oil shocks. This was only inevitable because by then the proven reserves found throughout the world were much larger than in the early 1970s, and oil was in over supply.

Giant platforms and improved drilling machines permitted deep-sea extraction of oil, making available giant oil deposits in the sea. Computers, geological equipment, and more advanced boring machines permitted horizontal land drilling, which allowed for a much higher rate of extraction even in oil fields that had already been abundantly vertically drilled. Ice-breaking instruments and superior equipment allowed also for the exploration and extraction of oil in frozen territories such as Alaska.

Perhaps the most impressive innovation was that which permitted the transformation of non-oil minerals into oil. Black hydrocarbon-bearing sands cover large parts of the Canadian territory. The possibility of extracting oil from them was long seen as possible, but the technology for such a feat did not exist, and if it ever came into being, most concluded that it would not be cost-effective.

The first oil sand enterprise saw its birth in 1967 when the provincial government of Alberta began to supply large subsidies to the private sector. In 1978, the government of Ontario created its own oil-sand company. Factories were created near the sand deposits that crushed, processed, and distilled the oil from the sand. The technology to separate the bitumen from the sand and to upgrade it into various blends of crude oil appeared almost as soon as strong government support was supplied. However, up to the late 1970s, the costs of separation were too high for it to be commercially viable. Since the 1980s, however, superior machines considerably reduced the costs of production, and by the 1990s even more advanced equipment brought the costs even lower. By then, costs were so low that profit margins were even higher than in regular liquid-oil operations, and drove Canada's largest oil company to stop exploring for conventional oil to concentrate on extracting it from sand. As for the quantities of oil that could be extracted from the sand, they were almost unlimited. By 1995, just the province of Alberta could claim to have more oil reserves than the whole of Saudi Arabia.[115]

Another way in which the technology-creating capacities of manufacturing found themselves reflected was in the course of investment. Since technology requires vast amounts of resources for its materialization, it is only inevitable that when the government promotes this sector enthusiastically, investment rises. The empirical data corroborates such an interpretation. It was during 1950 to 1969 when investment as a share of GDP in the United States was the highest in the second half of the twentieth century, and it was then that Washington promoted the sector the most. In most other OECD countries, factory output also grew the fastest during these decades and they too attained their highest levels of investment.

Savings followed a similar pattern. In the United States and in practically all other OECD nations, they accounted for a larger share of GDP when manufacturing was promoted the most, and as soon as support began to drop in the 1970s, their share began to fall. Since technology is the fundamental creator of wealth, and the speed by which wealth grows is the fundamental variable for determining how much is saved, it was inevitable that when factory output grew rapidly, as between 1950 and 1969, savings as a share of GDP were high.[116]

Corporate profits also behaved in a way that further substantiates the thesis of manufacturing. They were larger when support for the sector was strong and smaller when factory output grew at a slower pace. During 1950 to 1969, profits as a share of GDP accounted for about 13 percent, during 1970 to 1989 they accounted for only about 9 percent, and during the 1990s the figure fell to 8 percent. Since technology is the prime generator of wealth and profits ultimately depend on how fast wealth gets created, it was only inevitable that when factory output grew quickly the economy performed well and business did likewise.[117]

Economic growth during the second half of the twentieth century in the United States was fast by standards of the eighteenth century but it was just modest by standards of East Asia during the 1950 to 1997 period. As a result, living conditions improved at only a relatively slow pace. Much suggests that this period, as well as the previous ones, were of missed opportunities in which growth could have been much faster if support for manufacturing had been stronger. Had Washington at least promoted this sector twice as much as it did during the 1950 to 1997 period, it is very likely that by the turn of the century the technology for a definite cure against cancer, heart disease, and numerous other ailments would have already been available.

6
The USSR–Russia During the 1950 to 1997 Period

War and Economic Growth

The 1950 to 1997 period was considerably less convoluted than the previous half-century, but it was nonetheless marked by great change. The late twentieth century in particular was marked by such radical changes that the USSR actually ceased to exist and divided into fifteen independent republics. Central planning was also abandoned and there was a renunciation of authoritarian rule.

Such a far-reaching change took place largely without violence, and the second half of the twentieth century was probably the most peaceful in all of the country's history. Practically all economists believe that independent of the economic system applied in a country, the potential for fast growth is always higher when a nation is at peace and the government can concentrate on developing the civilian economy. Although such an assertion sounds reasonable at first glance, the fact is that the empirical evidence has frequently not supported it.

The case of the USSR is illustrative of this. The second half of the century was the most peaceful period in Russian history, but it was not then that the fastest rates of economic growth were attained. Economic growth during the preceding fifty years was about twice as fast, even though there were numerous armed conflicts. During that time, in fact, war-induced devastation reached levels far above anything previously experienced. The devastation of World War II alone was probably much greater than the accumulated destruction of all the wars Russia endured during the preceding two centuries.

If peace is indispensable for positive economic performance, as so many analysts have asserted, the macroeconomic figures should have been significantly different. The fact, however, is that real GDP (once adjusted to the levels of efficiency attained in the most competitive market economies) averaged about 3.1 percent annually in the 1900 to 1949 period and only 1.6 percent for the second half of the century. (See tables in appendix.)

Also worth noting is that between 1950 and 1997 the probability that the country would engage in war progressively decreased. With every decade that passed, tensions with the historic enemies of Russia, such as Germany, Japan, or any European country, rapidly declined. Even though tensions with China rose during the 1960s, they progressively diminished in the following decades. A similar situation took place with the United States. Since the end of the Cuban missile crisis in the early 1960s, tensions between the superpowers decreased. Since the mid-1980s, the USSR's relations with capitalist countries started to become even somewhat friendly, and in the early 1990s tensions disappeared completely as capitalism and democracy were embraced by Moscow.

It was during the 1950s that the largest share of GDP was allocated for defense, coinciding with the fastest GDP figures. On the other hand, during the 1990s when the armed forces absorbed the smallest share of the nation's resources, the economy performed the worst. It was not just that growth was absent during these years, there was actually a terrible contraction, of such immense proportions that it was actually steeper than the one experienced during World War II.[1]

How is it possible that the most destructive war in world history and the country that was most destroyed by this war managed to perform better than during the most peaceful years of the twentieth century? The 1990s not only were the most peaceful years, they were also years in which foreign aid and assistance flowed in the largest amounts during the whole century. In addition, the 1940s were the years in which market forces were distorted the most, while during the 1990s market forces were more operational than at probably any other time during the twentieth century.

There is not the slightest doubt that defense expenditures deliver a large amount of goods that cannot be consumed, and it is also beyond any doubt that market distortions create inefficiencies. It is as well beyond debate that the larger the expenditures in armaments, the larger the creation of goods that cannot be consumed, and that the larger the market distortions, the larger the inefficiencies. However, the degree to which these policies affect the economy negatively has not been yet established. Much suggests that their negative effect is considerable, but there are also many indications that even their compounded effect is not determinant for the overall performance of the economy.

If any of these variables, or even their compounded effect, were determinant for growth, then the development of events in the USSR–Russia during the second half of the twentieth century should have been completely different. The fact that it was not, clearly indicates that some other variable was the determinant one. The empirical data actually reveals that this other variable is so uniquely endowed with impressive growth-generating powers that even with extreme levels of market distortions and a vast allocation of funds for the production of armaments, it still manages to supersede those barriers and deliver growth.

The only variable that in the history of Russia and numerous other countries systematically has given signs of possessing those impressive powers is manufacturing. The growth of this variable is not fundamentally dependent on whether an economy functions under a market system. It is fundamentally dependent on whether a government decides to promote its growth. Since the Soviet government decided to promote its growth during the 1940s and the 1950s, the sector expanded. Since there was a massive subtraction of government support during the 1990s, the sector contracted. Weapons accounted for much of what this sector produced during the 1940s and 1950s. Notwithstanding their incapacity to be consumed, weapons are manufactured goods, which is why their production did not hamper economic growth.

The 1950s and the 1960s

Seen from this perspective, it becomes understandable why history took this course. A closer look at the development of events since the 1950s is very revealing of the impressive growth-generating powers of this sector. The 1950s started with Stalin still at the helm of power, and his determination to allocate priority to manufacturing was by then almost as strong as during the late 1920s when he took control of policy making. Stalin's determination to support this sector was anchored by a number of ideological motivations.

There was on the one hand the effort to be self-sufficient. The socialist autarkic model demanded that everything that was consumed also had to be produced domestically. Since most consumable goods were factory goods, the efforts to produce everything domestically ended fundamentally channeled into this sector. Another motivation was the attempt to demonstrate the superiority of the socialist system. Stalin and most of his comrades in the Politburo were extremely fixated on the idea of surpassing capitalist countries developmentally. Since most of what the most advanced capitalist countries produced were factory goods and these goods also carried the most prestige, Moscow decided to promote this sector on a large scale.[2]

Also, the events of World War II had shocked the Soviet leadership and convinced them that an extensive production of weapons was constantly

convinced them that an extensive production of weapons was constantly required even when there was no apparent threat. Reinforcing this idea was the rapid growth of tensions with capitalist countries and in particular with the United States in the aftermath of World War II.

So as a result of the strong ideological motivations of Stalin and the particular circumstances of the time, support for manufacturing was very strong in the early 1950s. This coincided with a very fast rate of factory output and a fast rate of economic growth.

However, factory output and economic growth were not as fast as during the 1930s and the 1920s, which coalesced with the fact that even though Stalin and the rest of the leadership were still strongly motivated, their enthusiasm had declined relative to the preceding decades. In the aftermath of the Bolshevik revolution, the country was lagging considerably behind capitalist nations and the desire to catch up was therefore stronger. Since during the 1920s and 1930s the USSR had attained the fastest rates of real economic growth in the world and therefore considerably narrowed the developmental gap with the future OECD countries, the goal of matching and overtaking capitalism became somewhat of a reality. As by the early 1950s, this goal began to look attainable, Stalin decided there was no longer a need to invest as much in manufacturing.

Since most of the Party leadership was convinced that the large investments in this sector during the preceding decades had hindered the development of agriculture and overall consumption, Stalin saw it as logical to relent a little on that front. Also, despite the Cold War tensions, Germany had been almost completely neutralized and a third of it placed under Soviet control, Japan had been disarmed, and the United States was too far away to represent much of a threat. The production of armaments, therefore, did not need to be that large. From Stalin's perspective, there were signals from all directions leading him to allocate a slightly smaller share of the nation's resources to this sector. That is why in the 1952 five-year plan, he made official what he had already begun to do a few years earlier. From the late 1940s, when the economy recuperated to its prewar levels, resources began to be transferred to nonmanufacturing sectors.[3]

In 1953, the dictator finally died from a stroke and his designated successor, Malenkov, took power. Malenkov had been instructed by his mentor not to deviate at all from the policies of the previous years, and he largely followed those instructions. However, during his brief reign up to 1955, he de-emphasized somewhat more the support for manufacturing, which correlated with a small deceleration of factory output and GDP. He was also convinced that by transferring resources from this sector to agriculture and the other sectors, overall levels of consumption would rise faster. That, however, did not occur.[4]

In 1955, the liberals within the Party ousted Malenkov and installed Nikita Krushchev in his place. They immediately decided to reconsider the level of priorities on all fronts. There was much economic liberalization across-the-board as rampant inefficiencies plagued the whole system and were the source of massive waste. The excessive centralism of the Stalinist planners was seen as the main cause of the poor quality of Soviet goods and in consequence more freedom was given to production entities. The reformers were also convinced that despite the Cold War and the lessons from World War II, Stalin had been spending too much on weapons, and they immediately cut defense expenditures. Even before the reformers came to power, the defense budget had begun to shrink as soon as Stalin died. However, from 1955 onward allocations for weapons were cut even more.[5]

Stalin had been convinced that the production of consumer goods had to be postponed during the initial stages of development because economic growth was ultimately dependent on the production of producer goods. The reformers, however, even though they shared much of this belief, were convinced that the distribution of resources had been too out of balance and that excessive preference for producer goods had hurt living conditions because it curtailed consumption. Even Stalin had begun to shift some resources for the production of consumer goods in 1952. His successor made more transfers, and Khrushchev many more.[6]

Khrushchev thought that by transferring resources from producer to consumer goods he would increase overall consumption and actually accelerate the rate of economic growth. Even Western economists saw things largely from this perspective and were convinced that consumption and economic performance would improve if resources were distributed more equally between the two domains. However, this did not occur and the economy actually decelerated.

A large share of transferred resources were used to produce consumer factory goods, such as clothing, footwear, radios, toys, automobiles, and household appliances. However, a large share of the resources that formerly financed the production of metals, machine tools, equipment, working instruments, electricity, and the like were utilized to promote agriculture, housing construction and welfare services. Of all these domains, it was agriculture that was promoted most. While Stalin had been largely disinterested in agriculture, Khrushchev was almost obsessed with it. The death of his first wife from the famine that assailed the country during the early 1920s was a lasting trauma for him, and throughout his political career he intervened in most aspects of agricultural policy. He fancied himself an agricultural expert, even though he had no formal education on the subject.[7]

As a result of this vision of the world, the majority of the reductions in defense and producer goods were transferred to agriculture and to other nonfactory domains. This resulted in an overall decrease in the share of the nation's resources allocated to manufacturing, and this ran parallel to a deceleration in the rate of factory output and in the rate of GDP. Agriculture, which was the prime beneficiary of this reallocation effort, did not experience an acceleration in its rate of production, and other fields that were also given a larger share of resources (such as housing) did not accelerate.[8]

Trying to increase consumption and to improve living conditions was clearly the most rational goal that the government could pursue, but this was evidently not the way to achieve it. Economic growth is what ultimately improves both, and much suggests that manufacturing ultimately lies behind growth. As support for the sector decreased, the rate of factory output had to decelerate and along with it the rest of the economy.

Despite the decrease in support during the second half of the 1950s, the sector still received a very large share of the nation's resources and overall support was strong during the whole decade. This correlated with an average growth of manufacturing of about 12 percent annually and a rate of GDP in real terms of about 4 percent. The official Soviet figure was much higher but with a large discount to compensate for the terrible quality of most goods, the figure falls by more than half.[9]

Even under Stalin, some measures to liberalize the economy had already begun. Malenkov undertook more, and after 1955 Khrushchev instituted many more. In 1956, for example, the tight regulation of labor in which workers were assigned a fixed job without the possibility of changing it was repealed, and after that labor mobility increased considerably. On average, in the 1950s, the economy was considerably less centrally planned than during the 1930s. However, it was during the 1930s that the economy expanded at a faster pace (growing by about 6 percent in real terms).[10]

As long as things are not analyzed from the perspective of manufacturing, such a situation remains a paradox; but, when seen from this perspective, the facts no longer present contradictions. Inefficiencies were indeed higher during the 1930s but support for the sector was much stronger, which delivered a much faster rate of factory output (about 17 percent per year). It was inevitable that the economy expanded at a faster pace considering that this was the essential variable.[11]

Khrushchev was almost obsessed with agriculture and was also highly interested in promoting consumer activities, but like most of his colleagues, he was also largely driven to demonstrate that communism was superior to capitalism. Since superiority, in whichever way it might be understood, is basically demonstrated by a more developed technology and since technology is mainly exhibited in factory goods, this motivation ended up channeling plentiful resources into that sector. The most noteworthy of these efforts to demonstrate superiority was the space program that Khrushchev launched almost as soon as he took power. The Soviets succeeded in demonstrating that in this field they were ahead of capitalist countries, for in 1957 they launched the first artificial satellite.[12]

Space goods were not the only factory goods promoted for the sake of demonstrating superiority. There were several others; nuclear-powered ships were the second most evident example. In 1959, the world's first civil nuclear-powered ship, the icebreaker *Lenin,* was completed and launched into the sea. The startled West concluded that such technological sophistication was largely the result of the huge investments the Soviets made in education, and they responded by allocating more funds for education.

Washington, which felt the most humiliated, forgot that during the 1950s Soviet factory output expanded about three times faster than in the United States. Since the creation and reproduction of technology is not fundamentally dependent on possessing an entrepreneurially driven culture or having a market economy, but on attaining factory growth, it was only natural that the Soviets made numerous technological breakthroughs. However, the West was right to condemn the distortions of market forces because without them the rate of technological growth would surely have been much faster. Much suggests that with the same strong support for the sector that Moscow supplied, technology would have probably expanded about three times faster under a capitalist system. The United States for example, with a rate of factory output about one-third as fast, attained almost a similar rate of technological development, which was reflected in the GDP figures. With a factory output of a little more than 4 percent per year during the 1950s, the American economy attained an economic growth of almost 4 percent. While the U.S. economy needed only a little more than a unit of manufacturing output in order to create a unit of economic growth, the Soviet economy needed about three units.[13]

As the 1960s began, Khrushchev continued to dominate policy making and to be convinced that the economy would perform better if resources were transferred from armaments and producer goods to agriculture and other consumer activities. Since agricultural production and other consumer activities had not grown faster during the second half of the 1950s, Khrushchev concluded that there was a need to transfer even more sources from the factories to agriculture, construction, and services. In the early 1960s, there was also further liberalization of the economy on all fronts. Production entities in all sectors of the economy were given more autonomy, there was more foreign trade, workers were given more liberty to choose their jobs, and prices were more deregulated.[14]

Khrushchev's fixation with agriculture drove him to attempt all sort of efforts to attain faster farm output. One of these efforts was to administratively reorganize the large collective farms. The farms were significantly consolidated in the hope that larger economies of scale would improve efficiency. There was also the Virgin Lands program, which targeted the historically neglected vast territories of southern Siberia and Central Asia, and very large investments were made to convert them into arable land. Crop diversification was also pursued enthusiastically and corn abundantly was promoted. Until then, corn had been little grown and Khrushchev saw it as a multipurpose plant that could largely boost Soviet agriculture.

However, notwithstanding all of these efforts, the rate of output of agriculture decelerated considerably. In 1962 to 1963, production was so low that it could not even meet the minimum demand. As a result, there were hunger riots and looting in several cities. Grain had to be imported from capitalist countries and the leadership thus endured terrible humiliation. Khrushchev never recovered from this failure and in 1964 he was removed from his post and supplanted by Brezhnev. A poor agricultural performance had also contributed to the ousting of Malenkov.[15]

Even though Khrushchev transferred many resources from weapons to civilian fields, he still presided over very large expenditures in armaments, and many of them were seen as unnecessary by many in the Party. His mismanagement of the Cuban missile crisis in the early 1960s was criticized by many in the Politburo as an unnecessary provocation of the United States. Without this provocation, tensions with Washington would have been lower, which would have led to a lower expenditure on weapons. Since most were convinced that spending on armaments was a hindrance to growth, they concluded that this resource misallocation was largely behind the deceleration of agriculture and the economy. So as soon as Brezhnev began to dictate policy, he transferred even more resources from the defense budget to agriculture and other civilian domains.[16]

Khrushchev was also accused of endorsing costly useless projects such as the space program. Many within the leadership echoed to a great extent the criticism of those in the West who asserted that the bold housing program initiated in the mid-1950s was held back by heavy spending in the space program.

In 1961, the Soviets were the first to put a man in space, which was a great political victory for a leadership intent on demonstrating the superiority of communism over capitalism. However, many in Moscow argued that increasing the Soviet population's level of consumption was a more important priority. For them, such a goal was being hindered by the vast expenditures that were made to create rockets, spacecrafts, and satellites. So as soon as Brezhnev initiated his mandate, he began to make noticeable reductions in the share of overall resources that were allocated to the space program.[17]

By the time Khrushchev fell from grace, most of the leadership was also convinced that there was a need to further liberalize the economy in order to improve efficiency, so this too was immediately undertaken by the new leader. The main emphasis of the reforms initiated in 1965 was decentralization by allowing a larger degree of responsiveness of production units to market forces. Liberalization, however, did not progress as fast as during the Khrushchev years, but policy making continued to move further in that direction.[18]

There was also a growing belief in the leadership that there was a need for a more professional and technical management of the economy. It was thought that people with little education, like most of those who had held the top positions in government until then, tended to make wrong decisions, which ultimately hampered the economy. Along with Brezhnev, Aleksei Kosygin was chosen to take direct charge of the economy. He was the first leader in Soviet history with considerable economic knowledge.

Even before the arrival of Kosygin, there was an increasing effort to give policy making a professional touch. In 1962, for example, Soviet economists started arguing that efficiency could be improved by the use of the new economics of linear programming and input-output analysis. Assisted by computers, these mathematical techniques, which had their origin in the West, were seen as productivity boosters that could help propel the economy. With the coming of Kosygin, these efforts were endorsed even more enthusiastically.[19]

However, despite the reduction in defense expenditures, the decrease in allocations for the space program and other prestige projects, the decentralization and overall liberalization of the economy, the professionalization of policy making, and the utilization of Western productivity-enhancing techniques, the economy continued to decelerate.

The 1960s was therefore a decade when, in spite of numerous efforts to accelerate the pace of economic growth, the GDP figures went in the opposite direction. While real economic growth (by OECD standards) had averaged about 4 percent annually during the 1950s, the figure dropped to about 3 percent in the 1960s. In this decade, a much larger share of budgetary allocations was assigned to agriculture in the hope that it would grow faster, and this was actually a period in which farm output grew at a much slower pace than in the 1950s.[20]

While agriculture grew by almost 5 percent per year in the 1950s, in the following decade the figure dropped to just 3 percent. These are Soviet official farm figures that must be significantly discounted in order to be comparable with those of the West, but the big difference between the two decades remains largely the same after the discount.

Throughout practically all of the country's history, a similar phenomenon had taken place. Agriculture tended to grow faster when a larger share of the nation's resources was allocated to manufacturing. Independent of whether the country was under a centrally planned system, a capitalist structure of production, or a feudal one, there was almost always an acceleration in the rate of farm output when Moscow increased support for manufacturing. When support was reduced, agriculture almost always decelerated its rate of output.

While manufacturing averaged about 12 percent in the 1950s, in the following decade the figure dropped to just 9 percent annually. Since the output of metals and machine tools as well as tractors, harvesters, irrigation equipment, fertilizers, and pesticides was slower, these goods became relatively scarce. As the overall rate of output of these essential farm materials slowed from the preceding decade, Soviet farmers had less means to incorporate technology into their work. With a lower technological input the farms had no alternative but to produce at a slower pace. Agriculture was almost helpless without these factory goods, for these goods embody technology and technology is the fundamental means by which agriculture improves its productivity.[21]

The same situation occurred with housing and all the other civilian fields that were more decisively promoted with the goal of seeing their output grow faster. This did not happen, and actually the opposite took place. Throughout history, it was almost always when support for manufacturing increased that housing and construction in general experienced an expansion. The other primary-sector activities and most services tended also to grow faster whenever support for manufacturing increased.

The situation was indeed a puzzling paradox, and it remains forever a paradox as long as the premises that common sense dictates are given priority. Common sense tends to be weak on logic and rich in first-hand experiences. Common sense systematically led communists and capitalists to see expenditures on weapons and space goods as subtractions of wealth, and it was also common sense that led them to see investments in agriculture, housing, and services as creators of wealth. However, when history is analyzed strictly logically, the inevitable conclusion is that actually the opposite took place.

It is also common sense that drove Westerners to the idea that market liberalization was essential for accelerating the rate of economic growth in the USSR. Even some within the Soviet leadership agreed. A logical analysis of history, however, reveals that liberalization was essential for improving quality but not for generating growth. It was in manufacturing that the bottom line for the attainment of growth resided, which is why as soon as support for this sector declined, the whole economy began to do likewise.

The 1970s and the 1980s

Since common sense continued to prevail in Moscow during the 1970s, and Brezhnev and Kosygin continued to rule, the policies of the preceding years were further purshed. The Kremlin pursued further the policy of transferring resources for the production of weapons to agriculture and other civilian fields. To achieve this without threatening national security, Brezhnev proclaimed in 1971 the policy of peaceful coexistence with capitalism, and several weapons reduction agreements with the United States were signed in order to set the foundations of this new era of détente.[22]

Almost everybody in Moscow thought that the economy was going to gain speed, as resources would be considerably less wasted in armaments. However, a reduction in the share of GDP that was allocated for the production of weapons once again coincided with a deceleration of the economy. Logic suggests that if those resources had been transferred to civilian manufacturing, the economy would have indeed performed better, but since the main goal of the reductions in defense was to increase the expenditures in agriculture, the overall rate of factory output inevitably decelerated. After having averaged about 9 percent in the 1960s, factory output expanded by just 6 percent during the 1970s.[23]

This coalesced once again with decelerated GDP figures, which, after having averaged in real terms about 3.0 percent, dropped in the 1970s to just 2.3 percent per year. There was also a further liberalization of the economy on all fronts during this decade, more reductions for the space program and for producer goods, and further improvement in the professional ranks of the bureaucracy and the Party. The concomitant effect of all of these policies, which common sense asserted were essential for improving economic performance, once again failed to deliver the desired results.[24]

Agriculture, which was the main beneficiary of all of these policies, illustrates how common sense rarely delivers positive results. The majority of the reductions in defense, the space program, and producer goods were transferred to this domain, and agriculture was liberalized the most. The most brilliant intellectuals and politicians were also assigned to this domain to determine how to improve its performance. During the 1970s, investment in agriculture increased by about 50 percent over the preceding decade. However, notwithstanding all of these efforts, the rate of output of this domain decelerated considerably. After having expanded by about 3 percent annually (Soviet figures) during the 1960s, the figure dropped to just 2 percent during the following decade.[25]

The additional efforts undertaken to produce more of other consumer goods and services ended up not materializing. The goal of generating a faster pace of consumption was also not attained, nor did living conditions improve at a faster rate. Quite to the contrary, rates of consumption grew at a slower pace than in the two preceding decades and living conditions improved more slowly.

The situation was only inevitable when analyzed from the perspective of manufacturing and the premise that this is the determinant sector for the generation of technology and therefore wealth.

In 1979, the Soviet Politburo launched an invasion over neighboring Afghanistan. This action raised tensions with the United States but by then the two superpowers had learned to coexist and the threat of a large-scale war had become very small. In addition, the fighting in Afghanistan was of low intensity and demanded only a fraction of the existing arsenals. In 1981, however, Washington decided to take action against what was seen as Soviet military expansionism and decided therefore to significantly increase defense expenditures. Although Moscow responded in kind and weapon production grew faster, the increases in defense expenditures of both countries were not large and were way below those of the 1950s and 1960s.[26]

The first half of the 1980s was marked by much discontinuity at the helm of the Party as one secretary general after another fell ill and died. Brezhnev died in 1982 and his successor, Andropov, died two years later. Only about a year after beginning to rule, Chernenko also died. Even though overall defense expenditures rose during the first half of the 1980s and the output of weapons accelerated some, the leadership during those years continued to be almost obsessed with agriculture and consumer activities. Thus, they continued to transfer resources from producer goods to agriculture and other fields.[27]

Allocations for iron, steel, other metals, machinery, equipment, and the like were cut more and transferred to the farms and to other consumer fields. Since the cuts in the producer domain were about the same as the increases in weapons, the end result was an overall similar allocation of resources to manufacturing. This situation was reflected in a similar pace of factory output that coincided with a similar rate of GDP.

Andropov and Chernenko were even more fixated than Brezhnev on seeing agriculture grow at a faster pace, which is why they transferred even more resources from producer goods to the farms. This is also why they instituted much liberalization in this domain. During the first half of the 1980s, there was an across-the-board liberalization of the economy, but by far the most liberalized domain was agriculture. In spite of these efforts, neither the economy nor agriculture improved.[28]

Agriculture was the nemesis of Soviet leaders because of the cold war rivalry with the United States and the comparisons that inevitably emerged from that rivalry. Agricultural productivity in the United States was at least two times higher than in the USSR. By 1985, for example, the USSR employed about 20 percent of its labor force in agriculture while the United States employed just 2 percent, and in spite of that the USSR could only produce about 85 percent of what America produced. Since by then food consumption was still relatively low and some malnutrition still endured, agriculture was seen as a priority for a socialist peace economy.[29]

Therefore when the Party chose a new leader in early 1985, they opted for Mikhail Gorbachev, who had concentrated on agriculture during most of his career and had proven to be a relatively successful productivity booster. During Chernenko's rule, for example, he had been largely in control of economic policy and had significantly promoted the cause of agriculture.[30]

As soon as he assumed power, he began to push further in that direction. He decreed a decisive liberalization of the whole economy, but with more emphasis in agriculture. He was also a strong believer in transferring resources from weapons and producer fields to the farms, and he immediately began to cut defense expenditures and reduce budgetary allocations for heavy manufacturing.[31]

From the start, however, Gorbachev gave clear signals of being interested in liberalizing and reforming the Soviet system in a much more decisive way than his predecessors. In 1987, for example, a new NEP (New Economic Policy) was introduced giving for the first time since the 1920s official recognition that the country needed the private sector to attain better economic performance.[32]

His plans for increasing investment in agriculture and other civilian domains were also much more ambitious than those of his predecessors. Like the former general secretaries, he thought that defense could provide the most funds for his plans. In order not to compromise the national security of the country, he began first by reducing tensions with OECD countries. He took unilateral actions that sent clear signals of his peaceful intentions and made very conciliatory proposals to capitalist countries (in particular to the United States). He began to pull the troops from Afghanistan; he reduced military and economic aid to Cuba, Vietnam, and other communist countries; he expanded economic cooperation with Western Europe; and above all he signed numerous arms reduction agreements with the United States.[33]

Defense expenditures as a share of GDP dropped by a large margin and most in the Kremlin and in the capitalist world became convinced that the economy was going to improve significantly as the "military burden" was considerably reduced. Almost everyone in the West was also convinced that an acceleration of the economy would take place because of the reintroduction of the private sector and the other market measures. However, the end result was exactly the opposite and the economy experienced a deceleration.

Since the reductions in defense were not transferred to civilian manufacturing and the reductions in producer goods were only partially reallocated into consumer factory fields, the result was a reduction in support for manufacturing. This was reflected in a decelerated rate of factory output for the second half of the 1980s, which correlated with a decelerated rate of economic growth.[34]

The 1980s was therefore another decade in which reality once again ignored common sense, because a decrease in defense expenditures failed to be accompanied by an improvement in the economy. This decade also experienced an increase in liberalization which failed to deliver faster growth and faster rates of productivity. A more professional and technocratic management of government was also incapable of delivering the desired results. It was evident that not even the combined effect of these efforts was essential for the attainment of growth.

What once again showed signs of being essential was manufacturing. The reduced support for the sector translated into a slower rate of factory output that averaged about 4.3 percent annually after having averaged about 6.2 percent in the 1970s. This in its turn coincided with a decelerated rate of GDP that averaged only 1.4 percent after having averaged about 2.3 percent (the GDP figures are adjusted to capitalist measurements).[35]

Defense expenditures as a share of GDP during the 1980s were actually at the lowest level in Soviet history. Also at this time, market forces were allowed to play a larger role and the policy making apparatus was at its most upgraded with professionals. Only the 1920s experienced lower defense expenditures and about a similar level of market liberty.

Not even the war-torn 1940s attained as low a rate. During the 1940s, real GDP averaged about 2.2 percent annually, which coincided with a rate of factory output much higher than in the 1980s. The 1940s was the most market-distorted decade in Soviet history and it was also when the largest share of GDP was utilized to produce armaments. How was it possible that growth was faster than in the 1980s? A rational response is possible only if it is assumed that manufacturing is the bottom line for the generation of economic growth.[36]

Not only did the numerous efforts to improve of the economy during the 1980s fail to achieve that goal, but the prime target of the reforms also failed to improve. Agriculture was the main focus of the reforms, and farm output, instead of growing faster, actually experienced a deceleration. As in the past, a slower rate of factory output coincided with a slower rate of agricultural output. Farm output in the 1980s averaged about 1 percent annually after having expanded by about 2 percent in the preceding decade.

With a lower output of tractors, irrigation equipment, and fertilizers, Soviet peasants had fewer means to apply technology to the land. Since technology is fundamentally responsible for improving agricultural productivity and it is embodied in factory goods, a slower output of these goods inevitably translated into a slower rate of implementing technology on the land.

Since the 1950s, many in the Kremlin became convinced that the fundamental mechanism for improving productivity in agriculture and in the rest of the economy was liberalization. As a result, with every decade that went by, market forces were allowed to play a larger role. However, the results actually went in the opposite direction. With every passing decade, productivity decelerated more and more. By the 1980s, it was at the lowest point since the end of World War II.[37]

By the 1980s, the Soviet economy was considerably less inefficient than in the 1950s and it is evident that such progress was the result of liberalization, but rates of productivity were considerably below those of the 1950s. It is clear that liberalization played a role in enhancing productivity, but it is also clear that it did not play a determinant role. The only variable with strong credentials for playing a determinant role is manufacturing—not just because the rates of this sector varied in direct proportion to the productivity rates, but also because this sector constantly attained the fastest rates of productivity. Factory productivity grew much faster than in agriculture, construction, and services.

As in capitalist countries, factories attained the fastest productivity figures almost permanently. The fact that capitalist and communist economies alike, notwithstanding their enormous differences, attained similar results on this matter strongly suggests that something very particular about this sector was behind that phenomenon. It suggests that something about the inherent nature of manufacturing managed to singlehandedly be responsible for the majority of productivity growth.

The fact that, under both systems of production technology found itself tightly bonded to the manufacturing sector supplies strong evidence that this sector is fundamentally responsible for the creation and reproduction of technology. Since technology is the main factor behind the growth of productivity, it becomes clear why the events in the Soviet Union took the course they did.

Manufacturing output grew at a slower pace during the 1980s than at any other time in Soviet history. Despite the slow rate, it was still relatively fast, which coincided with a relatively strong support for the sector. Defense absorbed about 16 percent of GDP, the large majority of which was used to finance armament production. It is precisely because support was relatively strong that technology continued to grow and productivity continued to expand.[38]

As with weapons, expenditures for space goods were also cut but they were still large. Consequently, the Soviets continued to attain impressive technological breakthroughs, which gave further evidence of the tight linkages between manufacturing and technology. During the 1980s, for example, the USSR put into orbit the first space station in world history. The share of GDP that was allocated to other fields of heavy manufacturing was smaller than before, but it was still large and the share that went to light manufacturing was also relatively large. By 1989, the overall share of factory output from GDP was almost 50 percent.[39]

Most analysts in OECD countries and many in the USSR saw the large size of this sector, as well as the strong support for it, as a terrible misallocation of resources. Manufacturing was seen as being too extensive, and thus as depriving the other sectors of resources and hampering economic growth. Many argued that this sector needed to shrink. The events of the 1990s, however, clearly demonstrated the invalidity of this idea and it also showed that numerous other ideas of the sort were inconsistent.

Among the other ideas was the assertion that the deceleration of the Soviet economy between 1950 and 1989 was largely the result of rising labor costs and the exhaustion of cheap natural resources. The explanations for the events of those decades from this perspective were largely similar to those concerning capitalist countries with decelerating rates of GDP. However, the fact is that the rising-labor-cost argument had never managed to add up consistently in capitalist countries, where competition with other countries is the norm and where foreign trade plays an important role in the economy.[40]

If numerous capitalist countries have succeeded in retaining fast rates of growth in spite of rising labor costs, then the more reason for a centrally planned economy closed to foreign trade. Singapore and Hong Kong, for example, experienced a much faster rate of labor-cost growth during the 1950 to 1989 period than the USSR, and they had the highest level of trade openness in the world. In spite of that, they managed to maintain a fast rate of growth during the 1980s.

As for a scarcity of natural resources, here again the argument fails to add up consistently with the facts. The case of the two city-states debunks this idea because they had probably the lowest per-capita availability of natural resources in the world during the whole period. The Soviet Union, on the other hand, had one of the highest. If it was possible to grow impressively while lacking natural resources almost completely, there was evidently no excuse for the USSR not to grow quickly independent of how exhausted certain natural resources seemed to be.

Many argued that the reason these resourceless city-states managed to grow so fast was that they were the most market-oriented economies in the world. However, the fact is that the growing market liberalization of the

USSR had coincided with deceleration during four decades, and when in the 1990s Russia and the ex-Soviet republics embraced capitalism, the economy not only decelerated, it actually plunged into a terrible depression.[41]

Soviet distortions were indeed great and exacted a heavy load on the economy. Western estimates calculate that even by the late 1980s only 7 to 18 percent of Soviet goods were able to meet the minimum quality and cost standards of OECD countries. However, during the 1990s, the majority of those market distortions were eliminated and inefficiencies were therefore significantly reduced.[42]

During the 1990s, Russia became almost as market-driven as the average OECD country, yet it fell into the worst economic contraction in its history. Perhaps only the contraction experienced between 1914 and 1920 was steeper than that of 1990 to 1997. It is worth noting that during the 1990s labor costs plunged by real international standards and the country proved to possess huge amounts of natural resources as it exported vast amounts of numerous commodities. All of those advantages, however, could not deliver a positive performance. The thorough endorsement of capitalism, the abundance of raw materials, and the low labor costs of the 1990s made it very evident that none of these variables, individually or as a group, could coherently explain the persistent economic deceleration of the 1950 to 1989 period.[43]

The 1990s

A closer look at the 1990s simultaneously reveals the inconsistency of orthodox arguments and the very strong logical tenets of manufacturism. The reforms that Gorbachev initiated in 1985 were not limited to the economy. He also fostered political reform so liberally that in no time the whole country and the whole Soviet-dominated bloc of countries began to experience political activity that led to radical transformations. In 1989, for example, several countries in Eastern Europe broke away from the Warsaw Pact and from the Council of Mutual Economic Assistance (CMEA). In that year the Democratic Republic of Germany disintegrated and Gorbachev gave the seal of approval for a reunification with the Federal Republic of Germany. Some time later, the Warsaw Pact and the CMEA were fully dissolved and most of its ex-members opted for taking the path of capitalism and democracy. Finally in 1991, the Soviet Union imploded into fifteen separate and independent states. The end of the USSR also marked the end of central planning and of an authoritarian regime because practically all of the new republics opted immediately for capitalism and democracy.[44]

Even before the breakup of the USSR, the republics had begun to behave largely like independent economies. By 1990, quasi-economic independence among the republics drove them to raise trade barriers against each other. The central planners' old policy of concentrating the production of particular items in one or two provinces in order to promote regional interdependence had some undesirable effects once those provinces became autonomous. As it no longer became possible to mobilize basic inputs for factory production among the provinces, this sector began to contract. The contraction of manufacturing in 1990 coincided with the first year of economic contraction.[45]

Many argued that it was the trade protection among the republics that caused the contraction, but the fact is also that the provinces began to trade very liberally with the rest of the world. Protectionism among the fifteen provinces effectively appeared, but liberal trade practices with the rest of the world did also. This more than compensated for the negative effects of this trade distortion and brought higher economic benefits. The economy therefore should have actually grown faster because of the improved trade picture.

In 1990, there was also much economic liberalization officially and unofficially. Market reforms were decreed that went much further than in the preceding years, and because the country began to fall into a semichaotic situation in which everybody did as they pleased, market forces became considerably more operational than what was officially sanctioned. However, notwithstanding such a productivity-enhancing situation, the economy contracted.

The next year, the country formally disintegrated and all fifteen newly independent states decided to overturn socialism and convert the economy into one governed by market forces. The new governments initiated policies to make the conversion as fast as possible. One of the most ardent reformers was the government of Russia, which inherited about 50 percent of the population, the majority of the territory, and about 60 percent of the wealth of the Soviet Union.[46]

The new government immediately began to privatize, liberalize, and deregulate the economy. The first stage of Russian privatization began in 1991 with the selloff of small businesses such as shops and restaurants. As soon as that was completed, a new stage began in early 1993 with the auctioning in exchange for vouchers of about 15,000 medium and large enterprises which employed the majority of the workforce. By the end of the next year, all of those companies had been transferred to the private sector.[47]

Once this phase of the privatization was completed, the economic face of Russia became almost unrecognizable. By then the private sector accounted for almost 60 percent of GDP, after having accounted for just 3 percent only a few years earlier. More than half of the population labored in firms that were in the hands of capitalists. By then, the private sector had even become larger than in some OECD countries such as Italy, which had a large number of state companies.[48]

Privatization of the urban economy was what most attracted the attention of Russians and foreigners because of its resolute break with the past, but reforms were undertaken in all areas and fields of the economy. Reforms in agriculture, for example, fully overhauled the Soviet farm system. By 1994, Russian agriculture was almost in another dimension with respect to a few years earlier: subsidies had been slashed; low-cost state loans had been eliminated; government-guaranteed purchases were significantly reduced; production quotas had been dropped; centralized pricing was scrapped; the state trading monopoly had been considerably broken up; and the bulk of state and collective farms had been quasi privatized.[49]

Privatization and liberalization were decisively pushed forward in every other primary activity and in every sector of the economy. In mining, fishing, forestry, construction, tourism, banking, and practically all other service activities, market forces became the factor deciding economic activity. Foreign trade was greatly liberalized and by 1994 Russia had a very unobstructed trade regime that was similar to that of numerous capitalist economies.

However, in spite of all of these positive policy undertakings, the economy contracted considerably during those years, unemployment and underemployment rose to unprecedented levels, and living conditions deteriorated significantly.[50]

In the other fourteen ex-Soviet republics, a similar phenomenon took place. Notwithstanding the encompassing liberalization and privatization in practically all of them, they all plunged into a depression that seriously deteriorated living conditions. The degree of interest with which each of the new governments tried to transform the economy into a capitalist one varied significantly, but all of them pushed decisively in that direction. Even in places like the Ukraine, where reform stalled in comparison to Russia, there was nonetheless a large transformation relative to the Soviet past. A capitalist economy also took root. The contractions, however, were not homogeneous. In places like the Ukraine, the contraction was worse than in all the others and in places like the Baltic republics the depression was less pronounced.[51]

In the countries of Eastern Europe that had formerly practiced a centrally planned system, whole-scale liberalization and privatization were also accompanied by negative GDP figures. During the first half of the 1990s, these countries experienced negative average GDP figures. As with the fifteen ex-Soviet republics, the degree of market liberalization varied from country to country but the governments in all of these nations decisively endorsed capitalism. The contractions also varied considerably. In Poland, the downturn was less steep, while in Bulgaria and Romania the depression was more pronounced.[52]

After half a decade of wholehearted liberalization and half a decade of the worst depression in the history of most of these ex-communist nations, many policy makers began to doubt the efficacy of the reforms. Opportunist politicians from left and right launched aggressive attacks against the reforms and vowed to stop them or even reverse them. These politicians had no credible alternative program for reactivating the economies of any of these countries, which together totaled almost thirty, but they were evidently right when they pointed out the shortcomings of the reforms.[53]

After decades of dismal levels of efficiency and productivity, it was beyond doubt that the high levels of competition that the reforms introduced were badly needed. The quality and cost-effectiveness of goods and services immediately experienced an impressive improvement across-the-board. However, the fact that the economy contracted, and in a very pronounced way, left little doubt that the capitalist reforms were not the bottom line for the generation of growth. The evidence clearly indicated that it was some other variable that played the determinant role.

Many economists and policy makers in these countries and in OECD nations concluded, therefore, that the determinant variable was inflation. Since very high levels of inflation were experienced in practically all of these nations during the first half of the 1990s, it was argued that financial instability was the cause of the contraction.[54]

During the decades of communism fiscal imbalances were maintained in most of these countries (much of the time at relatively low levels). In the Soviet Union, for example, during the 1966 to 1985 years budget deficits hovered at about 2 to 3 percent of GDP and the money supply was allowed to expand at a moderate pace. However, during 1986 and 1990, budget deficits rose significantly to average about 7 percent of GDP.[55]

Since the arrival of Gorbachev to the Kremlin, a loose monetary policy also began to be applied in an effort to stimulate the economy. That goal was not achieved, for the economy actually decelerated, but the money supply rose rapidly. Currency in circulation, which had grown by just 6 percent in 1985, rose by 12 percent in 1988 and by 28 percent in 1990. Since enterprises were allowed to pay their workers more but prices were not freed, inflationary pressures built up. When prices were suddenly set free in the early 1990s, inflation shot up to very high levels.[56]

To make matters worse, the fiscal imbalances and the loose monetary policy were continued during the first half of the 1990s, and inflation grew on average by about 770 percent annually (in Russia). It was therefore asserted that growth was impossible under those circumstances because low inflation was a precondition for growth.

Nobody can seriously question the merits of low inflation and it is also beyond any doubt that the best mechanisms for harnessing prices are sound fiscal and monetary policies. However, if it is true that low inflation is a precondition for growth, then Argentina and Brazil should have also experienced contractions. During these same years, those two South American countries also experienced hyperinflation, yet they managed to grow. In the case of Argentina the economy even grew relatively fast, averaging about 6 percent annually while enduring an average inflation of about 468 percent. Brazil, which endured a level of inflation almost double that of Russia during 1990 to 1994, managed nonetheless to grow by about 2 percent per year.[57]

A rate of 2 percent was not an impressive performance, but it was far superior to the extensive contraction in all ex-Soviet republics and Eastern European countries. It was evident that the inflation variable was not determinant for the attainment of economic growth. Throughout history there were several other examples in different countries where growth, even fast growth, was attained in spite of high inflation. There are also numerous examples in which, despite a very low inflation climate, growth was not attained. It is simply impossible to find a consistent correlation between inflation and growth.

The case of Kyrgyzstan, which was an ex-Soviet province in Central Asia, further reinforced the idea that there is no causality linkage between inflation and growth. When the USSR dissolved, it was the republic that embraced market reforms most enthusiastically. It not only privatized and liberalized at a breathtaking pace, it also fully endorsed the structural adjustment program of the International Monetary Fund (IMF). By 1994, the government was running healthy public finances, tight monetary policies were being applied, and monetary stability had been attained, yet the economy continued to fall. Inflation had been brought under control, but the economy continued to contract significantly way.[58]

The absence of a correlation, however, did not stop many analysts from arguing that the reason Argentina and Brazil had grown in spite of hyperinflation was that they had always been market economies, while the ex-CMEA countries had just made a drastic structural transformation which was inevitably traumatic. That argument, however, failed to explain why nations such as China, Vietnam, and Laos, which were also in the midst of transforming their economies from communism to capitalism, managed to attain speedy rates of economic growth. It was argued that China ran a cautious and relatively sound monetary policy that kept inflation low, which allowed for a smooth transition. Although China indeed kept prices at relatively low levels, Vietnam, did not and the country still managed to grow quickly. During 1987 to 1992, inflation in Vietnam was very high, but that was not a hindrance for the attainment of fast growth.[59]

Since the argument about inflation did not add up consistently, many economists therefore argued that the reason nations such as China, Vietnam, and Laos had grown rapidly was that they were less centrally planned than the CMEA nations at the moment when they started their reforms. If the more a nation was centrally planned, the harder the attainment of growth once it starts to embrace capitalism, China should have been the one that performed the worst. Among the three East Asian communist economies, China went furthest on matters of state centralization. However, exactly the opposite took place. China, which was the most centrally planned of the three by the time it started reforming (in 1979), grew the fastest. Laos, which was actually the least centralized of the three by the time it started reforming in the late 1980s, grew the least fast.[60]

It was also argued that when they started their reforms, the ex-CMEA nations had higher levels of foreign debt than China, Vietnam, and Laos when they started theirs. China in the late 1970s and Vietnam and Laos in the late 1980s indeed had very low levels of foreign debt as a share of GDP, while the levels in CMEA countries by the early 1990s were much higher. However, if high levels of foreign debt were an obstacle for growth, history would have not seen a large number of nations grow quickly while carrying a heavy load of foreign debt. Examples of the above are Germany during the 1920s and again after World War II; Japan after World War II; South Korea during the 1960s, 1970s, and 1980s; and to a lesser extent Argentina during the 1990s.[61]

Analysts argued that these were capitalist countries, which made it easy to grow despite the debt. However, if being capitalist was so important for attaining growth, why is it that China, which even by 1997 was still officially a socialist nation, managed to grow faster than practically all capitalist countries in the world during the 1990s. During these years, the private sector in China accounted for only 10 percent of GDP. Almost the same phenomenon was appreciated in Vietnam and Laos. Privatization was only barely carried out and the private sector accounted for only a fraction of GDP. In spite of that, China averaged about 8 percent of real economic growth during 1990 to 1997, Vietnam had about 5 to 6 percent, and Laos had about 4 to 5 percent.[62]

It is also worth noting that if having a large foreign debt in a centrally planned economy inhibited growth once that economy began to transform itself into a capitalist one, then it should have been Poland that performed the worst. Among the ex-CMEA countries, by 1990 Poland was the most in debt. By then, Poland's external debt as a share of GDP was about 82 percent. Not only was it way above that of the Soviet Union or any other CMEA country, it was actually one of the highest in the world, exceeding even that of Argentina, Brazil, or Mexico. However, during the 1990s it was Poland, the ex-CMEA nation, that performed best. Independent of how these arguments are presented, the fact is that they systematically failed to add up consistently with the empirical evidence.[63]

Misunderstanding Causality

China attracted much international attention during the 1990s because of its very fast rates of growth. The stark contrast in performance with the ex-USSR drove many analysts to conclude that foreign direct investment (FDI) had played an important role. This was apparently a variable that could explain why, in spite of its timid liberalization, China had performed so much better than all ex-CMEA countries. FDI indeed flowed in vast amounts into China during the 1990s while in the ex-USSR it came in small amounts. Since the majority of FDI came from offshore Chinese, such as those from Hong Kong and Taiwan, it was argued that it was these blood linkages that made the exceptional flows of capital possible.[64]

This argument, however, had too many weak points to be acceptable. It is first worth noting that FDI in China in the 1990s as a share of GDP was small, accounting only for about a tenth of the whole economy—too small, therefore, to have acted as a propeller. There was also the fact that in the 1980s, FDI flowed in just a tiny fraction compared with the 1990s, yet China still grew very fast.

Also worth noting is that there were numerous examples of countries that had no wealthy expatriate communities yet they attracted even more FDI than China did. Mexico attracted much higher amounts of FDI in per capita terms during the 1990s. Notwithstanding those advantages Mexico grew more slowly than China. The cases of Vietnam and Laos also debunked the argument of the offshore Chinese advantage. None of these two Southeast Asian countries had large and wealthy expatriate populations that invested abundantly in them in the 1990s, but they nonetheless managed to attract much FDI and grow fast. Also interesting to note is that despite attracting much larger per capita amounts of FDI, Mexico grew less than half as fast as China, yet Mexico had the advantage of being a capitalist country.

Not only was it a weak argument that the ex-Soviet republics had not attracted much FDI because they lacked a large and wealthy expatriate community, it was also false that FDI was determinant for the attainment of growth.

The majority of mainstream economists have always advised developing and developed countries with market economies to privatize, liberalize and deregulate as soon as possible. Up to the late 1980s, they had also argued that if CMEA countries would privatize and liberalize at a very fast pace, a positive economic performance would immediately take place. Soon afterward, however, practically all CMEA countries did precisely that and what they got in return was probably their worst ever economic performance. Since China liberalized at a slow pace, some economists and policy makers began therefore to argue that gradualism supplied better results than shock therapy.

This argument obviously forgot that countries such as Hungary and Poland began with gradual reforms in the early 1970s. However, even though there was economic growth during the 1970s and 1980s, it was never fast like that of China in the 1980s and 1990s. Also worth noting is that since the 1950s, economic reforms in the USSR followed the gradualist approach and what the country got in exchange was constantly decelerating GDP figures. Gorbachev's reforms in the second half of the 1980s tried precisely to imitate China's gradual efforts, and the economy, instead of growing faster, actually decelerated further.[65]

It was also argued that countries performed better when they gave priority to economic reform over political reform. China's example inspired many to assert that. However, it is worth remembering that from the 1950s to the 1980s in the Soviet Union and most of Eastern Europe, emphasis was put on economic reform while political reform was relegated to a secondary role. During those decades, however, GDP figures for most of these countries decelerated almost constantly. Therefore this could not have been the cause of the large gap in GDP figures between China and the ex-CMEA countries during the 1990s.

Still others purported that China's success resided largely in having started its reforms in agriculture. In the ex-USSR and most of Eastern Europe reform in the 1990s indeed gave priority to the nonagricultural sectors, while in China from 1979 liberalization went always further in agriculture. Things were not always like that, however. From the 1950s to the 1980s in CMEA countries, agriculture was the main priority of the liberalization efforts, and in spite of that the economy decelerated almost constantly.

Associated with this argument was another stating that China, Vietnam, and Laos found it easier to grow because they were developing countries with a very large agricultural sector. Transferring workers and resources from agriculture to the other sectors, according to this argument, was much easier than transferring them from inefficient urban enterprises to efficient ones. The latter was the challenge that the ex-CMEA countries confronted during the 1990s.[66]

If classic economic development is so easy that even nations with socialist economies can manage to grow quickly then nations even less developed and with market economies should have been able to grow much faster. The bulk of sub-Saharan Africa, for example, during the 1990s had economies that had never been centrally planned and which were considerably more market-driven than those of China, Vietnam, and Laos. They were also economies where the share of GDP and of the population that was in agriculture was larger than that of the three East Asian communist countries. However, economic growth was much slower. While real growth averaged about 7 percent in China during the 1980s, in sub-Saharan Africa it was just about 2 percent per year. In the 1990s,

China grew by about 8 percent and the countries south of the Sahara expanded by less than 2 percent annually.[67]

It is also worth noting that of these three Asian socialist economies, China was the least agricultural but the one that grew the fastest. It was Laos that had a larger share of the population working the land at the time when the capitalist reforms began, and therefore the one that should have moved more easily on the road of classic economic development. However, of the three it was Laos that grew most slowly.

By the time CMEA countries dumped central planning, the majority of the jobs had been covered by an extensive social welfare system. When communism ended, high social welfare expenditures remained and burdened the newly privatized companies with excessive costs. The high-cost production structure that prevailed during the 1990s was, according to many analysts, a major cause behind the poor economic performance of these nations. In China, on the contrary, when the communes were dismantled in 1979, nearly three-fourths of the workforce found itself outside the socialized economy with little social welfare protection. By 1991, the aggregate expenditure on social security in China equaled 6 percent of GDP, while in the bulk of ex-CMEA countries it was twice as high or even higher. That difference, according to many analysts, is what basically explained the great difference in economic performance between China and the ex-CMEA nations.[68]

China, Vietnam, and Laos were indeed considerably less burdened by social welfare expenditures. However, the difference does not seem large enough to explain why in the 1990s China attained one of the fastest rates of growth in the world while the ex-Soviet republics attained the worst and the nations in Eastern Europe attained the second worst. The social-welfare-expenditure difference was not proportionate to the GDP difference.

Also worth noting is that if a large social welfare expenditure were determinant or at least important in affecting negatively the economy, the Scandinavian countries should have experienced a contraction or at least stagnation, and not just during the 1990s. Since the mid-twentieth century, these countries have been the largest social welfare spenders in the world as a share of GDP. In the 1990s, their social welfare expenditures were much higher than in the ex-CMEA countries, but growth was nonetheless observed.

Many, however, argued that because Scandinavian countries were already very market-driven by 1990, it was possible for them to grow despite the burden of the large welfare expenditures. However, if having a market economy helps so much, then sub-Saharan countries should have grown very rapidly because by 1990 they were already very market-driven and they also had the lowest social welfare expenditures in the world as a share of GDP. Their rates of growth, however, were actually among the slowest in the world.

While none of these arguments manages to add up consistently with the facts, the thesis of manufacturing succeeds very well. During the first half of the 1990s, there was a very large subtraction of government support for manufacturing in all of the ex-CMEA countries and there was also a very large reduction of factory output. In addition, the contraction of the economy was proportionate to the reduction of manufacturing. However, the subtraction was not homogeneous. It was least pronounced in Eastern Europe compared with the ex-USSR and it was in Eastern Europe that factory output and GDP contracted the least. Within Eastern Europe, the situation was also not homogeneous. In countries such as Poland, Hungary, and the Czech Republic, the economy contracted the least and it was also in these nations where the new governments took the most actions to counterbalance the free fall of factory production.[69]

In the ex-USSR, the performance of the new fifteen states was also not homogenous. Some performed worst than others. In the Baltic states the economic downturn was less pronounced and the subtraction of government support for this sector was of a lower magnitude. Factory output did not drop as much as in the other republics.

The subtraction of support for the sector was driven by several ideological motivations. The dissolution of the Warsaw Pact and the end of the cold war drove the governments in Eastern Europe and the Soviet Union to extensively cut defense expenditures. There was absolutely nothing wrong in reducing weapon production drastically, but much suggests that it was a major error not to have transferred those resources to civilian manufacturing. Numerous armories were immediately ordered or driven to produce civilian goods, but under a greatly reduced budget.[70]

Nobody seriously considered transferring the resources to civilian manufacturing because there was nothing in the theories behind communism and capitalism that asserted that government support for this sector per se was essential or even important for the generation of growth. Practically all of the new governments of the ex-CMEA nations were eager to embrace capitalism, and Western analysts had for long argued that these countries had too much manufacturing. These analysts asserted that these nations needed to reduce the size of this sector, and the new policy makers were willing to follow their advice. Thus, the reductions in defense expenditures were seen as a wonderful opportunity to deconstruct this sector.[71]

Since the reductions in defense were accompanied by a catastrophic economic contraction, many economists began to argue that the large reductions and the transformation to capitalism were part of an inevitably traumatic process and some time needed to pass before these changes delivered positive results. However, by 1997 the majority of ex-CMEA nations were still waiting for those positive results. By then, their economies were still contracting, stagnating, or just growing slowly.[72]

The direct subtraction of support for the production of armaments translated also into a subtraction of support for related fields. Since iron, steel, machinery, equipment, and numerous other inputs for the fabrication of weapons were needed in lower amounts, budgetary allocations were significantly reduced for these fields.

The renunciation of the closed economy structure of production also delivered a subtraction of support. Since these countries no longer wanted to be autarkic, the need to produce everything domestically was no longer seen by the new governments as necessary. As a result, numerous fields that were not even indirectly associated with weapons were no longer supported or were considerably less promoted.

There was also a reduction of budgetary allocations for prestige projects that had originated when it was a priority to demonstrate that communism was superior to capitalism. The space program was the largest of these projects and as a result the production of space goods fell significantly.

With all this subtraction of support for the sector, it was inevitable that overall output would contract by a large margin. Many policy makers lamented the closure of so many factories, but most were also convinced that their countries had overinvested in this sector during the preceding decades and that it needed to shrink. Also, the majority of these factory goods were of very low quality and many therefore concluded that these countries were actually better off without them.[73]

These countries indeed obtained very few benefits from those low-quality goods, and such a large share of the factories were so plagued by obsolete machinery and equipment, decrepit buildings, and overstaffed work forces that they deserved to close down for good. It was also rational to suspend budgetary allocations and all other forms of subsidies to those enterprises, but there is much to suggest that it was a major error not to have utilized those resources and many others for the promotion of new and efficient private-sector factories. History and logic suggest that the resources should have been kept for the support of the sector.

The same was true for defense expenditures. The reductions were a positive undertaking and it had been still better if much more had been cut, but all those resources should have been utilized to finance new and efficient private-sector civilian factories. If the governments of these nations had endorsed an enthusiastic policy of support for this sector, probably not a single year of contraction would have occurred. That is what China, Vietnam, and Laos did and they effectively grew quickly from the very first year of transition to capitalism. In these three countries, support was strong from the start, factory output grew rapidly, as did their economies. It was Peking that promoted the sector the strongest and it was China that attained the fastest rates of economic growth.[74]

These three East Asian nations neglected privatization almost completely and while it was evident that this was an error, they nonetheless did what was fundamental for the generation of growth. The ex-CMEA governments, on the other hand, made a rational decision by privatizing and liberalizing rapidly, but they did not do what was essential for growth.

Much suggests that if they had rechanneled all the resources utilized until the end of the 1980s to produce weapons, weapons-related goods, and civilian factory goods into new and private civilian factories, things would have been completely different. Most likely, not a single year of recession would have occurred. Had policy makers allocated an even larger share of overall resources by supplying an abundance of fiscal, financial, and nonfinancial incentives to foreign and domestic manufacturers, it is highly likely that very fast rates of economic growth would have happened from the first year of reform. However, this was not done and the economy plunged massively.

Although the privatization and liberalization programs were wide-ranging, they still left a large share of enterprises in state hands and a considerable degree of regulation in place. Much suggests that it was an error not to have gone further. If every firm had been privatized and practically all remaining regulations would have been phased out, the ex-CMEA countries had significantly profited from the higher levels of efficiency that would have inevitably resulted. However, had these productivity-enhancing measures been undertaken without the endorsement of strong support for manufacturing, the performance of the economy would have only improved marginally. As the 1995 to 1997 years demonstrated, more privatization and liberalization for most of the ex-CMEA countries (in particular the ex-Soviet republics) did not coincide with economic growth.

Reforms, Change, and Weak Results

It was thought that one of the great failures of central planning was failing to develop the service sector and many believed that liberalization would bring its rapid development. The transformation to capitalism indeed delivered impressive development of the service sector, but it did not coincide with growth. Western analysts have long argued that services are very important and perhaps even essential for the generation of growth, in particular for developed economies.[75]

Since CMEA countries were not classified as developing nations, it was assumed that this idea would also apply to these economies; but that idea never added up consistently with the empirical evidence of OECD countries. The share of services in the GDPs of OECD countries relentlessly increased during the second half of the twentieth century, and as it did the rate of these nations' economic growth systematically decelerated.

The liberalization effectively delivered a huge growth of services in the ex-CMEA countries and their share of GDP expanded exponentially, but this was accompanied by massive contraction. Most services grew at an impressive pace (the legitimate ones as well as the illegitimate ones). By 1993, for example, the Czech Republic had the highest per-capita ratio of casinos in the world, while four years earlier not a single roulette wheel spun in the former Czechoslovakia. Criminal services and prostitution in the Czech Republic and in all the other ex-CMEA countries also grew at an exponential pace.[76]

Legitimate service activities also grew quickly. Those that expanded the most were wholesale and retail trade. Trade, in particular short-term trade, became the area that the newly privatized banks in these countries financed the most. In Russia, for example, the approximately 3,000 privatized banks began almost immediately to finance primarily mainline commercial activities such as retailing, where the returns are quick and the amounts of capital needed are relatively small. Mafia lenders also showed a strong preference for short-term credit to retailers and trading companies. Other activities that attracted the interest of bankers were currency dealings and stock trading, which also have very fast rates of return.[77]

Banks in these countries, once they were privatized, did what banks in every other country have always done when left to do as they pleased. They opted for concentrating on activities that promised to deliver profits in the shortest amount of time. Since investment in manufacturing takes by far the longest to show a profit, the inherent instincts of East European and ex-Soviet bankers was to cut practically all lines of credit to this sector.

This does not mean that they should have continued to finance obsolete, decrepit, inefficient, and overstaffed factories, but they could have financed new and efficient private factories, as well as industrial parks where foreign and domestic private manufacturers would have set operations in ready-made installations.

However, it is not realistic to expect private bankers to finance manufacturing without the backing of the state. In practically every country in the West and East Asia throughout almost all of their history, commercial banks supplied a large share of their capital to this sector only when the government (by means of incentives and regulations) drove them to adopt such a lending policy.

The vast amounts of capital that manufacturing requires compared to all the other sectors and the much longer periods of time needed to materialize a profit drove bankers instinctively to distance themselves from this sector. This behavior actually ended up undermining the profitability and the very existence of banks, because the well-being of banks ultimately depends on the strength of the economy; and the performance of the economy depends on how much is invested in manufacturing. By early 1996, for example, more than a fifth of Russia's banks had gone bankrupt or were almost insolvent, and most of the rest had everything but outstanding balance sheets. The government confronted the problem by creating deposit insurance, better supervision, and higher reserve requirements.[78]

This was indeed a very positive policy undertaking and it should have been adopted earlier, but as 1996 and 1997 demonstrated, this undertaking was not essential for the attainment of growth or for the well-being of the banks. During these last years, the economy continued to perform poorly and the banks did similarly.

During 1995 to 1997, services continued to grow rapidly and once again it became evident that this sector was not essential for the generation of growth. Tourism and numerous other service fields grew at a double-digit pace in Russia and in most of the ex-Soviet republics, but it did not translate into positive economic performance.[79]

Since by the end of the first half of the 1990s results had been catastrophic in practically all ex-CMEA countries, many argued that such a situation was the result of not having pushed the reforms all the way. So privatization, liberalization, and deregulation were pushed some more in practically all of these countries. By 1997, most of these countries had a private sector that was larger than that of several OECD countries and numerous developing nations. In that year, for example, Russia's private sector had grown to account for about 80 percent of GDP.[80]

However this was little help for most of these nations as the economy continued to contract, or at best grew only modestly. This situation once again demonstrated that even though positive, those undertakings were not essential for the generation of growth. Russia, for example, continued to significantly contract during those years. However, not all ex-CMEA countries continued to experience negative GDP figures. Among the ex-Soviet republics, a few, like the Baltic states, managed to grow. In Eastern Europe, the situation was much better as most managed to grow some and a few like Poland even managed to grow quickly.[81]

That diverse situation was paralleled again by proportionate levels of support for manufacturing and concomitant rates of factory output. While support for the sector continued to be taken away in most of the ex-USSR, in most of Eastern Europe the subtraction ended and policies promoting the sector were initiated. Among the ex-USSR republics, it was the Baltic states that attained the best factory performance, and these were among the only ex-Soviet economies that saw growth. In most of Eastern Europe, there was some factory growth and the faster producer of manufactures (Poland) also had the fastest growing economy.[82]

The differing performance had largely to do with the fact that the USSR had been the reason behind the creation of the Warsaw Pact, and in consequence spent the largest share of its GDP in weapons. As the pact was dissolved and the national security concerns behind it vanished, it was inevitable that cuts in defense would be larger in the ex-USSR. In Russia defense as a share of GDP averaged about 16 percent in the 1980s while in 1990 to 1997, the figure fell to about 8 percent. Since numerous civilian factory fields were directly linked to armaments, these fields also had to fall further in the ex-Soviet Union. As the leader of the communist bloc, it was in the Soviet Union where the bulk of efforts intended to demonstrate the superiority of this system over capitalism were undertaken. When the system collapsed, it was there that manufacturing of this sort had the biggest fall.[83]

This does not mean that the manufacturing and economic contraction were inevitable. It only means that the factory promotion efforts of the state of a strictly civilian and private nature had to be extremely strong in order to compensate for the decline of armaments and inefficient manufacturing. The government of Poland for example got more or less to this vision of things by the mid-1990s. Inspired by the example of China and numerous other East Asian nations, it began to promote the sector enthusiastically. By 1996, for example, there were five large special economic zones (SEZs) offering tax holidays for up to twenty years to foreign and domestic manufacturers, as well as financial assistance, free land for the factory, and other nonfinancial incentives. Outside these SEZs, there were also several fiscal, financial, and nonfinancial incentives that were offered to foreign and domestic manufacturers, although less abundant.

In most ex-CMEA nations, the respective governments were only barely motivated to try to imitate the example of China and other East Asian nations on matters involving manufacturing. Priority continued to be given to liberalization and to the promotion of primary-sector activities and services. The huge amount of natural resources that the ex-USSR possessed drove analysts and policy makers to the conclusion that these countries were particularly well endowed for the production of these goods. The fascination with services among most economists also drove policy makers in the wrong direction. To make matters worse, the idea that manufacturing needed to be further reduced in its share of GDP was also strong. With such a vision of the world, it was inevitable that factory output among most ex-Soviet republics continued to decrease during 1995 to 1997; and that coincided with a further contraction of GDP.[84]

That does not mean that liberalization should not have been continued.

More liberalization and privatization should have actually been undertaken, but much suggests that together with these reforms, a very decisive policy in support of manufacturing should have been endorsed.

Extremely high levels of inflation were experienced during the first half of the 1990s, coalescing with large budget deficits and a lax monetary policy. The International Monetary Fund was absolutely right in pressing these countries to eliminate their fiscal imbalances and to tighten monetary policy, but together with those undertakings there was a simultaneous need to promote factory production in a very decisive manner.[85]

One thing did not hinder the other. Many argued that fast inflation was one of the main factors hindering growth, but when Russia (for example) finally applied sound fiscal and monetary policies in 1997 and inflation fell to just 10 percent, the economy only managed to attain a rate of 0 percent. It was not inflation that was hindering growth, but the absence of support for manufacturing. During certain years of the 1990s, Vietnam and China endured high inflation but they simultaneously promoted factories decisively, which coincided with fast economic growth. In other years, they applied sound fiscal and monetary policies, and inflation was low, but they simultaneously supported the sector enthusiastically, and once again fast GDP figures were observed. In 1997, for example, China had a 0 percent level of inflation but the economy grew in real terms by about 6 to 7 percent. Some years earlier, inflation reached 25 percent and the economy grew by 10 percent.[86]

The distortion of market forces had long been held accountable by Western analysts as the fundamental reason agriculture performed badly in the USSR. During the 1990s, agriculture was largely left to be sculpted by market forces, and instead of improving, it did worse than during practically any other decade of the twentieth century. Only during the 1910 to 1919 decade was the performance worse. During 1990 to 1997, Russian agriculture contracted on average by about 1 percent annually.[87]

It was beyond any doubt that liberalization brought productivity-enhancing benefits to agriculture, but the empirical evidence also made it clear that this was not the determinant variable for the generation of growth in this domain. The manufacturing thesis, however, had no problem adding up consistently with the facts. In the twentieth century, it was only during the chaotic and governmentless 1910 to 1919 years that there was also a large subtraction in support for manufacturing. Factory output also contracted by a wide margin during this war decade.

As during the 1910s, the contraction in factory output during the 1990s also ultimately translated into less farm machinery, fertilizer, and other basic inputs. Since these were the goods that embodied technology, the newly market-driven agriculture was left without the capacity to add technology to its production efforts, and technology is the fundamental means for enhancing productivity. It was therefore inevitable that production dropped. During 1990 to 1997, manufacturing in Russia contracted on an average annual basis by about 9 percent and farm output by about 1 percent.[88]

Foreign trade presented a similar paradox. During Soviet times, Moscow was not interested in trading with the rest of the world and Western analysts were convinced that if the country would open to foreign trade, commerce would increase manifold. Some time later, their liberalization wish became reality. During the 1990s, foreign and domestic trade were thoroughly liberalized, but instead of expanding rapidly, they ended up contracting.

When seen from the perspective of manufacturing, the paradox is resolved. Since factory goods are the most tradable of goods and account for the bulk of international trade, a contraction of this sector inevitably delivered fewer goods that could be traded. Fewer factory goods also translated into a lower output of primary sector goods, which diminished even more the amount of goods that could be traded.[89]

A similar situation was observed in the other ex-Soviet republics and most nations in Eastern Europe. Notwithstanding their wide-ranging liberalization of agriculture and foreign trade, the average figures for these two domains during 1990 to 1997 were negative, stagnant, or just slightly positive. Practically all of the preceding history of these countries presented a similar development of events. Agriculture, the other primary activities, and trade tended to grow faster whenever government support for manufacturing increased. It was almost always when the state transferred resources from these domains to manufacturing that these domains grew at a faster pace. That was inevitable, because that was when technology expanded at a faster pace, and technology is ultimately responsible for the expansion of any field or domain.

With the significant subtraction of support for manufacturing during the 1990s, it was unavoidable that just about everything would contract. In Russia, for example, factory output contracted by about 9 percent per year during 1990 to 1997 and the official GDP figures fell by about 6 percent annually. However, just as the official figures during Soviet times needed to be readjusted, so too did the statistics during the 1990s. The readjustment of figures was justified on the ground that so much of what was no longer produced was of such bad quality that it was not strongly missed. A percent drop in the output of automobiles, electronics, apparel, footwear, tractors, pharmaceuticals, grains, or medical services in Russia, for example, was a considerably less important loss than the same percentage drop of those goods in Japan.[90]

The figure therefore most likely to reflect the real economic contraction of Russia during the 1990s is about 3 percent annually. In the other ex-

Soviet republics and nations of Eastern Europe, the readjusted statistics for the years of contraction should also be discounted by about 30 to 50 percent of the official figures.

However, notwithstanding the readjusted figures, up to the last of the ex-CMEA nations endured steep contractions in the early 1990s and their average performance between 1990 and 1997 was at best modest. Much suggests that such an economic crisis and the suffering that it brought to the inhabitants of those nations could have been easily avoided if a decisive policy in support of manufacturing had been endorsed.[91]

Manufacturing and Investment

The evidence substantiating a very strong correlation between manufacturing and economic growth is abundant. The evidence demonstrating a very strong parallelism between this sector and technology is also abundant. This situation inevitably leads to the suspicion that this sector is fundamentally responsible for the generation of technology.

The case of the USSR, more perhaps than any other, made that suspicion transform into causality. The centrally planned system was full of inefficiencies and the authoritarianism that accompanied it hampered creativity. It is evident that the only way so many inventions came about was by means of a factor that by its very nature always succeeds in creating technology. Western analysts argued that it was not possible for the Soviet Union to be inventive, but over and over again the country made outstanding discoveries that put it in the world's lead in particular fields. All of that was accompanied by a very decisive support of manufacturing.[92]

It is most likely not chance that the rate of technological development ran in paralleled to the level of support for the sector during the second half of the twentieth century. It was during the 1950s when the factory promotion efforts of the state were the strongest, and it was then inventions materialized at the fastest pace. From then onward, support progressively decreased and the rate of factory output progressively decelerated, and that coincided with a decelerating rate of technological development.

It is interesting to note that the deceleration in the rate at which inventions appeared took place regardless of the considerable improvement in several factors that most analysts saw as determinant for the development of technology. During the 1950 to 1989 period, there was on the one hand a growing economic liberalization, which was supposedly essential for better allocating resources and therefore for attaining a higher technological output of those resources. There was, on the other hand, a growing political freedom that allowed scientists and technicians to better express their creativity without risking demotion, jail, or death.[93]

There were also growing expenditures in research and development (R&D) as a share of GDP, which were explicitly attempting to accelerate the pace of innovation. While in the 1950s R&D expenditures as a share of GDP averaged about 1.8 percent, by the 1980s they had grown to account for about 3.4 percent. As time passed, the importance of scientists in society and even in politics progressively grew, and they increasingly influenced policy making for the sake of serving the interests of science. In addition, education significantly improved from several perspectives. On the one hand it became less ideological and more scientific and, on the other hand, the share of the population that completed secondary and university education relentlessly grew.[94]

From every perspective considered by orthodox interpretations as essential or important for technological development, the situation of the USSR should have been completely different. With the growing share of R&D expenditures, the growing educational levels of the population, the increasing numbers of scientists, and the progressive liberalization of the economy, it should have been during the 1950s when technology grew at the slowest pace and the 1980s when it grew the fastest.

The fact that it was not strongly suggests that a variable so inherently endowed with a strong capacity to generate technology, and which was not among the ones identified by orthodox interpretations, was the one that made that situation possible; and the only variable that fits that description and which correlated with the development of technology is manufacturing.

The 1990s went a step further to demonstrate that relationship. During these years, there was a massive subtraction of government support for this sector and factory output contracted substantially. This drop coincided with a complete absence of innovation and technological progress. There was actually a leap backward on that front. Technology is what fundamentally improves living conditions and living conditions dropped considerably between 1990 and 1997 in Russia and the other ex-Soviet republics. Morbidity rose dramatically, life expectancy fell to very low levels, and malnutrition became endemic. After having averaged about sixty-nine years in 1989, life expectancy by 1997 had fallen to fifty-nine years.[95]

Even diseases that had been largely eliminated, such as tuberculosis, made a large-scale comeback. Since the availability of factory goods dropped extensively and it was these goods that were the depositories of technology, a contraction in the sector's output inevitably translated into a contraction in the nation's supply of technology. Under those circumstances, a decline in living conditions was impossible to avoid.[96]

Independent of the low quality of Soviet tractors, they still managed to create more agricultural goods than when none were available. Food processing machines were also of a miserable quality but with a much lower number of them, the food supply had to decline. And a lower food supply had to translate into malnutrition, which opened the way to opportunistic diseases such as tuberculosis. Soviet pharmaceuticals were also of a much lower therapeutic efficacy than those of capitalist countries, but these medicines were still better than none at all.

Manufacturing systematically showed signs of transcending the differences imposed by opposing economic systems such as capitalism and communism. In capitalist countries, for example, this sector constantly absorbed practically all R&D expenditures and in centrally planned economies the same phenomenon was observed. From the 1930s until the breakup of the USSR, about three-quarters of R&D expenditures were utilized for the sake of making new and more powerful weapons. The largest share of R&D expenditures in the United States from the 1940s to the 1990s was also for the sake of developing new armaments.[97]

Due to World War II and the Cold War in both countries, weapons were by far the area of manufacturing that needed to be improved as fast as possible. As a result, policy makers in both countries supported this field more than any other, which inevitably meant it absorbed the most R&D expenditures. Also in both countries, the fields of manufacturing that were promoted the most made the most technological progress. In both nations during these periods, weapon technology made the most technical breakthroughs and claimed the largest number of patents.

The 1990s also presented a paradox from the perspective of orthodox interpretations of reality. The thorough liberalization of the economy allowed for a much better allocation of technological resources, and the almost complete political freedom that prevailed allowed scientists and technicians the utmost ability to fully exploit their creativity. On top of that, ideology was completely eliminated from the educational system, and defense expenditures were largely cut, supposedly allowing for increased investments in civilian technology. Western analysts had long claimed that if the USSR would undertake such measures, technological progress would accelerate its pace considerably. Once again, reality did not move in tandem with such a vision of the world and what ended up taking place was exactly the opposite.

Government R&D expenditures dropped by about three-fourths during 1990 to 1997 in Russia as well as in practically all of the ex-Soviet republics. Since the bulk of R&D was in weapons, most analysts thought at the start of the reforms that the newly created private sector would rehire the scientists and technicians that used to work in the armories and would buy the machinery and equipment to make and create new civilian goods. That did not occur. Most thought that there would be a wide-scale conversion of the defense plants into civilian factories, but even by 1997 that had still not taken place. The factories were just closed down, the machinery and equipment rusted away, and the scientists and technicians became unemployed or ended up in menial service activities.[98]

It is evident that the factories and the machinery were of a lower quality than those of OECD countries, but it is worth remembering that defense companies were by far the most modern and best equipped (at times even with OECD machinery). They also had the best scientists and technicians. Even if the installations could not match the cost and quality that the large flow of foreign imports imposed on the market, it was evident that the scientists and technicians were highly valuable assets, especially when seen from the perspective of the low salaries they were prepared to accept. Most analysts assumed that the new manufacturing private sector would rehire the highly qualified technical personnel. That did not occur.

Others thought that the service sector would absorb the scientists and give them meaningful jobs. That did not occur either. That was highly unlikely when it is considered that history has systematically recorded that the manufacturing sector has always given employment to the bulk of scientists and technicians.

There is much to suggest that it did not occur because the government did not promote manufacturing decisively. In the history of Russia, the private sector never took the initiative to invest in the sector and during the 1990s such a pattern of behavior did not change. In the West and East Asia, the private sector also systematically refused to invest in the sector unless there was first support from the government.

This is only understandable considering that this sector is fundamentally responsible for the creation and reproduction of technology, and the generation of technology is the most investment-intensive activity in the world. Without support from the government, the risks are simply too high and capitalists instinctively opt for the other sectors that demand much lower investments and which can attain a profitable return in a much shorter time.

Not by coincidence is it that during capitalist and socialist times, it was always manufacturing that absorbed the bulk of overall investment. Before the arrival of the Bolsheviks, this sector absorbed the vast majority of investment, and during their more than seven-decade rule, it absorbed almost three-fourths of overall investment.

The geographic distribution of factories also reflected the high investment nature of this sector. Between 1950 and 1989, for example, the province of Russia absorbed about 10 percent more investment than its share of the total Soviet population. That is, it absorbed 62 percent of overall investment while accounting for about 52 percent of the total population. That correlated with the fact that Moscow targeted the province of Russia as the main recipient of factory investment because of national security concerns. Defense plants were particularly concentrated in this province.[99]

The differing levels of support for manufacturing also correlated with the different levels of investment over time. During the 1950 to 1989 period, for example, investment as a share of GDP constantly decreased, paralleling the decreasing levels of government support for the sector. Factory output grew at the fastest pace during the 1950s, and it was then that the share of investment reached its largest size, accounting for about 30 percent of GDP. From then on, it declined, and by the 1980s it averaged just about 11 percent of GDP, coinciding with a much decelerated rate of factory output that was only one-third that of the 1950s.[100]

Western analysts expected that as the economy was progressively liberalized between 1950 and 1989, and the rewards for profit increased, investment would also rise, but exactly the opposite took place. Many therefore argued that since the system remained centrally planned, the small liberalization was practically meaningless. However, when the system was totally dismantled in the 1990s and market forces became fully operational, the same paradoxical situation continued to prevail. On this occasion, investment actually fell precipitously and there was even massive capital flight. Despite the economic plight of most ex-Soviets, savings remained high, which according to most economists is essential for the materialization of investment. However, investment never materialized. During 1990 to 1997, savings in Russia as a share of GDP were of about 30 percent. There was evidently enough domestic capital for investment, but it was never utilized for that purpose.[101]

That the economy was under capitalism or communism just did not make any difference. As long as government support for manufacturing was missing, investment would remain low or nonexistent. Without guarantees from the state that the vast investments that this sector requires would be profitably recuperated, the new capitalists just did not commit their capital to factories. Since this sector is by far the most investment-intensive, an absence of investment in manufacturing inevitably translated into an overall absence of investment throughout the economy.

In previous times, the same situation was observed. For centuries, Russian capitalists refused to invest in the sector and it was not until the late nineteenth century that Moscow for the first time decided to supply strong support for manufacturing. Immediately, investment rose for the first time to relatively high levels. In other capitalist countries, the same thing took place. During the second half of the twentieth century in most OECD countries investment as a share of GDP declined almost constantly, which coincided with decreasing levels of government support for the sector. During this period, the government that promoted factories the most among this group of nations was Japan's, which also coalesced with the fact that Japan was the OECD country that invested the largest share of its GDP.[102]

By the 1990s, Russia, the other ex-Soviet republics, and Eastern Europe were in much need of investment. Numerous measures were taken to stimulate it, but in most of these countries it never materialized. History strongly suggests that investment is most stimulated by endorsing a policy of support for manufacturing. The higher the incentives the state supplies to the producers of factory goods, the larger the investments in this sector and the higher the overall levels of investment.

History also strongly suggests that endorsing a decisive factory promotion policy is the best means for achieving numerous other goals that seem very disconnected with the economy, such as protecting the environment, increasing political freedom, and improving the lot of women.

Ecological consciousness was completely absent in Russia during practically all of its history. This coincided with a history full of famines and malnutrition in which the desperate situation of the population drove people to be indifferent to the needs of the environment. This also was paralleled by centuries in which there was no government support for manufacturing. By the mid- to late nineteenth century, the sector began for the first time to be promoted in a relatively strong way and famines suddenly and significantly diminished. Support was not always strong thereafter and at times there were even subtractions of support, but in broad general terms Moscow continued to promote factories rather decisively. That ran parallel with a constant improvement of living conditions.

By the mid-twentieth century, living conditions had risen considerably and the Soviet population as well as their leaders began to give importance to things that formerly were seen as almost irrelevant. Concern for the environment first surfaced in 1960 when work began on building two pulp and paper mills on the shores of Lake Baikal, the world's largest natural reservoir of fresh water. Many intellectuals complained and there was also some street protest. From then onward the continuation of a relatively fast factory output and the progressive improvement in living conditions coincided with a growing environmental awareness from the population and the leadership.[103]

There was growing political freedom during the second half of the twentieth century, notwithstanding the authoritarian system that prevailed during most of this period. Forced labor camps were rapidly dismantled after Stalin's death, and numerous forms of freedoms were increasingly allowed after that.[104]

The rule of law also developed considerably during this period. Until 1953, the USSR had only one law, that of Stalin, but after that the judicial system progressively evolved in a rather expeditious way. As wealth was continuously accumulated and the population became increasingly satisfied in its basic demands, its behavior became less violent and unpredictable and the leadership could no longer justify maintaining a totalitarian system. Under those circumstances, it was inevitable that freedom and the rule of law developed rapidly. The evidence very strongly suggests that manufacturing was fundamentally responsible for the creation of wealth, thus ultimately responsible for the development of more civilized forms of government.[105]

During the Soviet period, women continued to be treated as second-class citizens. Their average wages were considerably below those of men and they were given employment in fields with lower prestige, power, and financial rewards. Marital violence was endemic and the judicial system was deliberately structured to favor men. However, the mistreatment that Russian women endured in former times was even worse. Until the mid-nineteenth century when the economy remained perennially in stagnation, the abuses they endured were considerably worse. Only when the economy began to grow at a relatively fast pace did their condition begin to improve.[106]

As the economy continued to grow, the treatment of women progressively improved. Since economic growth was relatively strong during Soviet rule, the situation of women improved considerably, notwithstanding the authoritarian nature of the regime that tended to disregard civil rights. Wealth is almost exclusively responsible for the satisfaction of human needs, and only when basic needs are satisfied does a population begin to behave in a civilized way.

Manufacturing is what a logical analysis of history designates as the fundamental factor behind the creation of wealth. It follows, therefore, that the best policy a government can undertake in order to protect the environment, promote political freedom, foster the rule of law, and ameliorate the treatment of women consists in supporting this sector as much as possible.

Manufacturing during the second half of the twentieth century in the ex-USSR grew by about 4.5 percent per year, and the economy in real terms expanded by about 1.6 percent. There is much to suggest that if market distortions would have been as small as in Hong Kong, the same rate of factory output would have delivered an economic growth more than twice as fast as the one that was attained. Much also suggests that it was an error not to have promoted manufacturing more decisively. If Moscow would have promoted this sector as much as it did in the 1920s, and factories would have increased their output at a 21 percent annual rate while simultaneously endorsing a free market policy, then the 1950 to 1977 period would have witnessed an economic growth of about 17 percent per year. With such a rate, Russia would have been, by the late 1990s, the most developed nation in the world, with a very developed democratic and judicial system.

7
Germany: The Postwar Era

A Peace Economy and the Size of the State

The second half of the twentieth century was by far the most peaceful period in Germany's history. Not even the slightest border skirmish took place with any of its neighbors nor did the country participate in any military intervention in a faraway country. By the end of the century, the German Democratic Republic (GDR), which after the end of World War II had been created as a separate country, disintegrated and was absorbed by the Federal Republic of Germany (FRG). This event of very large historical proportions took place without the slightest form of violence.

The whole 1950 to 1997 period was therefore the only moment in Germany's history in which the government could wholeheartedly concentrate on the civilian economy. Considering that in all of the preceding half-century periods wars considerably disrupted normal economic activity, it would be logical to expect that the fastest rates of economic growth were attained between 1950 and 1997 years. Growth was indeed faster during this period, but the difference was so small relative to the two preceding war-torn periods that it becomes evident that peace in itself is not a determinant variable for the attainment of economic growth (see charts at the end of the book).

Logic suggests that all governments of the world should work hard to maintain peaceful relations with all other countries and to maintain peace within their borders, but the empirical evidence strongly indicates that peace in itself is not an important factor affecting growth. Economic growth in the FRG during the second half of the twentieth century averaged about 3.8 percent annually. In the preceding fifty years, the country was seriously disrupted by war and actually experienced defeat in the two largest armed conflicts in history. Great devastation was caused by these conflicts (in particular World War II) and the tremendous war preparations that preceded both conflicts delivered extremely large economic distortions. However, growth averaged about 3.5 percent annually during the years 1900 to 1949.

During the 1850 to 1999 period, when Germany engaged in three large-scale wars and several minor military operations, and when the country was constantly making preparations for war, the economy grew on average by about 3.7 percent per year. The second half of the twentieth century clearly demonstrated that economic growth was not dependent on militaristic government policies, as many people of a political rightist orientation came to believe during the period from the mid-nineteenth century to the mid-twentieth century. However, these two fifty-year periods also demonstrated that peace was not essential for growth.

The historical evidence clearly demonstrated that on this matter, all schools of thought had missed the bottom line. However, this was not the only misinterpretation. The second half of the twentieth century was a period in which government expenditure in West Germany as a share of GDP reached huge proportions. On average, the combined expenditure of federal, state, and local government during this period was about 41 percent. This was a very large increase compared with the previous half-century and, in addition, Germany's state expenditure became one of the largest among OECD countries.[1]

During the preceding half-century, government expenditure had averaged only half as much, and during the 1850 to 1999 period the combined weight of central, state, and local government was less than 10 percent of GDP. These facts largely invalidate ideas from left and right concerning the ideal size of the state. Those who argued that a large state was desirable could not explain why during the second half of the nineteenth century, when the state was not even one-fourth as large as a century later, the economy grew at the same rate as during the 1950 to 1997 period.[2]

Those who asserted that a small government was essential could not explain either why during the second half of the twentieth century, at a time when government expenditure had grown to gigantic proportions, it was possible for Germany to grow faster than it had in all of its preceding history. Reinforcing further the idea that a correlation was missing was the fact that during the first half of the nineteenth century the overall expenditure of all the German states was even smaller than during the following fifty years, and yet economic growth was largely stagnant.[3]

The compounded evidence of Germany's history as well as that of numerous other countries strongly suggests that a large government expenditure tends to hamper efficiency by allocating resources inadequately, but the data clearly indicates also that a small state in itself is not determinant for growth. The empirical information clearly reveals that neither is essential for growth.

Once again, it was evident that orthodox interpretations of reality were incapable of adding up consistently with the facts. It was also evident that a variable different from the ones traditionally considered by all sides of the debate had acted as the prime cause of growth. The capacity of this variable to generate growth was so vast that it could still overcome major obstacles and hindrances such as a large government expenditure during the second half of the twentieth century and a large production of weapons during the two preceding half centuries.

Manufacturing was the only variable that correlated consistently with the long-term empirical evidence. For example, growth was weak during the first half of the nineteenth century, notwithstanding the extremely small share of overall government expenditure, and this coalesced with a weak state support for manufacturing. This sector grew by about 1.2 percent annually and GDP by about 0.9 percent. In the following fifty years, the economy accelerated impressively, averaging about 3.7 percent per year, notwithstanding the large intensification of war and the huge increase in weapons production. The acceleration, however, coincided with a very large increase in the factory-promotion efforts of the state and this sector expanded by about 4.4 percent annually.

During the 1900 to 1949 period, German governments once again allocated a very large share of the nation's resources for the production of weapons, and once again the economy grew rapidly. This coalesced with extensive subsidies for manufacturing, with a rate of factory output that averaged about 4.7 percent, and with a rate of GDP that averaged about 3.5 percent annually.

Even though during the second half of the twentieth century German governments stopped spending on armaments, they began to enlarge by a significant margin the size of state expenditure. Worse still was that the large majority of the increase was for the sake of enlarging the social welfare budget. However, notwithstanding the very large share of the economy that was allocated for purely consumption purposes, the economy grew slightly faster than in the two preceding fifty-year periods and much faster than in the first half of the nineteenth century, when social welfare expenditures were nonexistent.

This, however, coincided with a strong factory promotion policy of the government, which was probably the most decisive in the country's history. Factory output expanded on an average annual basis by about 4.6 percent and the economy once again followed and grew by about 3.8 percent.

During the whole 1800 to 1997 period, as well as before, manufacturing constantly gave strong evidence of being intrinsically associated with the creation and reproduction of technology. As the sector that was the prime generator of technology, it had the power to create an abundance of wealth even when an abundance of policy errors were being committed.

That would explain why the large production of armaments during the 1850 to 1949 period did not hinder fast growth and actually correlated with a strong acceleration relative to all of the preceding history. Since armaments are manufactured goods, their production created technology. That technology was soon used for civilian purposes and in such a way that the civilian economy benefited and living conditions rose.

That would also explain why the large rise in the size of the state, and more particularly in social welfare expenditures during the second half of the twentieth century, did not hinder growth. Since government support for manufacturing during this period was as strong as during the two preceding fifty-year periods, the economy continued to grow at about the same pace.

A Divided Country During the 1950s

A closer look at the development of events during the 1950 to 1997 period illustrates the strong causal linkages between manufacturing and wealth creation. From the end of World War II until 1989, Germany stopped existing as a unified country, as the largest share of it adopted a market economy and a democratic political system, and the rest of it adopted an undemocratic system and a centrally planned economy. Officially, both Germanys were born in 1949.[4]

During the whole period in which both Germanys existed, rates of economic growth were not identical. The official statistics of the GDR frequently stated that economic growth was stronger in the east. However, once the official figures of the GDR are reconstructed to be consistent with the considerably lower quality of the goods and services produced there, the situation changes dramatically. Measured in this way, the FRG systematically attained the fastest rates of economic growth.

Many have argued that the reason the FRG performed better was that it practiced a capitalist system. The poor quality of the goods and services in the GDR left very little doubt about the inherent shortcomings of the system. However, it is highly unlikely that the slower rates of the GDR were the result of the centrally planned system. Real growth in East Germany during the 1950s was faster than in the FRG during the 1960s, 1970s, 1980s, and 1990s. Such a situation clearly demonstrates that despite the inefficiencies of the system, it was still capable of generating rates of growth that were faster than those attained in capitalist countries.[5]

It was like the USSR during the 1920s and 1930s, which in spite of its highly centrally planned system, still managed to grow faster in real terms than any capitalist country in the world during those decades. The case of the USSR, East Germany, and several other communist countries which at times managed to grow rapidly, strongly suggests that notwithstanding their numerous shortcomings, there was something about the policies undertaken by the governments of these countries that was capable of generating fast growth. There was evidently something with such impressive growth-generating powers that it managed to overcome all the growth-hindering distortions of the centrally planned system.[6]

Also worth noting is that even though the GDR's long-term average rate of growth was slower than in the FRG, the fact that growth in itself took place was very revealing. During the 1950 to 1989 period, GDP averaged on an annual basis about 4.3 percent for West Germany and about 2.8 percent for East Germany. If possessing a market-driven economy was essential for the generation of growth, then the GDR should have attained absolutely no growth during the whole period. However, it did grow, and actually faster than numerous capitalist economies in Africa, Asia, and Latin America.[7]

The fact that several other communist countries during that same period attained growth and that their rates of growth were faster than those attained by several capitalist countries goes a step further in substantiating the idea that the essential variable determining growth was not any of the ones traditionally presented.

In all of these countries, economic growth coincided with decisive policies in support of manufacturing and with fast factory output. During this period, rates of GDP were not homogeneous. Some communist countries grew faster than others and that in its turn coincided with different levels of support for manufacturing. Over time, most countries experienced decelerating rates of GDP and that coalesced with decreasing levels of support for the sector and decelerating rates of factory output.

In all communist countries, as well as in capitalist nations, the development of technology was constantly bonded intimately with manufacturing, which suggests a particular relationship of this sector with innovation. The fact that such a tight bond was appreciated throughout history in practically all corners of the world indicates that this sector has the intrinsic capacity to generate technology. Since technology is fundamentally responsible for the creation of wealth, it is understandable why the GDR and other communist countries managed to grow regardless of their numerous policy errors. Only like that is it that the empirical data makes sense.

Analyzed from this perspective, it becomes evident why the GDR systematically grew more slowly than West Germany. Support for manufacturing was less strong. During that time, Berlin could have given much more support to the sector, but it did not because the Soviets largely vetoed such a possibility.

The USSR endured such heavy losses at the hands of the Nazis that the idea of seeing a prosperous Germany was intolerable to the Soviet leadership in the decades that followed the end of World War II. Only under the pressure of a fast-growing West Germany after 1948 did Moscow reluctantly acquiesce to a policy that allowed the East German authorities to promote manufacturing. Subsequently, Moscow allowed only a rate of factory output similar to the one attained in the FRG.[8]

During practically the whole period in which the GDR was under Soviet control, Moscow followed a policy of having East Germany shadow the FRG's economic development. After the war, the Kremlin wanted nothing less than to see Germany dwell in material hardship, but since West Germany began to grow rapidly in the late 1940s, it felt compelled to let the eastern part do the same. So it stopped blocking East German authorities' desires to reactivate the economy. Since by the time the war ended the largest sector was manufacturing, the change of attitude translated into letting Berlin allocate a relatively large share of resources into factories.

Moscow, however, never liked seeing its ex-oppressor prosper. It therefore put limits on how fast the economy could grow, and the limits were largely determined by the rates West Germany attained. Being manufacturing the largest sector by the time the country was partitioned, it was this sector that seemed the most important for economic recovery but also where limits were most imposed to its expansion. The fact that this sector had produced the weapons that had caused the death of almost 25 million Soviets was perhaps the strongest reason behind Moscow's reluctance to see it grow very fast.[9]

Not that the East German authorities were willing to promote manufacturing in a extremely decisive way, but they would probably have supplied a higher level of support had the Kremlin not imposed limits. Under those circumstances, rates of factory output would have been faster, as would rates of real economic growth. The inherent inefficiencies of the system would have continued to deliver much waste and low-quality goods, but growth would have been faster. That is what happened during the 1950s, compared with the 1980s. During both decades, the waste and inefficiency in East Germany was almost identical but in the 1950s real economic growth was more than 5 percent annually while during the 1980s it was less than 2 percent. This coincided with a rate of factory output of about 12 percent in the 1950s and of only 4 percent in the 1980s.[10]

Interesting to note is that even though the Soviets were not interested after the war in seeing either of the Germanys recuperate economically, it was actually they who ultimately led the United States and its allies to let western Germany recuperate. Once that started to take place, Moscow felt forceded to let Eastern Germany do the same thing. In early 1948, the Soviets invaded Czechoslovakia and the Western allies immediately began to take decisive measures to counterbalance what seemed like the initiation of a large-scale effort to convert by force all Europe to communism.[11]

Until then, the allies had been blocking practically all efforts to produce any heavy manufactured good in West Germany because of the linkage of this domain with weapons. With the invasion of Czechoslovakia, the allies changed their position immediately and decided that the FRG from then on would be treated as an ally so that it could assist in the containment of communism. They lifted restrictions on heavy manufacturing, which by the end of the war accounted for the largest share of factory output, and allowed Bonn to promote this sector as much as it wanted, as long as weapons were not produced. The United States also disbursed an abundance of grants in the form of the Marshall Plan, which was used mostly for the financing of factories.[12]

Economic recovery was seen by Washington as the best mechanism to contain communism in Germany and the rest of Europe, because it was poverty and material hardship that were increasing the number of communist sympathizers. Because manufacturing was the largest sector of the economy, the economic future of West Germany was seen by Washington and Bonn as largely dependent on this sector. A large share of the nation's resources was therefore allocated to this sector by means of fiscal, financial, and nonfinancial incentives and other mechanisms. Manufacturing output immediately began to grow at an extremely fast pace and the economy did likewise.[13]

The abnormally high levels of support supplied to this sector during the 1950s stemmed not just from the large size of this sector, but basically from the vast destruction caused by the war. There was above all a very strong desire to reconstruct the numerous buildings and factories that had been destroyed by the bombardments.

Bonn allocated a large share of the nation's resources for the promotion of factories, because these had been particularly targeted by the aerial attacks. Investment in the production capacity of the country took priority and since this basically meant manufacturing, the government provided factory producers with low taxation, ample grants, abundant soft loans, free land for the production installations, guaranteed government purchases at prices that assured a profit, and other incentives. The country was also short of foreign exchange, which was needed to finance the importation of numerous essential goods. Since factory goods had accounted for the bulk of exports before the war, it was evident to Bonn that the vast majority of its export promotion efforts had to concentrate on manufacturing.[14]

By then also there was a long tradition of strong government support for manufacturing, which had started about a century earlier and which had constantly coincided with fast economic growth and a positive trade account. Tradition therefore predisposed policy makers to promote factory production even if there had been nothing to reconstruct and if foreign exchange had been plentiful. Since many factories had been destroyed and reserves were extremely low, levels of support for the sector rose to far higher levels than anything formerly experienced, which coincided with unprecedented rates of factory output and unprecedented rates of economic growth.

Manufacturing output during the 1950s averaged about 9.3 percent annually and economic growth averaged about 8.0 percent. Another situation that drove Bonn to allocate a larger share of the nation's resources than ever before to manufacturing was unemployment. West Germany emerged from the war with malnutrition and very high levels of unemployment, and even with the fast growth that took place in the late 1940s, by 1950 unemployment was still in double digits.[15]

By 1950 agriculture had already shrunk to such a small size that nobody seriously considered this domain as capable of generating many employment possibilities, while the service sector was not that large and very few people thought that this sector could absorb the unemployed effectively. So that left only manufacturing with the potential for providing employment to the millions in Germany without a job and to the millions who emigrated from Eastern Europe. In the aftermath of the war, about ten million emigrated to the FRG.[16]

The emigration was more intense in the immediate years after the war, but it was still strong during the 1950s and contributed to one of the fastest rates of population growth in the Germany of the twentieth century. Under the heavy pressure resulting from the high unemployment, plus the exceptional reconstruction efforts, the shortage of foreign exchange, and the traditional support for manufacturing, Bonn decreed the strongest level of support for the sector in Germany's history. Such an effort coincided with the fastest GDP figures in the country's history.

Orthodox Explanations of the Fast Growth

Many argued that the exceptional rates of growth were the result of the concomitant effect of a number of factors that were unique of those years. It was asserted that interest groups were in a very weak position during those years and therefore were not an obstacle to the implementation of rational economic policies. Since the late nineteenth century, labor unions were the group that had increased its demands the most and had increasingly forced policy makers to favor them at the expense of the rest of the economy. It was claimed that the Nazis had crushed worker organizations, that the war had dislocated them, and that the Allies had deprived them of their power, and as a result they were incapable of interfering in policy making.[17]

The history of the twentieth century has indeed demonstrated how in numerous countries labor unions have made such excessive and unrealistic demands that they have hampered the application of rational economic policies, and helped to undermine competitiveness. However, if the absence of such interference would be essential for the attainment of fast economic growth, then Germany and all other countries should have attained fast growth during the first half of the nineteenth century and during the preceding centuries. In those times, labor unions were nonexistent and the mass of the population tended to make few demands. However, it was only until the late nineteenth century that Germany grew rapidly for the first time and that actually coincided with the emergence of labor unions and with rapidly growing labor demands.

It was also argued that the Marshall aid acted as a spark or propeller on the economy, not just because it provided badly needed financing, but because it pushed German firms to import American machinery and equipment that possessed the most developed technology in the world. Marshall aid was indeed significant, but only when analyzed from the perspective that before the Plan started, the United States was providing absolutely no assistance. In reality, it accounted for just 3 percent of the GDP of the seventeen West European countries that benefited from it during the five years in which it was disbursed.[18]

Marshall aid was therefore too small to have acted as a propeller of growth for Germany or for the other nations in the region. It also ended in 1952 but in the following years the economy of Germany and most of its Western neighbors continued to grow quickly. In some countries such as France, growth was even faster during the 1960s than in the 1950s. Also, the country that received the most aid per capita was Britain, but grew at the slowest pace during the late 1940s and during the following two decades. Germany and Italy, which received a considerably lower amount of aid per capita, were the fastest growers in Western Europe. While Britain averaged a GDP rate of less than 3 percent during the 1950s, the FRG averaged about 8 percent and Italy about 6 percent.[19]

Also, the Soviet Union and the countries of Eastern Europe received absolutely no aid or any form of capital inflow from the United States. They also did not benefit from the importation of American machinery and equipment, and to make matters worse they practiced a very distorted economic system. However, in spite of all of these disadvantages, these countries managed to grow rapidly. The USSR, for example, grew by about 4 percent annually in real terms during the 1950s, which was faster than Britain. From every perspective, it becomes evident that Marshall aid could not have possibly been responsible for the fast growth of the FRG. It did have a positive effect, but the empirical evidence clearly reveals that it was just marginal.

It was also argued that the destruction of the war was so great and unprecedented that the reconstruction efforts ended up being more extensive than anything experienced before. Since a very large share of Germany's production capacity had been paralyzed by the war but not destroyed, it was claimed that bringing the installed capacity back into production was a relatively easy task.[20]

The problem with this argument is that it fails to explain why Switzerland and Sweden, which did not experience any destruction during the war because of their neutrality, managed to have fast growth afterward even though they did not have anything to reconstruct. Britain, which was one of the most destroyed countries, attained less than 3 percent annual growth in the 1950s while Switzerland, which barely saw a couple of bombs fall over its territory, averaged a GDP growth of almost 5 percent.[21]

The argument on reactivation of the installed capacity is also weak because Britain found itself in a similar situation, yet did not attain fast growth. By the end of the war, numerous British firms were paralyzed only because they lacked inputs or for some other reason, and reactivating their full production capacity was relatively easy.

It was also asserted that the FRG's fast growth was largely the result of the currency reform that took place in 1948, in which the deutsch mark (Dm) replaced the reichs mark. There was also the fact that monetary policy was devolved to the central bank and its management was placed in the hands of competent technocrats. Since many economists believe that sound monetary policies and low inflation are determinant for growth, it was concluded that such a situation made the exceptional rates of growth possible.[22]

As a matter of fact, the FRG did experience very low levels of inflation during the 1950s, but this argument forgets that during the Nazi 1930s (1933–39), monetary policy was managed irresponsibly and inflation was relatively high, yet the economy actually grew faster than in the 1950s. Other countries, such as South Korea during the 1960s and 1970s, also experienced high inflation while simultaneously attaining extremely fast rates of economic growth. Much suggests that pursuing a policy of low inflation is a rational effort, but the historical evidence also demonstrates that is not fundamental for the generation of growth.[23]

Currency reform also does not present a consistent correlation with growth. At times it has coincided with fast growth, as during the unification of all of the currencies of the German states in 1871 and that of 1948. At times it has not, as during the one that took place in 1990. During 1990 to 1997, the substitution of the GDR's currency with that of the FRGs did not deliver fast growth to the eastern part of Germany nor to the western part. In Brazil, currency reform in the 1980s and 1990s also did not coincide with fast economic growth.[24]

It was also purported that the fast growth of the 1950s was the result of market liberalization. The Allied occupational forces in the years immediately after the war broke down cartels and liberalized foreign trade in a very significant way. In 1948, the visionary Conrad Adenauer became prime minister of West Germany and named as minister of economics, Louis Erhard, one of the most reputable economists in the country. They immediately got to the task of deregulating and liberalizing the economy even more.[25]

These undertakings very evidently improved the efficiency of the available resources and unleashed latent energies in the population, but if liberalization was so important for growth, it becomes hard to understand why economic growth was so strong during the Nazi 1930s. During these years, the economy was highly regulated. On top of that, foreign trade was greatly curtailed and resources were abundantly wasted on weapons. However, economic growth was faster than in the 1950s. Here again the correlation was missing.[26]

Another idea introduced as an explanation of the fast growth was that the initiation of an effort to integrate the economies of Western Europe rationalized resources among its members and thus improved efficiency for all. In 1950, for example, the European Coal and Steel Community was inaugurated and in 1957 the Treaty of Rome created a free trade zone among its several members.[27]

Since Germany was a founding member of these organizations, it evidently benefited from the higher efficiency resulting from the increased competition. However, if economic integration was determinant for fast growth, it would have been natural for the FRG and the other members of the European Economic Community (ECC) to have experienced an accelerating rate of GDP.

During the second half of the twentieth century, practically all nations in Western Europe progressively became more and more integrated, and by the 1990s practically all of the nations of the region were members of the European Union (EU). By then goods, services, capital, and labor moved freely between its members and even some form of monetary union had already taken place. However, with every decade that went by, the economy of Germany and of practically all EU members decelerated more and more. While growth in western Germany averaged about 8 percent in the 1950s, the figure was just 2 percent between 1990 and 1997. Much suggests that economic integration had numerous positive effects, but the evidence also clearly indicates that it was not essential for the attainment of growth.[28]

In the 1950s, there was a large influx of workers from Eastern Europe.

Many claimed that they had a high performance potential, and that their skills contributed significantly to the generation of fast growth. The reality, however, was not quite like that. Even though many of the immigrants were highly skilled, most were not. Most were simply expelled by the new governments in Eastern Europe for being Germans who had been transplanted there by the Nazis in their efforts at colonization, or for having sympathized with the Nazis. Others were just escaping communism. It is, therefore, unlikely that average individuals could have acted as a booster on the economy.[29]

Also, during the Nazi 1930s there was a considerable outflow of scientists, technicians, and other highly talented people from Germany because of their Jewish origins or because they clashed with the Nazis. However, despite this large brain drain, economic growth was even faster than during the 1950s. During 1933 to 1939, GDP grew by about 13 percent annually while during the 1950s it expanded by just 8 percent.

The situation of Nazi Germany suggests that even if the bulk of the immigrants in the 1950s had been highly skilled persons, it is unlikely that they could have noticeably affected growth. The fact that during the 1990s western Germany received a large number of people from East Germany, Eastern Europe, and the ex-Yugoslavia and grew by only 2 percent, further reinforces the idea that the immigrants of the 1950s were not determinant or even important for the generation of fast growth.[30]

It was also asserted that because of the scarcities of the war, there was a strong, pent-up demand, and consequently producers felt secure about rapidly increasing output. If pent-up demand from the scarcities of war deliver fast growth, than South Korea should have grown even faster than the FRG in the aftermath of its civil war with the North. The civil war was very destructive for South Korea and it was a considerably less developed economy than West Germany in the aftermath of World War II. However, from 1954 to 1961 South Korea averaged a GDP of less than 4 percent annually.[31]

It is also important to point out that by 1954 the FRG had already surpassed the best standards of living achieved during the Nazi years and in consequence demand had also surpassed all preceding levels. From then onward, there was no longer pent-up demand, but the economy continued to grow rapidly, clearly indicating that another variable was propelling growth.[32]

Also, during the 1980s and 1990s, sub-Saharan Africa had a much lower level of per capita consumption than the FRG during the 1950s. Among the nations of that region there was a very strong desire to increase consumption very rapidly, yet economic growth averaged only 2 percent annually. On the other hand, between 1990 and 1997, Singapore had a higher level of per capita consumption than western Germany at that same time, and had never experienced a destructive war. In spite of having demand even more satisfied than most OECD countries, it had no problem growing by more the 8 percent annually.[33]

The empirical evidence clearly demonstrates that having a large pent-up demand because of war or underdevelopment is in no way determinant for the generation of growth; nor is having a very satisfied demand a deterrent to fast growth. During the 1990s, Singapore actually grew faster than the FRG during the 1950s.

Others have claimed that the fast growth of world trade during the 1950s allowed for a fast growth of exports, which in their turn acted as a propeller on the economy. World trade indeed expanded rapidly during that decade and the FRG's exports grew faster than ever before, but there is little to suggest that this was the cause of the fast GDP figures.[34]

For Germany and for most other nations in the West and East Asia, it is not possible to establish a consistent correlation between these variables and growth. During most of history a parallelism has been observed, but frequently it broke down, clearly indicating that the relationship was not causal. During the 1930s, for example world trade collapsed, Germany's exports shrank, but the economy managed to grow rapidly. On the other hand, world trade during the 1990s expanded at a faster pace than in the 1980s, but for OECD countries and numerous other nations, that did not coincide with faster economic growth. There was actually a deceleration in the GDP figures.[35]

None of these arguments managed to convincingly demonstrate that they had been determinant or even important for the generation of growth. If individually they were irrelevant or marginal, then it is evident that as a group they could not have been determinant. On the other hand, the thesis of manufacturing did add up consistently with the facts. Not only could it explain why western Germany grew so fast during the 1950s, but also why most countries in Western Europe grew faster than Britain, and why CMEA nations grew relatively fast. It could also explain why the Nazis attained impressive rates of growth, why exports have at times not correlated with GDP figures, and numerous other apparent paradoxes.

Manufacturism is even capable of explaining the bigger paradox of the GDR. The economy of Nazi Germany was highly regulated and distorted, but it was still operating within the framework of a capitalist system. The economy of the GDR, however, deliberately attempted from the very beginning to destroy all forms of capitalism. Since 1945, nationalization started with everything that had belonged to the Nazi government, then moved rapidly beyond that, and by 1950 about 90 percent of the economy was in state hands. A centrally planned system of production was introduced also almost as soon as the Red Army took over the territory. During the 1950s, centralization and nationalization moved forward and well before the decade ended capitalism had been completely annihilated.[36]

Under these circumstances, it is hard to see how Eastern Germany could have experienced any growth at all. However, the fact is that it did, and the differing rates of GDP varied in direct proportion to the differing levels of government support for manufacturing. During 1946 and 1947, for example, growth was slow and almost identical to that of West Germany, coinciding with similar levels of support for the sector on both sides. The Soviets wanted the Germans to suffer but since the Allies allowed Bonn to promote manufacturing a little, Moscow decided that East Germany had to match the performance of the West in order to avoid a situation that would suggest that capitalism was superior to socialism. As a result, factory output grew slowly on both sides and so did the economy.[37]

When in early 1948 the Allies decided to let Bonn promote the sector as much as it wanted and even assisted the FRG in this goal, the Soviets were caught off-guard. They were convinced that the Allies wanted to see Germany dwell in economic hardship and that they would never allow her to recover. Once it became clear that they had wrongly evaluated the situation, policies were changed and a much larger promotion of the sector was immediately allowed. However, since their desires of retribution were so strong, they did not give Berlin as much liberty on that front as the Allies gave Bonn. In both Germanys, factory output began to grow exponentially faster (1948-1949), but the much stronger support of Bonn coincided with rates that were about twice as fast as in the East. This in turn correlated with rates of economic growth that were exponentially faster for both sides than in 1946 and 1947 but that were much faster for the FRG.[38]

Soon afterward Moscow came to terms with reality and decided that it would let Berlin promote manufacturing as much as Bonn did. It was concluded that the GDR would inevitably grow more slowly if such an approach was not adopted, and such a situation would bring worldwide disrepute to the cause of socialism. Since communists saw services as not very useful and agriculture by 1950 averaged only 14 percent of GDP, they saw manufacturing as the only sector capable of propelling the economy.[39]

There were also strong reconstruction needs pushing for strong support for the sector, but the main motivation for promoting the sector during the 1950s was the desire to match or surpass the FRG. Since Bonn allocated a large share of West Germany's resources to this sector and got more than 9 percent average factory growth, Berlin did likewise, attaining a 12 percent rate of manufacturing production. That coalesced with an economic growth of 10 percent, according to the official figures, which was above 8 percent for the FRG. The official statistics, however, need to be significantly discounted because the quality of East German goods and services from the start was considerably below that of capitalist countries with highly competitive markets.[40]

As a result, real economic growth in the GDR during the 1950s was about 5 percent per year. The inefficiencies of the centrally planned system immediately became evident. Since from the beginning of this decade the economy was already operating under this system, the whole decade was riddled with the production of low-quality goods and services. Under those circumstances, resources ended up being wasted on a grand scale. While the FRG needed only a little more than a unit of factory output to generate a unit of GDP, the GDR needed almost three units of manufacturing to deliver a unit of real GDP.

Many argued that the reason the GDR grew fast was that there was much to reconstruct and because a large share of the production capacity that had been created in capitalist times was simply brought back into production. However, the fact is that by 1955 the economy was already at parity with the previous peak level during Nazi rule. Since reconstruction had been already largely completed, it is hard to understand why the economy continued to grow at all, and why it even grew fast.[41]

Also worth noting is that the Soviets conducted a very large dismantling of the GDR's production capacity, which was sent to the USSR. Most of it took place in the years immediately after the war and amounted for more than one-quarter of the production capacity of 1939. In addition, the Soviets exacted war reparation payments from current production between 1945 and 1953, which amounted to about 15 percent of GDP.[42]

With such a massive subtraction of the existing machinery and equipment by the Soviets and with such high war reparations, it is hard to see how even a capitalist economy could have grown fast. However, the fact is that the economy was operating under a very distorted system that (according to Western analysts) lacked the capacity to generate any growth. Many Western economists predicted that under those circumstances a constant recession would take place or at best stagnation would occur. The fact is that fast growth took place in the late 1940s and the 1950s, and the only thing that can coherently explain such a phenomenon is the manufacturing thesis.[43]

The 1960s and the 1970s

During the 1960s, rates of GDP slowed considerably for both Germanys. In the FRG, the figure dropped from 8.0 percent to 4.6 percent, coinciding with a decrease in support for the sector and a deceleration in the rate of factory output, which went from 9.3 percent to 5.7 percent. Since the economy had recovered in the early 1950s to its best former peak level and by 1959 reconstruction in strictly statistical terms had been largely completed, Bonn decided that the incentives supplied to manufacturers should be reduced. Most German policy makers were convinced that market distortions should be kept to a minimum and only exceptional circumstances such as the devastation of World War II justified mobilizing resources on a large scale to favor this sector.[44]

Since nothing in their vision of the world stated that manufacturing was the determinant factor affecting growth, they saw no reason why in the 1960s they should maintain the same level of support of the preceding decade. As a result, factory output inevitably decelerated and the economy followed suit.[45]

The fast export growth of the late 1940s and the 1950s rapidly increased foreign exchange and the trade deficits began to turn into surpluses. As these balance-of-payments pressures were reduced, the government concluded that there was no longer a need for such large subsidies to exporters. Since the bulk of exports were factory goods, these subsidies fundamentally assisted the sector. As they were reduced, the sector's rate of output decelerated and exports did likewise. After having grown by about 15 percent annually during the 1950s, exports expanded by just 10 percent in the 1960s.[46]

Although statistically by 1959 the FRG had vastly exceeded the peak level attained by the Nazis, the ravages of the war were still largely visible in the thousands of destroyed buildings. The authorities therefore still felt psychologically pressured to support the sector decisively in order to overcome this situation and to promote exports. Balance-of-payments concerns had fallen considerably, but they were still significant and there was still a need to push exports. Keeping with a tradition of more than a century in which support for the sector had coincided with relatively fast growth and a positive trade balance, the authorities also promoted factories on the belief that something about this sector delivered positive results. Because of all of the above, support for the sector was still strong, which coalesced with a rapid growth of factory output and GDP.

Confronted with a decelerated but still rapidly growing economy in western Germany, the government of the GDR once again tried to match the FRG and promoted manufacturing less decisively than in the preceding decade. As a result, factory output decelerated to a rate similar to that of the FRG and averaged about 5 percent per year. The official figures state that the economy grew by about 4 percent annually during the 1960s. However, the real figures must have been about 2 percent considering the very low quality of East German goods and services.[47]

Once again, the economy of the GDR expanded even though the private sector and competition were almost totally nonexistent. Western economists asserted that growth under those circumstances was not possible and once again reality demonstrated that their theories had missed the mark. It was evident that something with impressive growth-generating powers had made that apparent paradox possible. The paradox increased under the reforming government of Ulbricht from 1963 to 1970, when even more decentralization efforts were made. A little liberalization of the economy was therefore undertaken during the 1960s and market forces were slightly more operational. However, instead of experiencing a small improvement, the economy went through a massive deceleration.[48]

Only the manufacturing thesis can make sense out of that apparently contradictory information. Since support for the sector largely decreased, there was a big deceleration of the rate of factory output, and when that happened the economy had to grow much more slowly because wealth was created at a much slower pace. Since manufacturing is the prime creator of technology and technology is the prime creator of wealth, these developments were inevitable.

Although the deceleration of the FRG during the 1960s was significant, Bonn did not worry much about it because a rate of almost 5 percent was still fast and unemployment during that decade was just 1 percent. During the 1970s, however, the economy grew by less than 3 percent per year and under those circumstances unemployment immediately began to rise, averaging about 3 percent during the decade. An unemployment rate of 3 percent was not high, but everyone would have preferred the situation of the preceding decade, in which real incomes also grew faster.[49]

Analysts concluded that the cause of such a deceleration resided in numerous factors. It was argued that responsibility resided on the one hand on a less flexible labor market, resulting from restrictions that were imposed on inflows of foreign labor, as well as on a rise in protection of labor-intensive industries. Other supposed causes were new competition from the NICs, the faster growth of wages over productivity, the oil shocks, and a fast rise of social security expenditures. Rising environmental costs, and the end of the Bretton Woods system which removed the undervaluation of the deutsch mark and hence the subsidy on exports, were also given responsibility for the slowdown.[50]

As unemployment fell to very low levels in the 1960s, the FRG began to import a considerable number of foreign workers, mostly from the Mediterranean. These workers made it possible for companies to continue smoothly with their expansion plans, and due to their low wages they also reduced production costs. In the 1970s, the authorities began restricting the inflow of foreign workers and many concluded that these measures instilled rigidities in the labor market, which ultimately hampered economic growth. There were indeed some benefits that came with the foreign workers, but the authorities began restricting their inflow after the economy began to slow down and not before. Since the slowdown of the early 1970s diminished the employment possibilities of the German population, the authorities decided to reduce the supply of foreign workers. The restrictions were an effect of the deceleration and not the other way around.[51]

Tied to this argument is the one that labor costs rose very rapidly during the 1950s and 1960s and that they reduced the competitiveness of German goods. They did grow fast, but the economy grew at a faster pace and in consequence the share of wages in GDP was almost constantly on the decline during that period. While in 1950 it was of about 66 percent, in 1960 it was of just 60 percent, and by 1970 it had fallen a little bit more. Since wages increasingly became a smaller share of overall production costs, it is highly unlikely that they could have reduced competitiveness. The lower share of wages should have actually had the inverse effect, but the fact is that exports and GDP decelerated.[52]

This was further confirmed by the fact that during the 1980s and 1990s, the share of wages from GDP continued to decline, and instead of accelerating, the economy went in the opposite direction and decelerated further. Wages once again grew more slowly during these decades than the economy and as a result they accounted for a smaller share of GDP. If the economy managed to grow impressively fast during the 1950s (8 percent) while wages accounted for about 63 percent of GDP, then faster economic growth should have been attained during the 1990s when wages accounted only for about 54 percent of GDP. The fact is, however, that the economy of western Germany grew on average by just 2 percent during 1990 to 1997. Wages accounted for a considerably lower share of production costs during these years, a fact that was supposed to have increased competitiveness, yet exports grew by just 2 percent, a fraction of the 15 percent rate of the 1950s.[53]

It was also argued that wages rose faster than productivity during the 1960s and that is what supposedly caused the deceleration in the following decade. Wages indeed outpaced productivity, but in the 1970s, 1980s, and 1990s it was productivity that grew faster than wages. If the 1970s endured a deceleration because of the high wage-to-productivity ratio of the 1960s, the 1980s and the 1990s should have experienced an acceleration of the GDP figures because of the inversion of the wage-productivity ratio. That, however, was not the case. During these last two decades of the twentieth century, the economy grew more slowly than in the 1970s.[54]

The 1970s witnessed the coming of age of the NICs. Some like Hong Kong and Taiwan had begun growing fast since the 1950s while South Korea and Singapore began to experience impressive GDP figures in the 1960s. Most became more export-oriented over time and in the 1970s, the average

GDP and export rate of growth of the four was faster than ever before. The FRG therefore experienced during the 1970s the strongest level of competition from the NICs to date. It was therefore argued that those new and efficient low-wage producers increased their exports so much that they reduced the world share of German exports. As a result, exports experienced a deceleration, which led the economy in the same direction.

If the entrance of the NICs into world markets had really affected Germany to the point of decelerating the economy, then Japan should have attained even slower rates of GDP and exports during the 1970s, because Japan was competing more directly with the four dragons. Not only did these four economies export much more to Japan than to the FRG, but the main export market of the NICs and Japan was the United States (for Germany it was Western Europe). However, Japan grew faster and exported at a faster pace than West Germany. While Japan's economy expanded by about 5.2 percent annually, that of the FRG grew by just 2.6 percent, and while exports averaged about 9 percent per year in Japan, in the FRG the figure was just 6 percent.[55]

In the 1980s, the same phenomenon took place. The NICs exported much larger amounts of goods to Japan than to the FRG and the majority of their non-Asian exports went to the United States. In spite of enduring the pressure of NIC competition much more intensively than Germany, Japan grew by about 4.0 percent annually while the FRG expanded by just 2.0 percent. Exports grew also faster in Japan averaging about 7.4 percent annually while in Germany the figure was just 4.4 percent. The NICs were indeed very competitive producers, but the evidence demonstrates that they could not possibly have played even a significant role in the deceleration the FRG experienced in the 1970s and in the 1980s.

The strong rise in competition from Asia during the 1970s drove Bonn and other governments in Western Europe to raise trade barriers, and the protection of labor-intensive fields was seen by many as the cause of the deceleration. The protection was justified on several grounds. Textiles and apparel, which were the largest East Asian exports, were protected on the grounds that the domestic producers contributed to the national security of West European countries because they provided uniforms for the armed forces. There were also employment preservation reasons and others of the sort.[56]

There is an abundance of historical evidence showing how trade protection in whichever form it comes reduces efficiency. However, history also demonstrates that trade distortions affect growth in only a marginal way. During the late nineteenth century, for example, a rise in protectionism in Germany coincided with an acceleration in the rate of economic growth and with fast GDP figures. During the Nazi 1930s, trade distortions were extensive, yet the economy grew at an impressive pace. Numerous other countries have also experienced a similar phenomenon. During the 1970s, Japan, Taiwan, and South Korea applied trade barriers that were much higher than those of the FRG. In spite of that, they attained GDP figures that were two to four times faster. The FRG and these East Asian countries very evidently committed a policy error by distorting trade, but it is also clear that the main variable determining growth was not trade.

It was also purported that the large increase in the price of oil was behind the economic slowdown of the 1970s. Had oil really been determinant or at least important in the generation of growth, then the economy should have accelerated in the 1980s, considering that the price of oil fell significantly. That was not the case and what actually happened was that the economy of the FRG slowed down even more. In the 1990s, the price of oil fell even more and once again the German economy refused to grow faster. The fact that numerous developed and developing countries experienced a similar situation substantiates further the idea that oil had practically no effect on the economy during the 1970s.[57]

The big increase in the price of oil contributed to a considerable acceleration in the rate of inflation. After having averaged about 2.7 percent annually during the 1960s, inflation accelerated to a rate of about 5.3 percent during the 1970s. It was, therefore, concluded that this situation was largely responsible for the deceleration of the GDP figures, considering that most economists believe that high inflation has a negative effect on the economy. A comparison with other countries rapidly reveals that inflation could not have possibly played a major role in the slowdown. In Japan, for example, inflation also accelerated considerably and averaged about 8.2 percent, yet notwithstanding a rate that was much faster than in the FRG, GDP grew at a much stronger pace, averaging about 5 percent per year.[58]

In South Korea, inflation during the 1970s averaged about 18 percent per year and not only did the economy grow by an impressive 10 percent per year, but it also experienced a noticeable acceleration relative to the 1960s. The case of South Korea was the ultimate piece of evidence demonstrating that high inflation, although unpleasant, does not have a determinant or even an important effect on growth.[59]

The fact that the FRG applied tight monetary policies during the 1980s and that inflation fell to the same low level of the 1960s, with no improvement in the GDP figures substantiates further the idea that there is no causality linkage between inflation and growth. Not only was there no improvement, the economy actually ended up decelerating further.

During the 1970s, there was a very large increase in social security expenditures in the FRG and many came to see this situation as a main cause for the deceleration. There was also a very large increase in government expenditure, mostly as a result of the enlargement of the social welfare budget. The combined expenses of federal, state, and local government averaged about 32 percent of GDP in the 1950s. In the following decade, the figure jumped to 36 percent and in the 1970s it went to 44 percent.[60]

During the following decade, a further enlargement of government expenditure once again coincided with a deceleration of the GDP figures. However, during 1990 to 1997, the share of government expenditure from GDP continued to grow, averaging about 49 percent after having been 47 percent during the 1980s. This time, however, the economy did not continue to decelerate and averaged the exact same rate of the 1980s. The breakdown of the correlation suggested an absence of a causality linkage.[61]

Also worth noting is that overall government expenditure increased considerably during the 1930s relative to the preceding decade and on this occasion there was a noticeable acceleration of the GDP figures, from about 4.0 percent annually to 5.0 percent. In the 1940s, the share of state expenditure from GDP declined and the economy instead of growing faster, actually decelerated to a rate of about 3.5 percent per year. During the 1950s, the expenditures of the state rose considerably and the economy experienced a large acceleration in its rate of growth, contradicting once again what most mainstream economists believe about the role and size of the state. A long-term correlation between government expenditure and growth was absent and so was any possibility of a causality linkage.

Even if social welfare expenditures are isolated, it is still hard to establish a correlation. During the 1930s, social welfare expenditures were significantly reduced, but expenditures on weapons rose significantly. Social welfare is frequently seen as reducing growth possibilities because it is purely a consumption activity. However, expenditures on weapons are seen in a worse way because they are accused of subtracting resources from the civilian economy. The 1930s, therefore, should have not experienced an acceleration of the economy, but the fact is that growth was faster. In the 1950s, social welfare expenditures increased by a very large margin relative to the preceding two decades, yet the economy grew much faster. The 1950s actually witnessed the largest share of GDP allocated to social welfare in the country's history up to that date, and notwithstanding such a situation economic growth was faster than ever before.[62]

The absence of a correlation clearly reveals that it was not the enlargement of the state or even the enlargement of social welfare expenditures that was fundamentally responsible for the economic deceleration of the 1970s. Also worth noting is that during the second half of the nineteenth century, government expenditure and social welfare expenditures as a share of GDP rose progressively. Such a rise, however, coincided with constantly accelerating rates of economic growth, giving further credence to the idea that none of these variables affected growth significantly. The cases of Hong Kong and Singapore suggest that a small government expenditure and a small expenditure on social welfare deliver better results than the opposite, but the benefits of such undertakings are evidently small.[63]

Others therefore argued that the main factor causing the deceleration of the 1970s was the rapid rise in environmental regulations that burdened firms with extra costs and reduced competitiveness. The FRG indeed began to experience a relatively large environmental consciousness during the 1970s, which translated into a considerable amount of legislation that forced companies to adopt environmental protection measures and which drove the government to make relatively large expenditures on the matter. It is also true that during the 1980s the share of GDP spent to protect the environment rose even more and that the economy experienced a further deceleration. However, during the 1990s the share of the nation's resources allocated to reduce pollution and protect ecosystems rose once again, but this time the economy did not decelerate and grew at exactly the same pace as during the preceding decade.[64]

The correlation not only lacked strength in the case of western Germany, but other countries demonstrated that increased expenditures in environmental protection did not coincide with a deceleration of the economy or with slow growth. Although most OECD countries showed a similar development of events as Germany did, Singapore revealed a very different scenario. Most OECD countries experienced rising environmental costs with decelerating GDP figures during the second half of the twentieth century, and at a time when they were spending the most on the environment (the 1990s) their rates of economic growth were the slowest.

Singapore's situation was completely different. The island's impressive economic growth during the second half of the twentieth century eventually converted this city-state into one of the most developed nations of the world. During the 1990s, the island's income per capita measured in purchasing-power parity terms was higher than that of western Germany. Its developed status and its high population density inevitably led the government to enact strict environmental protection legislation, which considerably increased the production costs of the firms installed in the island. In addition, the government allocated a large share of the budget for this purpose.[65]

During 1990 to 1997, Singapore had regulations on pollution control and related matters that were as strict as those of western Germany, and it also spent a similar share of its GDP on environmental protection as compared with developed OECD countries. However, while western Germany and the bulk of OECD countries averaged a weak growth of just 2 percent per year,

Singapore managed to attain more than 8 percent. Also worth noting is that the 1990s witnessed an acceleration in the island's rate of growth, because during the 1980s the GDP figures had averaged about 7 percent. In the 1980s, however, the share of the island's resources allocated to protect the environment was smaller than in the following decade.[66]

The case of Singapore clearly demonstrated that it was possible to attain impressive rates of economic growth for a developed nation while decisively protecting the environment, but it also showed that an increase in environmental protection was not incompatible with even an acceleration of an already fast rate of economic growth. From the empirical evidence, it becomes very clear that the rise in environmental costs in the FRG during the 1970s cannot be held responsible even in a minor way for the marked deceleration of the economy.

The early 1970s saw the end of the Bretton Woods system of fixed exchanged rates, which removed the undervaluation of the deutsch mark and hence the subsidy on exports. Since exports were believed by many to be determinant for growth, it was concluded that the revaluation of the currency reduced export competitiveness and thus economic growth.

When the FRG underwent currency reform in 1948, the exchange rate was fixed to the U.S. dollar at a rate that reflected its real market valuation. Since from then onward until the early 1970s, the FRG's economy systematically grew faster than that of the United States, the DM inevitably gained in strength relative to the American currency. Because most other OECD countries experienced a similar situation and progressively exerted pressure on the U.S. currency, the system was eventually dissolved.[67]

However, it is important to note that even though the exchange rate was the same during the 1950s and 1960s and the subsidy on exports was therefore the same, the FRG's rate of exports varied considerably. While during the first decade the rate was about 15 percent per year, during the 1960s the rate dropped to just 10 percent. Had the overvalued exchange rate been determinant for exports, the figures should not have varied so much. Once the overvaluation was removed, exports dropped to a pace of about 6 percent during the 1970s, a percentage reduction just a little bit larger than the one experienced in the 1960s. If the subsidy was so important for exports, then the end of the overvaluation should have brought a much steeper reduction of exports. The evidence suggests that the end of the Bretton Woods system was not fundamentally responsible for the drop in the rate of exports and neither was it determinant for the economic slow down. The evidence suggests that the exchange rate at best played a marginal role in the deceleration of the 1970s.

Substantiating further that idea is the case of Singapore, which during most of the second half of the twentieth century progressively experienced a constant appreciation of its currency. While this appreciation should have reduced export competitiveness, the economy systematically grew very fast and exports did likewise. During the 1980s, for example, despite much appreciation, exports grew by about 12 percent annually and the economy by about 7 percent. During 1990 to 1997, the currency strengthened still more and exports and the economy grew even faster, averaging about 16 percent and 8 percent respectively.[68]

None of the hypotheses that were presented as the cause of the economic deceleration of the FRG during the 1970s managed to even show a capacity to affect growth noticeably. It is therefore unlikely that even the collective effect of them all could have caused the slowdown of exports and of the economy. It was evidently some other variable that took the bulk of responsibility, and the only variable that managed to add up consistently with the facts was manufacturing.

By 1969, the FRG had greatly recovered from the destruction of the war and the physical scars of the conflict had largely disappeared. Production had abundantly exceeded the peak prewar levels and foreign exchange reserves had risen to relatively high levels. As a result, Bonn concluded that the manufacturing promotion efforts resulting from the effects of the war were practically concluded. In consequence, the government reduced by a large margin the incentives it supplied to factory producers. Tax exemptions and tax reductions were diminished, grants were reduced, subsidized financing was largely curtailed, land grants for the factory became fewer, subsidized utilities were reduced, guaranteed government purchases with relatively large profit margins were diminished, and other incentives were also cut.

With such a considerable reduction in support for the sector, the rate of factory output inevitably decelerated in a significant way, and along with it the rest of the economy. After having grown by 5.7 percent during the 1960s, factory output averaged only 3.2 percent in the 1970s, and the economy followed, moving from an annual rate of 4.6 percent to 2.6 percent. The government tried to fight the deceleration with Keynesian policies and increased expenditures in infrastructure, education, job training, and social welfare, but the economy never recovered the dynamism of the 1960s, much less of the 1950s.[69]

Because there were still a few war-reconstruction and balance-of-payments concerns, Bonn still supplied a certain amount of incentives to factory producers, and because of a relatively long tradition of promoting the sector dating to the mid-nineteenth century, the government gave an additional level of support. Because of this support, factory output still managed to expand at a modest rate.

The manufacturing thesis is also best able to explain the events of the GDR during the 1970s, which presented a terrible paradox for mainstream interpretations. During this decade, East Germany once again attained some economic growth even though market forces were almost totally nonexistent. Western economists were convinced that without the private sector, economic growth was impossible, but once again the economy of the GDR grew. It grew in real terms by about 2.4 percent annually.[70]

This phenomenon coalesced with a relatively strong promotion of manufacturing, which delivered a factory rate of output of about 6 percent per year. The official statistics claim that the economy grew by about 5 percent per year but because the system was riddled with inefficiencies, a very large discount is needed in order to render the GDP figures compatible with those of highly efficient capitalist economies.

Since Berlin was trying to outperform or at least match the FRG economically and technologically, and these two goals were fundamentally associated with manufacturing, the state's promotion efforts ended up fundamentally assisting this sector. Technology is the one thing that most clearly demonstrates superiority. Since Moscow and Berlin wanted to demonstrate that communism was superior to capitalism, they allocated a large share of the GDR's resources to the sector most associated with innovation. East German policy makers therefore decreed policies that generated a rate of factory output much faster than that of the FRG. Since in the preceding decades just matching West Germany's rates had proved to be insufficient in matching the pace of technology and wealth creation in the FRG, more was needed.[71]

There was also a significant amount of investment in the fabrication of weapons. At first, the Soviets did not want to allow absolutely any form of armament production in the country that was responsible for almost 25 million Soviet deaths during World War II. However, since the Allies made the FRG a member of NATO in 1954 and progressively allowed it to produce weapons, the Soviets felt compelled to do likewise. By the 1970s, the FRG's defense expenditures accounted for about 3 percent of GDP and those of the GDR were about 6 percent. Weapon production in this way helped to increase factory production over the rate that an exclusive civilian production would have delivered. Considering the much smaller size of the GDR, the Soviets thought it necessary to compensate by allowing for a much larger production of weapons.[72]

The 1980s

The 1980s witnessed once again a deceleration of the economy of the FRG. After having expanded by about 2.6 percent annually during the 1970s, the

GDP figures dropped to just 2.0 percent. This rate was insufficient to provide employment to all despite a population growth that was the lowest in the world. During this decade, the rate of population was 0.0 percent. However, despite the lack of growth of the workforce, unemployment doubled, going from about 3 percent during the 1970s to 6 percent in the 1980s.[73]

This time Bonn became very alarmed about such a development and endless debates over the possible causes took place. Much attention was given to fiscal imbalances as a main cause of the deceleration. Since the time Germany became a country in 1871, most governments remained regularly convinced about the merits of fiscal rectitude. Budget deficits were seen as a hindrance to growth. After World War II such a vision of things continued to prevail and during the 1950s a large budget surplus was attained.

Attention on this matter has regularly centered on the federal government. However, the fact is that maintaining fiscal rectitude at the regional and local government level is almost as important, especially in a decentralized country like the FRG where state and municipal government account for about half of overall government expenditure.[74]

During the 1950s, the consolidated account of the federal, state, and local governments managed to show a large surplus averaging about 4 percent of GDP. In the 1960s, a consolidated surplus was once again attained, although this time the surplus averaged only 1 percent of GDP. In the 1970s, however, the balance shifted and there was a deficit, which averaged about 2 percent of GDP. As the 1970s came to an end it, became evident that Keynesian stimulation efforts had failed to reactivate the economy, and the deficit that seemed to result from them began to be seen as the source of the deceleration.[75]

With a change of government in 1982 a new policy stance was introduced. The priority became reducing the deficit as well as reducing overall government expenditure. This would supposedly release resources from the public sector to the private sector, where they would be used more efficiently. Subsidies to all sectors were reduced and tax exemptions were diminished so as to reduce tax distortions.[76]

In spite of its commitment to these goals, Bonn failed to eliminate the deficit and to reduce government expenditure. It did not even manage to stabilize either one. It only succeeded in slowing down the growth of the deficit and of government expenditure. After having averaged 2 percent, the overall deficit rose to 3 percent of GDP in the 1980s and overall government expenditure rose from 44 percent to 47 percent of GDP. The deficit of the federal government fell somewhat, but that of regional and local governments rose considerably and the consolidated account of the three delivered a larger imbalance.[77]

The governments of the 1970s committed grave errors by liberally increasing expenditure, especially because so much of it was in social welfare. However, the situation of the 1980s clearly demonstrated that running balanced or surplus budgets did not exclusively depend on the desire to balance the accounts of the government.

History has supplied a considerable amount of information that suggests that it is much easier to run balanced or surplus budgets when fast economic growth takes place. The 1950s and the 1960s for the FRG were a clear example of that. The deficits of the following three decades all coincided with slow growth. Japan also ran surpluses during the 1950s and 1960s while economic growth was impressive, and ran mostly deficits in the following three decades when growth had decelerated by a large margin. Hong Kong and Singapore ran almost constantly balanced or surplus budgets during practically the whole second half of the twentieth century, coalescing with impressive economic growth during practically the whole period.[78]

Much also suggests that economic growth is not the result of having balanced or surplus budgets. The Nazi 1930s, for example, witnessed impressive economic growth while budget deficits were regularly experienced. The United States during 1940 to 1943 ran gigantic budget deficits and it nevertheless attained impressive GDP figures. South Korea during the 1960s, 1970s, and 1980s endured constantly budget deficits, but very fast economic growth was nonetheless attained.[79]

Maintaining balanced or surplus budgets is evidently the rational thing to do for any government, but the historical data very clearly demonstrates that it is not essential or even important for the generation of fast growth. The empirical evidence also demonstrates that once fast growth is attained, it is much easier to have fiscal rectitude.

What gives strong signs of being essential for the generation of growth is manufacturing. For example, the very strong support that Berlin provided to the sector during the Nazi 1930s is consistent with the double-digit factory and GDP growth of those years, giving a clear understanding why it was still possible to achieve that in spite of the budget deficits. The wide-ranging factory-promotion efforts of Washington between 1940 and 1943, can also explain why the U.S. economy grew extremely fast despite gigantic budget deficits. And the enthusiastic factory promotion efforts of Seoul during 1960 to 1989 can rationally justify the impressive economic growth notwithstanding the constant budget deficits.[80]

Even if Bonn had succeeded in eliminating the federal deficit during the 1980s and regional and local governments had done the same, much suggests that the economy nonetheless would have decelerated because support for manufacturing was decreased. The same thing is likely to have occurred even if overall government expenditure had been reduced, because history demonstrates that small government in itself is not essential or even important for the generation of growth. There are benefits in avoiding budget deficits and in having a small government, but it is evident that they are only marginal. The bottom line resides in manufacturing.

That would also explain why despite much deregulation during the 1980s, which improved efficiency, the privatization of numerous state-owned firms, which also improved efficiency, and measures to improve education and job training, which improved the working skills of the workforce, the economy still decelerated. It is evident that these efforts were not essential for the generation of growth.[81]

Another measure that was taken in an effort to improve efficiency and which also failed to reverse the decelerating trend was the liberalization of the labor market. A myriad of rigidities and disincentives affecting the labor market existed by the early 1980s, which affected labor productivity in a negative way. Laws that made it easier for employers to fire employees were enacted and unemployment benefits were somewhat reduced. In this way, employers were given some liberty to get rid of unnecessary or inefficient workers and workers were pushed some so that they would search for a job more actively. In spite of those productivity-enhancing measures, the economy decelerated.[82]

On matters referring to the amount of hours worked per week, the government also attempted to stop the rapid decrease in working hours that the preceding decades had witnessed. During the 1950s, for example, the average number of hours worked per week in the FRG was about 45 and by the 1970s the figure had dropped to about 38. The efforts of the government succeeded in arresting this trend during the 1980s and the figure remained largely the same. That, however, was also incapable of preventing the deceleration.[83]

The fact that the efforts on this matter failed to deliver the desired results suggests that the number of working hours is not a variable that plays a determinant role in the generation of growth. Substantiating this idea further is the fact that during the second half of the nineteenth century the number of hours worked per week constantly declined in a significant way, and in spite of that, the economy constantly accelerated its rate of growth. Important to note is that during the early nineteenth century the average German worked almost three times the hours per week that the average German in the 1980s worked, and in spite of that, growth was considerably slower. This fact strongly suggests that this variable at best has a marginal role in the generation of growth.

The manufacturing thesis has no problem making the historical data add up consistently. During the early nineteenth century, state support for this sector was very weak and the very slow rates of manufacturing output that resulted from such weak support had to deliver a slow rate of economic growth even if people worked more than ninety hours per week. During the period 1850 to 1999, the factory promotion efforts of the German governments became progressively more decisive, which is why, notwithstanding the continuous reduction in the number of hours worked, the economy experienced a constant acceleration.

This thesis is also the only one capable of explaining why economic growth took place in the GDR during the 1980s, even though the economy was almost totally centrally planned and the private sector accounted only for about 4 percent of GDP. If mainstream interpretations of reality were right; no growth should have taken place. The fact that the economy expanded in the fourth consecutive decade under a communist system clearly indicates that a variable different from the ones considered by orthodox interpretations as essential for growth made such a phenomenon possible.[84]

In the short history of the GDR, the 1980s was the decade when market forces were more operational, and this also presented a paradox. Following as always the guidelines set by Moscow, Berlin considerably liberalized the economy during these years. Since the Soviet economy was liberalized significantly, specially since the arrival of Gorbachev in 1985, East German policy makers decreed similar policies. Considering that market forces were more operational in this decade, Western economists predicted that the economy would perform better, but the fact is that there was a noticeable deceleration. Officially, the economy expanded by a little more than 3 percent annually after having grown by about 5 percent in the preceding decade.

Once again, the figure needs to be discounted so as to compensate for the very low quality of the goods and the services produced there. Once that factor was deducted, the figure expressing real economic growth dropped to about 1.7 percent annually. The growth in itself coincided again with Berlin's factory-promotion efforts, which happened to be considerably less pronounced than in the preceding decade. That in its turn coalesced with a decelerated rate of factory output that averaged about 4 percent annually after having been about 6 percent in the 1970s.[85]

The decreased support for the sector once again reflected the decreased support that took place in the FRG. There was also a considerable reduction in political tensions between Washington and Moscow, which immediately translated into much lower tensions between Bonn and Berlin, delivering a cut in defense expenditures in both sides. A slower rate of armament output inevitably contributed to a slower overall rate of factory output. It was not inevitable that the events took that course, but since the cuts in defense were not transferred to civilian manufacturing, factory output had to significantly decelerate.

Reunification and Eastern Germany

The configuration of both Germanys began to change dramatically on November 9, 1989, when the GDR's policy makers gave the order to open the until-then heavily guarded border with the FRG, and to begin dismantling the wall separating the two countries. Bonn immediately drafted a plan to unify the two countries, and on July 1, 1990, economic, monetary and political unification took place.[86]

The way unification took place generated much debate, in particular over exchange-rate policy. At the time the wall fell, a DM could buy many East German marks. For political reasons, Bonn decided to convert a large share of the existing currency in circulation at a one-to-one rate, which largely overvalued the East German mark and distorted market forces in the process.[87]

This overvaluation of about 440 percent was evidently an error, but it is worth remembering that not all of the currency was converted at that rate. A large share was converted at a more market-determined exchange rate. On top of that, the introduction of the DM in East Germany had the advantage of being the equivalent of a comprehensive price reform. The extensive distortions of the centrally planned system were phased out all at once.[88]

The introduction of the DM also immediately created an efficient capital market that allowed for the efficient financing of all enterprises within the East German territory. By 1991, West German banks were already providing the bulk of the financing in the former GDR. That translated not only into an efficient management of capital, but also into an abundance of capital for a region lacking it. It meant that from one day to another it was possible for all enterprises in the ex-GDR to import from any part of the world the best machinery, equipment, and any other input for their operations.[89]

There was also a comprehensive privatization program that began almost as soon as unification took place and which rapidly succeeded in selling to domestic and foreign capitalists the majority of the enterprises of the ex-GDR. Those that were overburdened by liabilities and did not arouse an interest from the private sector were simply liquidated. From 1990 to 1994, the Treuhandanstalt (Bonn's privatizing authority) sold or liquidated about 12,270 companies from a total of 12,370 that until 1989 constituted the totality of the GDR's companies. It was the fastest and most wide-ranging privatization in history.[90]

Although not as fast, there was also a comprehensive effort to clear up property claims that originated in the postwar years when the communists nationalized everything without compensation. Bonn took measures to have the courts settle these claims as fast as possible so that investment would not be deterred.[91]

With unification also came a professional management of monetary and fiscal policy. Since the GDR was simply absorbed by the FRG, its monetary and fiscal policy began immediately to be fully in control of Frankfurt and Bonn, and under those circumstances it fell into highly professional and experienced hands. That was particularly evident in monetary policy, because the FRG's central bank had by then earned a reputation for being one of the best, if not the best in the world.

With all the privatization, liberalization, deregulation, sound monetary policies, and adequate management of the budget that was undertaken, most economists predicted that Eastern Germany would grow at an impressive pace in the years immediately after reunification. Since the GDR's situation in 1990 largely resembled that of the FRG in 1948, when a currency reform was introduced and the economy was decisively liberalized, many concluded that East Germany would also grow at a double-digit pace just as West Germany did at that time.[92]

That, however, did not occur. The economy systematically failed to grow fast. At first there was a massive depression, then some growth and eventually even a few years of fast growth, but by the end of the 1990s the economy of Eastern Germany had slowed to very low levels.[93]

During 1990 to 1997, the economy averaged only a 2.6 percent annual rate of growth. It was argued that because trade collapsed with the CMEA countries (Eastern Europe and the USSR), a contraction was inevitable, but the fact also is that trade with communist countries was immediately supplanted by trade with capitalist countries, which offered many more benefits. Also worth noting is that when Vietnam lost its aid and its trading privileges with the USSR in the early 1990s, its economy did not collapse. On the contrary, it grew very fast throughout the 1990s, notwithstanding the fact that liberalization and privatization were undertaken only in a very timid-way. Also, fiscal and monetary policies were handled by Hanoi considerably less professionally, yet Vietnam attained a rate of real economic growth of almost 6 percent annually (the official figure was more than 8 percent).[94]

That situation was even more paradoxical when one considers that during the whole period Eastern Germany received massive transfers of capital from Bonn. They were much larger than the transfers the United States made to the FRG during 1948 to 1952 under the Marshall Plan. Marshall aid accounted only for about 2 percent of the FRG's GDP at the time, while the subsidies from Bonn to the East during 1990 to 1994, accounted for about 36 percent of the East's GDP.[95]

These huge unilateral transfers, which were given fundamentally in the form of grants, tax exemptions, and soft loans, absorbed about 5 percent of western Germany's GDP. Taxes were raised noticeably in the West in order to finance this massive aid program which continued in the rest of the 1990s. The overall transfers during 1990 to 1997, accounted for about one-third of the GDP of eastern Germany. At no other moment in the history of Germany or practically any other country had such a vast aid program been undertaken.[96]

By 1997, almost US$700 billion had been transferred to the East, yet the economy had averaged a rate of less than 3 percent annually during the 1990s. Vietnam, on the other hand, received no aid and even saw the aid that came from the USSR eliminated. It did not even have a large and wealthy expatriate community, as China did, that could invest abundantly in it, and it had not yet fully recovered from the vast destruction of the war that ended in the early 1970s. In spite of that, it managed to grow more than twice as fast as East Germany.[97]

Such a perplexing situation revealed a lot about the inability of orthodox efforts to understand the nature of economic growth. If the bottom line of economic growth resides in privatization, liberalization, fiscal rectitude and sound monetary policies, then eastern Germany should have attained impressive rates of economic growth and Vietnam should have done poorly. If to that is added the extensive aid that it received, the ex-GDR should have attained the fastest rates of growth in the world. The fact that none of this took place clearly demonstrates that the policies dictated by Bonn had erred on the most important matter concerning growth.[98]

Many insisted that the problem resided in the overvalued exchange rate at which the currency was set at the time of unification. This was evidently an error, but the rest of the reforms that most economists saw as essential were undertaken in the way most thought was ideal. Considering the massive aid, there is no excuse for not having grown rapidly, not to mention that numerous economies have managed to grow very rapidly while simultaneously having overvalued exchange rates. China's yuan during 1980 to 1993, for example, was significantly overvalued, yet the economy grew by about 7 percent annually in real terms.[99]

It is also worth noting that China during those years had a fixed exchange rate that did not allow for an adequate response to market signals. In the ex-GDR, however, the new currency constantly reflected its true market value.

Many, therefore, argued that the situation of the GDR was exceptional because it was transiting from communism to a completely different system and traumas were therefore inevitable. The fact, however, is that other countries went through a similar change and nonetheless did much better. China was also transiting from communism to capitalism and had an overvalued exchange rate. In addition, it only timidly liberalized its economy, it made practically no privatization, and it practiced a lax monetary and a lax fiscal policy relative to East Germany. If China managed to grow rapidly, there was absolutely no excuse for the ex-GDR not to have grown much faster.[100]

There was also the argument that wages in the East were too high relative to labor productivity, which was largely the result of the overvalued exchange rate and the decision of Bonn to set wages in the East as close as possible to the level of western Germany. Indeed, even by 1997 wages were still unrealistically high averaging about 75 percent of those in the West, but with a productivity of just 50 percent of that of West Germany. Setting wages so high was evidently an error, but that could not explain why Eastern Europe and in particular the ex-republics of the Soviet Union had performed much worse notwithstanding the fact that wages had not been subsidized.[101]

In these countries, wages plummeted and governments only barely made a few efforts to raise them. Market forces determined the exchange rate and wages, and in many countries they were evidently undervalued relative to labor productivity. If scientists and technicians, who had proven their inventiveness and resourcefulness during communist times, could be hired for a few hundred dollars per month, it was evident that their labor assets were being undervalued. However, notwithstanding this situation, all of these countries experienced an even worse economic performance than East Germany. Between 1990 and 1997, all of them attained slower GDP figures. Poland, which grew the fastest, expanded by about 2 percent annually and Russia, which was among the worst performers, contracted by about 3 percent annually.[102]

If the overvaluation of wages relative to productivity in eastern Germany had at least played an important role in growth, the situation of the ex-CMEA countries should have been very different. The fact that they performed even worse strongly suggested that although a policy error, this situation at most affected the economy marginally.

It was also asserted that a major factor responsible for the modest performance of the ex-GDR during the 1990s was the inadequate infrastructure that existed at the time of unification. That might have been true for 1990, but soon afterward the situation changed dramatically. A very large share of the massive aid that Bonn supplied was utilized to develop the infrastructure of the territory in order to bring it to parity with that of western Germany. There was therefore much investment in highways, roads, railroads, airports, telecommunications, sewage systems, water systems, electricity, and ports.[103]

Infrastructure improved at an impressive speed, yet growth was only modest. By 1997, for example, at a time when infrastructure had already almost reached parity with the West, the economy grew by just 2 percent. Also important to note is that at the time of unification the East's infrastructure was considerably more developed than that of China, Vietnam, and Laos. However, notwithstanding the considerable retardation of these three countries on that matter and the fact that they remained throughout the 1990s among the most distorted economies in the world, they all grew much faster than the ex-GDR. The retardation of infrastructure of the three actually increased during these years because they all invested on this field at a much slower pace than eastern Germany.[104]

How is it, then, that they grew much faster? Evidently, it was because infrastructure at best plays a minor role in the generation of growth, and because these three countries did promote abundantly whatever is it that is fundamentally responsible for growth. Although unaware of what exactly it is that causes growth, it is evident that these East Asian countries promoted it unconsciously during the 1990s.

Many argued that the "all at once" liberalization of the economy, and in particular the liberalization of trade, did not give the companies in the East time to adapt to the very high levels of competition of world markets. The fact that shock therapy also failed in Eastern Europe and the Soviet Union gave further credence to this idea. Also, the fact that China, Vietnam, and Laos retained high trade barriers while growing rapidly reinforced even more this vision of things. It was also remembered that in the years since 1948, relative high trade barriers were in place in the FRG.[105]

It is true that numerous countries throughout history attained fast economic growth while endorsing protectionism, but it is also a fact that protectionism on numerous occasions coincided with stagnation. A clear example of this last was the United States during the 1930s. Also worth noting is that during the second half of the twentieth century, the fastest rates of growth in the world were attained by Hong Kong. They were also the fastest that had ever been attained over a fifty-year period anywhere in the world. This achievement took place while the government in Hong Kong practiced the most liberal of trade regimes.[106]

The third-fastest GDP figures in world history over a half-century were achieved by Singapore during the period 1950 to 1997, and such a feat was also accomplished while a totally free trade system was in operation. The long-term sustainability of growth while free trade is practiced is further reinforced by the fact that when the whole twentieth century is analyzed, it was Hong Kong and Singapore that attained the fastest rates of growth in the world; and they both did it while constantly practicing free trade.[107]

It is simply impossible to find a consistent correlation between protectionism and growth. A causality linkage is therefore out of the question. History actually suggests that even though free trade is not essential for the generation of growth, its contribution is nonetheless positive. Much suggests, therefore, that the adoption of a liberal trade policy for the ex-GDR was a positive undertaking and the reason growth was weak was that the determinant variable behind growth was not promoted.

Manufacturing is indeed not dependent on trade for its growth. It is fundamentally dependent on the desire of the government to see it grow, and the fact is that during 1990 to 1997 Bonn was only modestly interested in seeing this sector grow in eastern Germany. By then, the bulk of economists and policy makers were convinced that services were the most growth-generating sector of a developed economy, and even though the GDR was considerably lagging behind the FRG in 1990, most in Bonn classified it as a developed economy. Services were therefore significantly promoted and they grew at an impressive pace.[108]

Since infrastructure was seriously lagging behind that of the West at the time of unification and the government thought that this was essential for the attainment of growth, it decided to allocate a vast amount of funds for this domain, and this domain grew at an impressive pace. Housing was in a relatively poor condition at the time of unification. Numerous buildings that had been significantly damaged during World War II were still standing in this bombarded form. Bonn therefore abundantly subsidized the construction of houses and other real estate projects.[109]

Construction grew in consequence, at an impressive double-digit pace, and as early as 1994 this sector accounted for about 18 percent of GDP while in West Germany it accounted for only 6 percent. The share of services in GDP also grew abundantly while that of manufacturing contracted significantly. An absence of support for the sector coincided in the first years with a massive contraction, although soon afterward factory production began to be modestly promoted.[110]

With the weak support that manufacturing received and the strong support that services and construction received, it was inevitable that rates on average were extremely weak during the 1990s for manufacturing and very fast for services and construction. Factory output from 1990 to 1997 expanded by about 1.4 percent per year and after having accounted for about 34 percent of the GDR's GDP in 1989, by 1997 it had shrunk to about 18 percent.

By the time the GDR ceased to exist, manufacturing accounted for a larger share of GDP than in the FRG and most economists had come to see this as an economic aberration resulting from the irrationality of the centrally planned system. The policy of systematically repressing services that communist governments endorsed was also seen as abominable by most Western economists. So when unification took place, Bonn utilized this opportunity to deconstruct manufacturing.[111]

By not being supplied with abundant incentives, the sector was incapacitated from growing rapidly. History has systematically demonstrated that when support is subtracted or when it is weak, even in a highly efficient capitalist economy, the sector always contracts or grows slowly. The more the reason, therefore, for that to have occurred in an economy riddled with inefficiencies. That does not mean that Bonn should have tried to keep inefficient and decrepit factories afloat. The bulk of the ex-GDR's factories did not have the installations, the machinery, the equipment or even the trained personnel to be competitive in world markets, and the bulk of them therefore deserved nothing less than to be shut down for good.[112]

Bonn actually kept many of these factories artificially alive for fear of seeing unemployment rise too high, and this was evidently a policy error. What the government should have done was to utilize the massive aid that it transferred to the East and use it all, or at least the vast majority of it, for the promotion of new and highly efficient private factories. By supplying an abundance of fiscal, financial, and nonfinancial incentives to domestic and foreign capitalists for investment in this sector, there is practically no doubt that independent of how much the GDR's factories contracted their output, the one coming from the new factories would have been far larger.

Eastern Germany could have easily avoided the massive contraction of the early 1990s and grown even at a double-digit pace throughout 1990 to 1997, had the massive capital transfers from western Germany been utilized to promote manufacturing. The aid of almost US$700 billion was so large that it could to have finance the fabrication of almost seven space stations of the sort that the fourteen countries with the most developed space technology in the world will complete by the year 2012. It was even enough to finance a manned expedition to Mars, which NASA estimates would cost about US$600 billion.[113]

Had the money been used for such space projects and all of the spacecraft needed to materialize them had been fabricated in Eastern Germany, the overall rate of factory output would have inevitably been much faster and the economy would have surely behaved similarly. However, there was not a need to allocate all of the resources in just one field. An across-the-board policy of large incentives to all manufacturing fields would have been an even better approach. With such an attractive investment environment, vast flows of FDI from western Germany and the rest of the world would have been guaranteed, assuring in this way the efficiency and quality of the output.[114]

That, however, did not occur because there was hardly anything in the vision of the world of most of Germany's policy makers that led them to the idea that manufacturing in itself was essential for the generation of economic growth. Most were fascinated with services, and the minority that thought that factory growth was important could not substantiate scientifically their gut feelings about this sector. So the end result was that about 70 percent of the huge net transfers went simply into consumption in the form of social welfare benefits, and much of the rest went to subsidize construction and services.[115]

Only a tiny fraction of those vast resources were utilized to promote manufacturing. Large sums were spent renovating town centers, in particular Berlin, which was designated as the capital of the reunified country, but only small sums were allocated for the promotion of factories. With such weak support, it was only inevitable that this sector expanded by just 1 percent annually during the 1990s, and that by 1997 the East produced only 6 percent of Germany's overall factory output and a similar small share of the reunified GDP. By then, the East's share of total exports was just 2 percent. Considering that manufactures are the most exportable of goods, such a situation was inevitable.[116]

Spending such a vast share of total transfers in social welfare benefits was a major error that achieved very few of its goals. The majority went to pay unemployment benefit and finance job-training programs, but by 1997 unemployment was still in double digits averaging about 17 percent. Only 20 percent of the people enrolled in job training programs landed jobs.[117]

The history of Germany and of numerous other nations has supplied many examples in which despite an absence of skilled workers, job creation was fast and unemployment was rapidly brought to low levels. History also supplies numerous examples of the opposite, in which an abundance of qualified workers was accompanied by poor economic performance and rising unemployment. These situations clearly revealed that independent of what common sense indicates, employment creation is not dependent on the education or skill levels of a population.

History suggests that it is dependent on the rate by which manufacturing grows, and for this sector to grow it needs support from the state. The stronger the support the faster the growth, and therefore the faster the creation of employment. There was practically no logic in allocating funds for job training and unemployment benefits because that contributed practically nothing in solving the unemployment problem. Those funds needed to be used for the promotion of factories, and by allocating them to this sector fast economic growth would have been assured, which would have rapidly eliminated unemployment. History has supplied enough evidence to conclude that a massive creation of employment would have taken place immediately after the resources had been reallocated, without even a small lag of time.

The United States, for example, by early 1939 was enduring an unemployment of about 19 percent and all of the social engineering efforts that had been undertaken to redress this situation since the early 1930s had failed.

World War II suddenly forced Washington to reallocate the unemployment-benefit resources, the ones destined for job training, the ones used for infrastructure, the ones that financed education, the ones that subsidized housing, agriculture, and the other primary activities, onto the factories that produced weapons. Manufacturing immediately began to grow at a double-digit pace, the economy did likewise, jobs were created at a breathtaking pace, and by 1943 unemployment had been completely eliminated and there was even overemployment.[118]

In the FRG after World War II a similar phenomenon took place. Very little was spent on unemployment benefits and job training programs notwithstanding the high joblessness of those years, but resources were abundantly channeled into the promotion of factories. Factory output was impressive, the economy did likewise, jobs were created at a very rapid pace, and during the 1960s unemployment was just 1 percent.

The case of Poland also suggests that what eastern Germany needed was to promote factory production much more enthusiastically. Even though on average during 1990 to 1997, the former GDR grew a little faster than Poland, once it stopped contracting in 1992 Poland grew faster. Worth noticing is that such a feat was achieved despite an almost complete absence of aid from abroad. Following the empirical example of China and numerous other East Asian nations, Warsaw began to promote somewhat enthusiastically the manufacturing sector by offering ample incentives to domestic and foreign capitalists. By 1996, there were five special economic zones (SEZs) offering tax holidays for up to twenty years, financial assistance, free land for the factory, and other incentives. Outside the SEZs, there were also numerous incentives for the sector, although in lower amounts. By 1997, FDI in Poland, of which the large majority was in manufacturing, accounted for about 4.5 percent of GDP. The automaker, Volvo, for example by then produced most of its worldwide output of truck parts in Poland.[119]

During 1992 to 1997 factory output in Poland was much faster than in eastern Germany and unemployment progressively declined while in the former GDR it remained fixed at very high levels. Bonn wanted to raise the standards of living in the East at a very fast pace so as to put them at parity with those of the West, but by 1997 the goal had largely failed. Much suggests that if the massive aid transfers had been used to promote manufacturing, impressive economic growth would have been achieved and the goal of achieving parity in standards of living would have been met.

The 1990s in Western Germany

Unification also brought numerous changes to western Germany although they were of a minor scale in comparison to the ones the East underwent during the 1990s. Taxes were raised noticeably to finance the development of the East. The same government that had been attempting to redress the fiscal imbalances since the early 1980s, had to relegate that goal to a secondary position in order to meet the vast demand for funds that the East required. Budget deficits rose significantly during 1990 to 1997. The federal deficit, which had averaged about 2 percent of GDP in the 1980s rose to 3 percent in the 1990s, and the consolidated deficit of federal, regional, and municipal governments rose from about 3 percent of GDP to 6 percent.[120]

The goal of reducing the size of the state also had to be temporarily abandoned for the sake of unification, and the consolidated expenditure of federal, regional, and municipal government rose from an average of about 47 percent of GDP in the 1980s to 49 percent in the 1990s. Public sector debt also increased noticeably. The increased expenditure, the larger budget deficits, and the expansion of public debt ended up affecting prices, and inflation rose slightly. After having averaged about 2.7 percent annually in the 1980s, inflation rose to 3.1 percent in 1990 to 1997.[121]

However, notwithstanding all of these negative side effects of unification plus the large transfer of resources to the East, which absorbed about 5 percent of the West's GDP, the economy did not deteriorate. Many economists and policy makers predicted a deceleration, in particular because of the large increase in budget deficits, but the fact is that the economy grew at exactly the same pace as in the 1980s. GDP once again averaged 2.0 percent annually.

Although this situation seemed somewhat contradictory, it became very consistent once things were analyzed from the perspective of manufacturing. During the 1990s, Bonn's support for this sector was almost identical to that of the preceding decade and factory output averaged a rate that was almost identical to that of the 1980s. After having averaged about 2.4 percent per year in the last cold war decade, manufacturing expanded by about 2.3 percent in the 1990s.[122]

By the end of the 1970s, reconstruction in all its forms had been fully completed. The destruction of World War II could no longer be observed in any form and the psychological pressure to assist manufacturing because of this reason therefore disappeared completely. Balance-of-payments concerns had also completely vanished as the FRG's current account surpluses had continued to grow and the foreign exchange reserves of the country had become very large. The support that Bonn supplied to manufacturing during the 1980s resulted mostly from the long tradition of support that had started around the mid-nineteenth century and which had systematically correlated with growth.

Policy makers could not understand why this correlation was so strong, in particular when most of economists increasingly asserted that services were the bottom line for the generation of growth in a developed economy.

However, the history of the country clearly revealed a strong parallelism between this sector and growth. Bonn therefore decreed during the 1980s a considerable number of fiscal, financial, and nonfinancial incentives to factory producers, but which were less onerous than those of the 1970s when the remnants of the reconstruction and balance-of-payment pressures forced the government to provide more subsidies.

In the 1990s, the motivations of the government to promote this sector remained unaltered. It was limited to the long tradition of favorable experiences which spoke in favor of support for the sector. Since the motivations were almost identical, the level of support was almost identical, which correlated with almost identical rates of factory output and GDP. The end of the cold war brought significant changes in defense expenditure and during the 1990s defense as a share of GDP fell to about 2 percent, while in the 1980s it had averaged about 3 percent. Armament production in consequence fell by a large margin, but most of the defense cutbacks were transferred to civilian manufacturing and the overall rate of factory output remained largely unaltered.[123]

State aid to manufacturing during the 1990s averaged a little more than 2 percent as a share of GDP in western Germany, which was about the same as in most other countries in western Europe. This in its turn coincided with similar rates of factory output and similar rates of GDP in the rest of western Europe. In the 1950s, direct state aid in the form of grants, subsidized lending, and tax exemptions accounted for about 9 percent of GDP, in the FRG which coalesced with a rate of factory output and a rate of GDP about four times faster than that of the 1990s. In most of western Europe, the same thing took place. Support for the sector was much stronger in the 1950s and 1960s and a much larger share of GDP was allocated to the factories, and for these countries also the economy grew much faster than in the 1990s.[124]

Support for civilian manufacturing during the 1990s as well as formerly, in western Germany and in the rest of Western Europe, was not constant and across-the-board. The level varied significantly over time and in the fields that received preference. It tended to be driven by a confluence of circumstances that pressured the government in one direction at one moment, and in another direction at another.

National security was, for example, one of these pressures. The producers of commercial sea vessels, for example, received incentives constantly, in the FRG and in the bulk of nations in western Europe, because of the strategic need to have a merchant navy that could back up its military counterpart. Textile, garment, and footwear producers also got subsidies on the ground that they were essential for the supply of uniforms and boots for the armed forces. Metal producers, machine tool producers, airplane makers, as well as the makers of electronics, computers, microchips, and numerous other goods that have a dual utility for military and civilian uses, were provided also with fiscal, financial, and nonfinancial incentives.[125]

In western Germany, the subsidization of civilian factory fields with linkages to the military was not as strong as in other countries of Western Europe such as France and Britain, which had a much larger defense expenditure. After the end of the war, the FRG took a passive role in defense, so its production of armaments as a share of GDP was considerably lower. As a result, the civilian factory fields that had a direct and indirect linkage with armaments were also subsidized less than in countries like Britain and France. In spite of the different levels of support among these countries for the production of weapons and weapons-related goods, overall it tended to decrease during the course of this half-century period. This correlated with decelerating rates of output of this sort of goods.

However, in Germany the century-old tradition of promoting manufacturing was much stronger than in countries like Britain and France, and the end result was an overall stronger level of support and a faster rate of factory output, during the second half of the twentieth century. That stronger tradition of support was channeled mostly into the areas that the particularities of Germany demanded. In Europe, during the second half of the twentieth century, western Germany was perhaps the most devoted protector of the environment, and as such it promoted more than any other country the development of environmental manufacturing.

Environmental consciousness rose over time and in consequence government subsidies for the production of factory goods that had pollution-reduction capabilities also rose. The problem of waste disposal, for example, grew rapidly in the decades after the war and there were rising predictions that Germany would be buried under a mountain of refuse. In the early 1980s, however, Bonn and more particularly the regional governments began to enthusiastically promote the production of machines that would convert waste into fertilizer, plastic, and electricity, as well as machines that would recycle paper, glass, and metal. Garbage tonnage, which had been growing very rapidly, immediately experienced a pronounced deceleration. In the 1990s, support for the production of this sort of goods increased even more as the environmental demands of the country rose, and output of environmental factory goods immediately began to grow at a much faster pace. This in turn coincided with a massive reduction in garbage tonnage, falling by almost half from the level of the 1980s. Most other European governments also subsidized this factory field much more than before and they too witnessed a rapid rise in the production of these goods and a decrease in garbage tonnage.[126]

Other environmental concerns, such as fear of possible fallout from nuclear plants or the global-warming effect of the heat-trapping gases emitted by oil and coal, drove the government to finance the production of machines that would reverse such a situation. A very large share of the research and development (R&D) costs and the overall production costs of factory goods that would offer alternative sources of energy was covered by the government.

In the 1990s, subsidies for the production of machinery and equipment that could create energy from biomass, tidal ocean waves, geothermal activity, from waste, wind, sunlight, and landfill gas, rose considerably and so did the rate of output of this sort of goods. By the late 1990s, these new and environmentally friendly forms of energy generation accounted for about 8 percent of western Germany's energy supply. Most OECD countries also increased their level of support for this sort of manufacturing and there too the rate of output of this kind of goods experienced a noticeable acceleration.[127]

A field that was strongly supported during the second half of the twentieth century was medical manufacturing because of its tight linkages with the overall health of the population. Pharmaceutical and medical equipment makers regularly received subsidies from the state, in particular for R&D. Bonn and regional governments paid for much of the R&D needed for the creation of new disease-fighting technologies, which practically always were embodied in a factory good.

In the 1990s, there was much support for the production of bio-engineered medicines, which was immediately followed by a massive acceleration in the rate of output of these factory goods. Until the 1980s, the fears of environmentalists about this new field inhibited any action from the government, but in the early 1990s restrictions were lifted and the government began granting about US$800 million annually to producers to cover R&D costs. Soft loans, tax exemptions, and nonfinancial incentives were also supplied so as to reduce costs even more. The direct subsidies of federal and state governments during 1990 to 1997 accounted for about 60 percent of the startup costs of these factories.[128]

Reinforcing the thesis that manufacturing practically only grows when the state supports it is the way the different factory fields evolved during the second half of the twentieth century. Their rate of expansion systematically paralleled the level of support that the government offered. In the 1950s and 1960s, for example, there was strong support across-the-board for all fields of manufacturing and there was also a rapid growth for all of them. In the next three decades, support fell considerably across-the-board and practically all grew at a much slower pace.

However, notwithstanding the overall deceleration of the last decades, the 1980s and 1990s witnessed a pronounced acceleration in the rate of environmental manufacturing, coinciding with a very large increase in the promotion of this field. In the 1990s, biotech manufacturing grew at an impressive pace relative to the past, which once again coalesced with an almost total absence of support up to the 1980s and with relatively strong support in the following years.

Other OECD countries experienced the same phenomenon. The factory-promotion efforts of the state decreased more or less across-the-board during the course of the 1950 to 1997 period and the rates of output of most fields progressively decelerated. However, some fields like environmental manufacturing and biotech manufacturing grew at a faster pace between 1980 to 1997 than before, which moved in tandem with an increase in the amount of subsidies that the large majority of OECD governments supplied to these fields.[129]

The growth of these fields in OECD countries, however, was not homogeneous. The fastest rates for environmental manufacturing were observed in western Europe, and that coincided with the region that subsidized this field, the most. In some particular areas of this field, however, western Europe did not take the lead. The production of solar-energy-generating equipment, for example, grew the fastest in Japan, which correlated with the fact that Tokyo provided the producers of photovoltaic equipment with much larger subsidies than any other OECD country. In biotechnology, however, it was the United States that attained the fastest output of genetically altered factory goods and this was paralleled by the fact that federal and state support for this field in the United States was stronger than in any other OECD country. The federal government alone budgeted more than US$3 billion in the early 1990s for R&D.[130]

Another field in which Germany attained a faster rate of output than most other OECD countries was trains. This situation once again was accompanied by a much larger flow of subsidies from the state to the producers of these goods during the 1990s and during most of the preceding decades. Only countries like France and Japan managed to grow at a pace that was faster, and that coincided with a level of support for this field from Paris and Tokyo that was somewhat stronger. Western Europe as a whole attained a much faster rate of output than the United States, and the fact is that Washington subsidized train production in a considerably less intense way. During the 1990s, western Europe's railways absorbed about US$31 billion annually in subsidies, of which a large share was used to reduce the expenses of train producers.[131]

Manufacturing's Particular Nature

The reason manufacturing so systematically (in Germany and elsewhere) tended to grow almost only when the state supported it seems to reside in the extremely high investment requirements of this sector. Over and over again, and not just during the second half of the twentieth century, the manufacturing sector proved to be by far the most investment-intensive. Not only did it require a gigantic investment per unit of output relative to the other three sectors, it also required much more time to see a profitable return on investment in comparison to the other sectors.

With these inherent characteristics, it was only inevitable that whenever support was weak and the intrinsic high-risk nature of this sector was not reduced with incentives, capitalists and entrepreneurs instinctively allocated their capital and energies into the other three sectors. Under those circumstances, manufacturing output had to be slow and likewise the economy.

Only the government, with its capacity to allocate resources at will, has the means to supply incentives to this sector in order to reduce its inherently high costs. And only in this way can the intrinsically risk-averse nature of the private sector toward this sector be overturned. The larger the incentives that are supplied, the greater the private sector's possibilities for profit and the larger therefore the generation of interest from capitalists to invest their resources in this sector.

Why would this sector be so different from the others with respect to its investment requirements? What would seem most likely to explain this phenomenon is technology. If this sector is causally linked to the process of technology creation, then it would be only natural that it is extremely investment-intensive because the generation of technology is by far the most investment-absorbing activity in the world. Even the reproduction of an existing technology is an effort that requires a huge investment.

The empirical evidence very conclusively demonstrates a strong parallelism between this sector and technology during the second half of the twentieth century. The fact that such a parallelism was observed in Germany, in western Europe, in Russia, in the United States, and in East Asia very strongly suggests that it was not chance. The fact that such a correlation was also observed during practically all of the preceding history of the West and East Asia adds further weight to the suspicion that manufacturing is intrinsically linked to the process of technology creation.

During the second half of the twentieth century, western Germany experienced a continuous reduction in the rate of technical progress, which was paralleled by a continuous deceleration in the rate of factory output. The bulk of the other OECD countries also experienced a deceleration in the rate of patent registrations and in the rate by which they imported technology, and this ran in tandem with decreasing levels of government support for the sector.[132]

During its short existence, the GDR underwent a similar process. In the 1949 to 1989 period, it was in the 1950s when inventions appeared at the fastest pace and it was also then that manufacturing grew at the fastest rate. It was during the 1980s that factory output grew at the slowest rate and it was then that technology in all its forms (created domestically or imported) was acquired at the slowest pace.

Eastern Europe and the Soviet Union also went through the same process. During the course of the 1950 to 1990 period, their promotion of the sector progressively decreased, their rates of factory output progressively decelerated, and the pace of technological development slowed down proportionately. The cases of the GDR and the rest of communist countries are particularly illustrative for substantiating the idea that manufacturing is naturally endowed with the capacity to create technology. Western economists constantly argued that wealth creation, economic growth, and the generation of technology were not possible under a centrally planned system, but over and over again wealth was created, growth was attained, and technology was developed. At times, the creation of technology was so outstanding in these communist countries that it was even superior to that from any capitalist country.

This situation ultimately leads to the conclusion that the only way that technology could have been created in a system riddled with inefficiencies was if the essential factor needed for its development resided in manufacturing. Since this sector is not dependent on a market economy for its growth, but on the desire of the government to see it grow, and since those communist governments wanted it to grow, the development of technology was inevitable. Since technology is the fundamental creator of wealth, it also became inevitable that growth would take place.

A market economy is essential for improving efficiency and maximizing resources, but not for the growth of factory output. Since there is no rationality in wasting resources, it follows that only the endorsement of a capitalist economy is logically justifiable. However, the paradox of communist economies can only be solved if it is assumed that manufacturing is the bottom line for the creation of technology and therefore for the creation of wealth.

Such a thesis finds itself substantiated by numerous facts. The figures for productivity are a case in point because they show a clear correlation with the differing levels of government support for the sector. Rates of productivity during the second half of the twentieth century were the highest during the 1950s (for the FRG and the GDR). This coincided with the decade when Bonn and Berlin allocated the largest share of their nation's resources to this sector and when rates of factory output were the fastest.[133]

From then onward rates of productivity decelerated for both Germanys correlating with decelerating rates of factory output. In the FRG, for example, total factor productivity averaged about 5 percent annually during the 1950s and factory output averaged about 9 percent. In the 1960s, factory production fell to about 6 percent and productivity dropped to about 4 percent. In the 1970s, the rate of factory output decelerated to about 3 percent and productivity fell to a pace of about 2 percent.[134]

Finally, factory production during the 1980s and 1990s averaged about 2 percent per year and productivity about 1 percent. Since productivity is fundamentally determined by technology and since technology is fundamentally determined by manufacturing, it was only inevitable that when this sector was strongly promoted and its growth was fast, productivity also expanded at a fast pace.[135]

Many have argued that productivity is fundamentally dependent on levels of competition and the degree by which market forces are operational. However, that vision of things cannot explain why the FRG and the GDR during the period 1950 to 1989 experienced decelerating rates of productivity while the economies of both were increasingly liberalized. Although to a considerably different degree, market forces in both Germanys were much more operational during the 1980s than in any of the preceding decades. However, it was in the 1980s when both Germanys attained their lowest productivity figures.

Such a vision of things cannot either explain why the rest of the OECD countries and the rest of the CMEA nations progressively experienced a deceleration in their productivity figures during the second half of the twentieth century, notwithstanding the progressive application of more and more liberalization measures. The manufacturing thesis has no problem adding up consistently with these facts because during that time the factory promotion efforts in practically all of these countries progressively decreased.

Also worth noting is that during the whole period manufacturing was the sector that attained the fastest rates of productivity. In the FRG and the GDR, as well as in the rest of OECD and CMEA nations, this sector was the one that attained the fastest rates of productivity. During the 1980s, for example, in OECD countries manufacturing productivity averaged about 3 percent per year while service productivity averaged only 1 percent. In the 1990s, this pattern remained largely unaltered.[136]

These were the only sectors large enough to have propelled the economy and therefore the ones whose productivity rates mattered. In the bulk of CMEA countries also, manufacturing systematically attained the fastest productivity rates. It is worth noting that countries that practiced such radically different economic systems as capitalism and communism constantly experienced the same development of events on such important matters as productivity and technology. Such a situation strongly suggests that a variable different than the ones considered essential by both systems was the one fundamentally responsible for the creation of wealth.

Manufacturing's inherent and perhaps unique capacity to create technology found itself expressed in numerous ways. The most unbelievable feats were constantly achieved as soon as government support was supplied for the manufacturing field related to that feat.

A case in point is the production of airplanes. As World War II came to an end, the United States emerged with the most sophisticated airplane technology in the world, and its lead was further reinforced with the capture of German and Japanese patents. As the undisputed leader of the anticommunist alliance, during the 1950s and 1960s Washington continued to invest much larger amounts of capital in the development of military airplane technology than any Western European country. As a result, it increased its lead even more in this field. As had frequently occurred in the past, military technology was soon utilized for civilian uses and, as a result, American commercial airplane technology was by far the most advanced in the world.

The very large technological lead of United States airplane makers drove numerous analysts in Western Europe to conclude that the Europeans would never catch up and that all efforts in that direction were doomed to fail. However, in 1970, four Western European governments (among them that of the FRG) decided to promote this field decisively, and for that purpose created Airbus Industrie. By 1994, the manufacturer had received about US$15 billion in state subsidies and had managed to become a highly competitive producer. It could match American plane makers in cost, quality, and technology. It was so competitive that it had even leapfrogged American companies by introducing a revolutionary technology of a computerized fly-by-wire system that reduced human pilot error. Its success was very clearly reflected in the fact that it had gone from zero in 1970 to capture about 30 percent of the world market of large commercial planes by the mid-1990s.[137]

Not only did the case of Airbus demonstrate that catching up technologically with a competitor with a large lead depended fundamentally on the degree to which a manufacturing field was promoted, it also gave clear evidence of the extremely investment-intensive nature of this sector. Boeing Corporation, for example, took about twenty years to turn a profit from the civil business it entered after World War II, and meanwhile had to live off military orders that guaranteed high profit margins. Airbus, on the other hand, spent the first quarter century of its existence without being able to report profits.

Not only does manufacturing demand huge investments because of its technology-intensive nature, it also demands an extremely long waiting time for the recuperation of the invested capital. It is precisely because of those characteristics that without government support the private sector instinctively opts not to invest in it. This is precisely the reason no European private company had dared challenge the monopoly of the Americans in that field before 1970, and why no private company in the remainder of the twentieth century dared invest in the field without the strong backing of the state. Without government support, they all concluded that the venture was just too risky.

Another example is space technology, which until the 1970s was almost monopolized in the noncommunist world by the United States, precisely because Washington had been the only one that had invested extensively in it. Washington's decisive support of this field (in particular during the 1960s) was driven almost exclusively by the political rivalry with the USSR. By the early 1970s, the United States had succeeded in developing a more sophisticated space technology than that of the USSR and was light years ahead of Western Europe. So most analysts concluded that Western Europe would never catch up and that all efforts in that direction would be a waste of resources.

However, in the 1970s, several European governments led by France decided to invest abundantly in space technology and founded Arianespace. They allocated vast amounts of capital to the project and soon this state-owned company (in which Bonn had a large stake) succeeded in fabricating space goods that could compete on price, quality, and technology with those of the Americans. Soon the Europeans managed even to outperform the Americans and in the late 1980s the rocket consortium began to dominate the world market for the launch of commercial spacecraft. During the 1990s, Arianespace launched about 60 percent of the world total.[138]

The catching up and the superseding of American technology coincided with a government promotion effort that was much more decisive than that of Washington. Since the Americans felt by the early 1970s that they had won the space race against the Soviets, their support for this field decreased considerably, and that was paralleled by a decelerated rate of technological progress in this field. The much stronger support of West European governments for space manufacturing in its turn coincided with a much faster pace of development of space technology. Under those circumstances, catching up was inevitable.

The case of Arianespace once again demonstrated that the rate of technological growth was fundamentally dependent on the degree to which governments supported manufacturing. It also showed that it had practically nothing to do with ideas such as the ones that assert that certain nations have a higher or a lower natural inclination for developing particular fields, domains, or sectors of the economy.

The development of environmental technology further reinforced this idea. The possibility of developing alternative renewable sources of energy in a cost-effective way was seen up to the mid-1980s in Germany and in most other countries of the world as an unattainable goal. However, as green parties increasingly gained political clout, the government felt forced to allocate much more funding for the production of environmental manufacturing. Immediately, the technology needed to reduce the costs of generating energy by means of wind, the sun, waste, biomass, landfill gas, and so on began to appear. By the late 1990s, about 8 percent of the generation of energy in Germany came from these sources and their costs by then were competitive with those of traditional sources of energy, notwithstanding the very low oil prices.[139]

Investment, Savings, and the Nonmanufacturing Sectors

Another way in which the technology-generating capacities of manufacturing found their expression was the way in which investment developed during the course of the years 1950 to 1997. If technology is fundamentally generated by this sector, and considering that technology demands a massive amount of investment, it would be logical to expect to find a correlation between levels of government support for this sector and overall levels of investment. That is precisely what this half-century period shows.

During the 1950s, for example, investment as a share of GDP in the FRG was by far the highest during this period, averaging almost 30 percent, coinciding with the most decisive support of manufacturing between 1950 and 1997. From then onward, Bonn progressively allocated a smaller share of the nation's resources to this sector, rates of factory output progressively decelerated, and levels of investment did likewise. By the 1990s, investment in western Germany accounted only for about 7 percent of GDP.[140]

Savings also followed a similar pattern of development. They rose to very high levels during the 1950s and from then they progressively declined, correlating with decelerating rates of factory output. In the 1950s, savings as a share of GDP averaged about a quarter of the whole economy and manufacturing grew by 9 percent annually. In the 1960s, the sector grew by about 6 percent and savings averaged just 20 percent of GDP. In the 1970s, factory output decelerated to just 3 percent per year and savings dropped to 15 percent. In the 1980s, manufacturing grew by more than 2 percent and savings fell to about 12 percent. In the 1990s, factory production grew at almost the same pace as in the preceding decade and savings accounted for about 11 percent of GDP.[141]

Since manufacturing is the prime creator of technology and therefore wealth, and since the possibilities for savings are always larger when wealth gets created at a fast pace, it was only inevitable that when the government promoted this sector in a decisive way, savings rose to higher levels.

In most of the OECD countries, the same phenomenon was experienced during the course of the second half of the twentieth century. Investment and savings progressively declined and despite numerous efforts to arrest such a decline, the trend continued in an almost uninterrupted way up to the 1990s. While none of the efforts to explain this situation managed to add up consistently with the facts, the one of manufacturing did. In practically all of these countries, government support for manufacturing progressively decreased during this period.

Many argued that it was the high levels of development of OECD countries that rendered high levels of savings unnecessary and that it was an inevitable consequence of being a wealthy nation. However, by the 1990s, Singapore was slightly more developed than most OECD countries and it nonetheless possessed an extremely high level of savings that was not only four times higher than the OECD average, but was actually the highest in the world. That coincided with a much stronger level of government support for manufacturing than the one supplied by OECD governments and it also coalesced with a rate of factory output that was almost four times faster.[142]

Things were not always like that. In the 1950s, OECD countries had higher levels of savings and investment than Singapore, and this was paralleled by stronger levels of support for the sector and faster rates of factory output in OECD countries.[143]

Another way the technology-creating powers of manufacturing found their expression was in the way the other sectors of the economy developed. They all tended to grow faster when the factory-promotion efforts of the state were strong. In agriculture, for example, the fastest rates for the FRG and the GDR were attained during the 1950s, correlating with the decade when Bonn and Berlin allocated the largest share of their nation's resources to manufacturing.[144]

From then onward, support for the sector decreased almost constantly, rates of factory output progressively decelerated, and rates of agricultural production did likewise. While in the 1950s rates of farm output averaged more than 4 percent in the FRG and less than 4 percent in the GDR, by the 1980s they averaged less than 2 percent in both countries.[145]

Services also grew at their fastest rate during the 1950s and from then onward they progressively decelerated. They grew by about 6 percent annually during the 1950s in the FRG, by about 4 percent in the 1960s, by about 2 percent in the 1970s, by about 3 percent in the 1980s, and by about 2 percent during 1990 to 1997. There was indeed an upsurge in the rate of growth of services during the 1980s as a result of deregulation. However, it is also a fact that deregulation moved forward in the 1990s and in spite of that the rate of services fell to a level that was practically the lowest in the second half of the twentieth century.[146]

A similar phenomenon took place in most of OECD countries during this period. The rate of services decelerated almost constantly, coinciding with the decreasing levels of government support for manufacturing. In the 1950s and 1960s, services expanded on average by about 5 percent annually, in the 1970s by more than 3 percent per year, in the 1980s by about 3 percent, and in the 1990s by a little more than 2 percent.

It was increasingly argued by economists that services were the determinant sector for the generation of economic growth in a mature economy, but practically all mature economies during the second half of the twentieth century experienced a progressive deceleration of their GDP figures as the share of services increased. The correlation actually went in the opposite direction. The larger the share of services in GDP, the slower the rates of economic growth. It was also asserted that services had a very large capacity to create jobs but the fact is that in the bulk of OECD countries the growing share of services coincided with rising levels of unemployment and/or underemployment. The fact that they systematically attained terribly low levels of productivity also made it highly unlikely that this service could have acted as a propeller on the economy.

The empirical evidence of Germany and of the bulk of OECD countries clearly indicated that services could not have possibly been responsible for the generation of growth and employment in mature economies. It also suggested that the bottom line of growth and employment resided in the manufacturing sector and that if a government wished a fast development of services and the other sectors, the only rational policy to follow was to promote factory production in the most wide-ranging way.

Throughout history, it regularly occurred that when governments subtracted resources from services, agriculture, and the other sectors in order to transfer them to manufacturing, the growth of services, agriculture, and the other sectors grew the fastest. At first glance such a situation presents itself as a paradox. The contradiction disappears if it is assumed that manufacturing is the fundamental creator of technology. As such it provides the other sectors with the technology that enables them to expand.

The technology-generating power of manufacturing found itself also reflected in the way overall profit margins behaved. They tended to be high when support for manufacturing was strong and low when the factory-promotion efforts of the state were low. During the Nazi 1930s, for example, profit margins were extremely high, which was paralleled by a double-digit growth of factory output. In the following decade, profit margins fell significantly and manufacturing output also decelerated in a significant way. In the 1950s, they greatly increased, paralleling the very large increase in the factory promotion efforts of the state. From then onward, profit margins progressively decreased and that coincided with progressively decreasing levels of subsidies for the sector.[147]

Since this sector is the prime creator of technology and since technology is the prime creator of wealth, it was only inevitable that when this sector grew rapidly and wealth was created at a fast pace, firms ended up capturing a larger share of that wealth.

Unemployment, Liberalization, Deregulation, Globalization, and Job Creation

During 1990 to 1997, western Germany endured its highest levels of unemployment during the second half of the twentieth century. Unemployment averaged more than 9 percent annually. Many argued that this situation was caused by the existence of a large public sector, high social welfare expenditures, high labor costs, and inflexible work practices. A considerable amount of evidence demonstrated that all of these aspects indeed had a negative effect on the economy and it was also clear that a contraction in the size of the state, a reduction in social welfare expenditures, wage moderation, and deregulation of the labor market would bring benefits.[148]

However, a long-term analysis of history clearly revealed that even the compounded effect of the four factors did not have the capacity to affect growth in a significant way. During the first half of the nineteenth century, for example, the size of the state as a share of GDP was only a tiny fraction of that of the late twentieth century. Social welfare spending was also almost completely nonexistent, labor costs were actually below a subsistence level, and regulations on labor and the economy were few. However, economic growth was much slower than from 1980 to 1997.

Also worth noting is that during the 1950s there was a considerable increase in the size of the state, in social welfare expenditures, in labor costs, in labor regulations, and in regulations on business. In spite of this, the economy grew to unprecedented rates, surpassing any preceding decade in the whole history of the country.

Although a causality linkage was missing, the evidence did suggest that the enlargement of these factors affected the economy in a negative way. Perhaps the one that had the largest of these marginal negative effect was social welfare expenditures. By the 1990s, they accounted for more than half of overall government expenditure and for about 30 percent of the GDP of western Germany. All of these expenditures were clearly of a purely consumption nature and were contributing in absolutely nothing to the generation of growth.[149]

The job training programs, the unemployment benefits, the job security regulations and even the expenditures in education were totally incapable of arresting the rise in unemployment. The goals that were pursued with the application of these programs were clearly not being attained; that was so because history demonstrates that growth is not dependent on the educational levels of a population, the skills of a workforce, or on job-protecting regulations. Growth (more particularly fast growth), which is the only real source of employment creation, has systematically given signs of being fundamentally dependent on manufacturing.

Other social welfare expenditures such as those allocated for health also showed an inability to deliver the goals that were pursued. By the 1990s, a growing share of the population could not be supplied with health services. History has repeatedly demonstrated that when economic growth is slow, it is practically impossible to supply a population with anything in an adequate way. Health services in Germany and elsewhere tended always to improve at the fastest pace when the economy grew at the fastest pace. The best way, therefore, of improving the health of the population was evidently by attaining the fastest possible GDP figures, and much suggests that the bottom line for the attainment of growth was in manufacturing.[150]

In 1950, a process of integration with several countries in western Europe began. This effort incessantly expanded the areas of the economy that became integrated. The number of countries that became members of this union also relentlessly increased over time. While during the 1950s the effort was limited to trade in some goods, by the 1990s it included the free flow of all sorts of goods, services, capital, labor, and even a considerable degree of currency integration; while in the 1950s the number of members was just six, by the 1990s, the figure had grown to fifteen.[151].

The project of further integration, which during the 1990s attracted most the attention of Germans and West Europeans, was currency unification. In the early part of this decade the governments of the region decided that the time had come for this idea to materialize. They therefore decided to establish a number of criteria for any nation that wished to qualify for membership in the creation of a single currency. It was demanded that countries maintain inflation rates at low levels, not exceeding more than 3 percent, that federal government deficits do not exceed an amount equaling 3 percent of GDP, that gross government debt not exceed a quantity equivalent to 60 percent of GDP, and a few other requirements.[152]

Most economists in Germany and abroad considered this criteria fundamental to a successful monetary integration. Most economists had actually urged European governments to apply almost identical policies well before the idea of currency unification appeared, supposedly because they were also fundamental for a positive economic performance. Although much suggested that these policies had positive effects, Bonn and the other EU governments increasingly saw them as essential for the generation of fast growth.

Others, however, took the reverse view and concluded that the efforts to rein in the budget, inflation, and government expenditure decreased the capacity to stimulate the economy with fiscal and monetary policy, and was therefore a fundamental reason for the slow growth and the high unemployment.

There was an abundance of evidence in the history of Germany and other countries however, that clearly demonstrated that fiscal and monetary policy have only a very small effect on the economy. The empirical data very clearly shows that even the concomitant effect of both has only a marginal effect on growth. During the 1930s, in the United States and the 1990s in Japan, for example, the governments of these countries significantly increased expenditures in an effort to stimulate the economy, and simultaneously lowered real interest rates by a wide margin. In both cases, however, the economy remained stagnant.

West Germany and numerous OECD countries during the 1950s and 1960s attained their fastest rates of economic growth in their entire history. It took place while budgets were balanced or in surplus, while government expenditure was much smaller than in the late twentieth century, while experiencing low inflation, and while having a small government debt. This situation clearly demonstrated that large government expenditure, budget deficits, high inflation, and public-sector debt were in no way helpful for the attainment of growth.

During the 1980s and 1990s, Singapore ran gigantic budget surpluses, government expenditure was but a fraction of that of western Germany, inflation was extremely low, and public debt was nonexistent. The fact that impressive GDP figures were nonetheless attained clearly showed that the criteria for monetary union was in no way hindering Western Europe from growing fast. The fact that by then Singapore was already a developed country made it even more evident that those accusing the Maastricht criteria of being responsible for the slow growth and high unemployment of Western Europe were simply missing the bottom line.

Much in history actually suggested that running balanced or surplus budgets, maintaining low inflation, having low or no public sector debt, and keeping government expenditure at low levels had positive effects on the economy. The fact that numerous countries at some moment in history attained fast growth while practicing the reverse of the above made it also plain that this was not essential for the generation of growth.

Those who were in favor of monetary union tended frequently to do the same as those who opposed it. They tended to overestimate the benefits that it carried with it. Benefits such as the elimination of the foreign exchange risk for those European Union companies that sell out of their home market into other EU countries were sure to act as a stimulant on the economy. Other aspects such as creating price transparency, the encouragement of competition within the EU, the exertion of a downward pressure on public spending, the lowering of transaction costs, and the increase in certainty for investors seemed also a sure by-product of a single currency.[153]

The reduction of inflation, the increase in labor mobility, and the decrease of financial speculation as a result of a reduction in currency speculation also seemed to become a sure effect of currency unification. Although all of these aspects seem capable of delivering positive effects on the economy, much also suggests that the boost on the economy will only be marginal. That boost is highly unlikely to accelerate the economy of western Germany and the EU to a rate of at least 3 percent on a sustained basis, which is the minimum needed for a reduction in unemployment and underemployment in developed economies.[154]

The evidence clearly indicates that economic and monetary unification, despite its numerous benefits, cannot solve Germany's troubles or those of Western Europe. During the second half of the twentieth century, economic integration rapidly and constantly moved forward. That actually coincided with a progressive deceleration of the GDP figures. The deceleration was not the result of the integration, but those facts clearly revealed that the benefits of integration were only marginal.[155]

During the 1990s, in particular, many argued that the numerous benefits of economic integration were neutralized by the phenomenon of globalization which created a very large outflow of German and West European capital to third countries. The end of communism drove many nations that had been formerly closed or semi-closed to open their borders. With low labor costs and a favorable investment climate, many ex-communist and developing countries began to attract vast amounts of capital from OECD nations.[156]

The case of the NICs, however, clearly demonstrated that the phenomenon of globalization could not possibly have been responsible, even in a minor way, for the slow growth of Germany and its EU partners. By the 1990s, nations like South Korea and Taiwan reached levels of development similar to those of Spain, while Hong Kong and Singapore had levels of income per capita in purchasing-power parity terms were similar to the most developed EU members. All of these four nations also experienced a massive emigration of domestic firms to countries with much lower labor costs. These outflows were actually larger (in relative terms) than those Western Europe endured. However, notwithstanding this huge outflow of capital, all of these economies grew much faster than western Germany and most EU nations. Singapore, for example, grew more than four times faster than western Germany.

This situation clearly revealed that globalization was in no way responsible for the slow growth, for Hong Kong and Singapore were actually the most globalized economies in the world. These city-states demonstrated that even with the most globalized economy, it was still possible to attain impressive GDP figures and bring unemployment to nonexistent levels. It is likely not chance that between 1990 and 1997, government support for manufacturing in the NICs was much stronger than in EU countries and rates of factory output were also much faster.

During the second half of the twentieth century, living conditions in Germany improved at a fast pace. Health, education, and overall incomes rose rapidly, and for the first time in its history (at least the Western part), it experienced a democratic form of government during the whole period. The country also experienced rising environmental consciousness, and plants and animals were increasingly given more protection.

It is evident that none of this would have taken place without the fast economic growth that the country experienced, and a very large amount of data suggests that this growth was fundamentally the result of the relatively strong factory promotion efforts of the state. Logic also suggests that living conditions would have improved much more, democracy would have developed more, and the protection of the environment would have gone further if government support for manufacturing had been much stronger.

Much points in the direction that the 8 percent annual GDP growth of the 1950s could have been easily sustained during the following four decades if manufacturing had continued to grow at more than 9 percent per year. That would have easily taken place if Bonn had continued to promote this sector as much as it did in the 1950s.

8
Great Britain During the Second Half of the Twentieth Century

Culture and Growth

During the second half of the twentieth century, Britain attained one of the slowest rates of economic growth among OECD countries. Not only was it one of the slowest, but the gap with many of these countries was considerable. The large majority of its Western European neighbors and practically all of its former colonies in North America and Oceania grew at a considerably faster pace. Worse still was the fact that the rates were so relatively weak that the British population was forced to very high levels of underemployment and unpleasant levels of unemployment during the whole period. Real incomes grew at a very slow pace and income distribution tended to worsen notwithstanding the numerous social welfare programs.

This was actually the second half-century period in which Britain attained slower GDP figures than the bulk of OECD countries. During the first half of the twentieth century, the large majority of the future OECD countries attained rates of economic growth that were faster than Britain's. Even during the second half of the nineteenth century, numerous nations in Western Europe, North America, Oceania, and Japan grew faster than Britain. (See tables at the end of the book.)

Such a persistent underperformance convinced numerous British policy makers and economists that there was something about the British culture that inhibited fast growth. It was asserted, in particular during the 1950s and 1960s when the growth gap with other developed countries was the widest, that the British were disinclined for work and that they were reluctant to invest. It was also asserted that they were resistant to innovation, that they lacked entrepreneurial drive, that they were noninventive, and that they were not strongly motivated toward education.[1]

Such a vision of the world represented a complete renunciation of ideas that for centuries had prevailed in Britain and which had also been believed by numerous other nations. From the early sixteenth century up to the mid-nineteenth century, Britain grew at a faster pace than any other country. During that time, it was actually asserted that the British were the most hard-working, the most entrepreneurial, the most willing to invest, the most inventive, the most creative, and the most committed to education.

Culture is a variable that over time remains largely constant. If it had really been responsible for Britain's success for about 350 years, it is very hard to understand why that same culture did not continue to deliver positive results from the mid-nineteenth century to the present. Logic finds it extremely hard to accept that after having been an asset for so long, culture suddenly turned into a liability. Until the mid-nineteenth century, it was also argued by many intellectuals of the time that it was culture that inhibited the nations of continental Europe and numerous other regions of the world from growing as fast as Britain. However, from then onward, those same growth-hindering cultures began to deliver, on a sustained basis, faster GDP rates than Britain.[2]

An untold number of nations have been classified at different moments in history as incapable of ever growing fast because of a culture that supposedly hampered investment, savings, and innovation. In the majority of cases, such a vision of things proved inconsistent, as at a later time fast growth was attained even though the culture of the country had not experienced any change. Japan is a case in point. Until the mid-nineteenth century most analysts thought that the country's culture was responsible for the perennial economic stagnation that had prevailed over the preceding centuries. However, from the last decades of this century up to the late twentieth century, Japan grew faster than all OECD countries.[3]

Until the mid-twentieth century, it was also argued by many Western analysts that China would forever remain in stagnation because of a culture that inhibited investment. In the 1950s, however, the country began to grow, and by the late twentieth century it attained one of the fastest rates of growth in the world. The same goes for most of the rest of East Asia. Up to the mid-twentieth century, many experts thought that stagnation was inevitable because of the culture of those countries. However, during the second half of the twentieth century, many of the countries in the region attained the fastest rates of growth in the whole world.

These are just the most outstanding examples that demonstrate the inability of the culture argument to add up consistently with the empirical evidence. The abundance of examples, as well as the abrupt changes in the rates of economic growth of numerous nations that did not see any change in their culture, leads inevitably to the conclusion that the culture variable could not have played an important role in the generation of growth. A long-term and logical analysis of history leads to the conclusion that culture at best plays a marginal role in growth.

It was evidently some other variable that was fundamentally responsible for Britain's weak performance during the period 1950 to 1997, as well as during the two preceding fifty-year periods. A variable with strong signs of having played the determinant role is manufacturing, because over the centuries it succeeded in correlating consistently with the empirical evidence. From the early sixteenth century until the mid-nineteenth century, Britain was the country that attained the fastest rates of manufacturing output in the world. Much suggests that this fast growth resulted not from chance, because during this long period London constantly subsidized this sector more than any other nation.

The evidence demonstrates also that the slower rates of factory output from the mid-nineteenth century onward, compared with most of the future OECD countries, were not the result of uncontrollable forces. The support that the British government gave to this sector during that period was systematically inferior to that supplied by most of the future OECD countries. During the second half of the twentieth century, the same phenomenon took place. Rates of factory output were slower than in most other OECD countries and London's promotion efforts were also weaker than those of most other developed nations.

Between 1950 and 1997, British factory output averaged about 2.6 percent annually and the economy expanded by about 2.3 percent. In the United States, factory output grew by about 3.5 percent and GDP by about 3.0 percent. In Germany, manufacturing grew by 4.6 percent and the economy by about 3.8 percent. In Japan, the sector expanded by about 7.5 percent per year and the economy by about 6.3 percent. That in turn coincided with the fact that it was Tokyo that subsidized this sector the most, and that it was the British government that promoted it the least.[4]

None of the governments of these countries at that time believed that manufacturing was determinant or even important for the generation of growth. However, because of a confluence of particular events that were related to national security concerns, balance-of-payments motivations, technological development goals, nationalistic self-sufficiency ideas, war reconstruction pressures, employment creation goals, and other ideological motivations, they ended up supporting the sector. Since these motivations were not homogeneous through time, levels of support for the sector ended up varying considerably and rates of factory output moved in tandem. Since these pressures varied also in intensity among nations, their levels of support were considerably different and their rates of factory output, too.[5]

The 1950s

During the second half of the twentieth century, Britain was pressured by most of these circumstances into giving support to the sector and it was also under a similar level of pressure as several other countries. However, because the political class was deeply convinced that distorting market forces was intrinsically wrong, it promoted the sector at the lowest possible level that those pressures allowed. During all this time, there was a persistent reluctance to break away from laissez-faire traditions. Notwithstanding the much stronger factory promotion efforts of its most important neighbors, such as France and Germany (in particular during the 1950s and 1960s), London remained convinced that subsidies were unnecessary. Even with the relatively low support that it supplied, many policy makers and economists regularly accused the government of undermining the private sector and the natural regenerative process of the free-market economy.[6]

Britain endured more physical destruction during World War II than France and Italy and its factories were more destroyed, damaged, and paralyzed than in these two countries. The pressure to reconstruct and rehabilitate these factories therefore forced London to supply an additional level of incentives to manufacturers than those provided during peacetimes. However, because distorting market forces was seen as a major policy error, it did not supply as many incentives as Paris and Rome, which were not fixated with this idea. Bonn was not obsessed with this idea either, and Germany had a relatively long tradition of distorting market forces for the sake of promoting manufacturing in a relatively strong way.[7]

After the war, Britain was enduring a scarcity of foreign exchange that was as acute as the one experienced by its Western European neighbors, and as a result it wanted to see its exports grow rapidly. Since the bulk of British exports were factory goods, such an additional desire to earn hard currency translated into a larger amount of incentives to the sector. However, since London was intrinsically against all forms of distortions, it subsidized exports less intensively than Bonn, Paris, Rome, and several other West European governments did.

The war also caused a scarcity of consumer goods and the British endured it almost as much as the Germans. There was therefore as much pent-up demand as in many of its war-affected neighbors, and since most of the demand was for manufactured goods, London's desire to see this demand satisfied translated into an increase in incentives for factory producers. However, since their noninterventionist vision of the world was so strong, British policy makers supplied considerably less support to manufacturers than their German, French, and Italian counterparts did.[8]

That is why, notwithstanding the fact that Britain received the most Marshall aid per capita in Europe, it was actually one of the countries with the slowest rates of factory output and GDP during the late 1940s and 1950s. Manufacturing tends to grow only when it receives government support and the weak support that British governments supplied to it at the central, regional, and municipal level compared with its neighbors had to deliver a much slower rate. In the 1950s, Britain's manufacturing production expanded by 3.1 percent annually and the economy grew by 2.7 percent.[9]

In West Germany, where support was the strongest in Western Europe, the sector grew by 9.3 percent and the economy by 8.0 percent. Italian policy makers, who were the second most enthusiastic promoters of manufacturing, managed to make the sector grow by about 7.1 percent per year and GDP shadowed that figure by growing at 5.9 percent. French policy makers were somewhat less motivated; factory output was about 5.8 percent per year and the economy grew by about 4.6 percent. In Tokyo, where the political class, as a result of a very strong nationalism that wanted to demonstrate that Japan was second to no one on economic matters, plus the motivations of war reconstruction, balance-of-payments, and pent-up demand, decided to promote manufacturing more decisively than anybody in Western Europe. This coincided with a rate of factory output of about 11.1 percent annually and a rate of economic growth of about 9.1 percent during the 1950s.[10]

There were some who argued in favor of imitating the much stronger factory promotion efforts of Britain's neighbors, but the majority of British policy makers were persistently convinced that the inferior British GDP rates were the result not of fewer machines, but of an inadequate organization of production. The much lower productivity rates of Britain during the 1950s relative to most of its neighbors drove most politicians and economists to the conclusion that an increase in productivity could be attained by means of a reorganization of production on more efficient lines. Since support for manufacturing was seen as intrinsically wrong, other explanations such as this one were systematically sought.[11]

Making more efficient use of the existing resources was nonetheless an idea that clearly had its merits. During 1946 to 1951, the Laborists nationalized numerous enterprises and once in state hands these firms systematically attained lower efficiency levels than before. Regulations on business were increased during these years and they also delivered a negative effect on efficiency. These were evidently policy errors, but later on they were partially eliminated and the GDP figures nevertheless remained largely unchanged. When the Conservatives regained power in the 1950s, they denationalized some of the state enterprises and reduced some of the regulations on business, but the growth figures were about the same as the ones attained during the late 1940s to early 1950s. In the 1960s, no more enterprises were nationalized and even though it was a decade when state companies were less numerous than in the 1950s, the GDP figures were practically identical during both decades.[12]

British policy making oscillated during most of the second half of the twentieth century in between years of nationalization and years of denationalization. During the mid-1970s, there was another round of nationalization and during the 1980s the bulk of the existing state companies were privatized. In the 1990s, practically all of the remaining state enterprises were handed over to the private sector. During that time, however, the Laborists, who were the ones who believed in nationalization, and the Tories, who were against it, had a very important common denominator in spite of their differences. Both believed that independent of whether a company was in state hands or in private hands, a large level of support for it could not be justified.

The Laborists were much less committed to the principle of letting market forces operate freely, but their desire to manipulate the market was circumscribed to their social redistributive goals. Their socialistic vision of the world drove them to the conclusion that the only thing that could justify market distortions was the goal of redistributing the wealth of the nation in a more equal way. Not all of the enterprises that they nationalized were in the manufacturing sector, but most were. Therefore, once these factories fell into the hands of the state, they were not provided with more subsidies than when they had been in the hands of capitalists. They were simply used to finance the enlarged social welfare programs of the Laborists.[13]

The nationalization effort was an error because it subtracted competitive pressure on these firms. However, if the government had supplied them with an abundance of incentives, then their rate of output would have considerably accelerated and the economy would have moved in tandem. That is what took place, for example, in Italy, where a very large number of state factories were created and regularly supplied with a much higher level of support than in Britain since the late 1940s. In France and in Germany, although state firms were much less numerous than in Italy, the government also supplied state and private manufacturers with a much higher level of subsidies than in Britain. There, too, rates of factory output and GDP were much faster since the late 1940s than in Britain.[14]

When the conservatives privatized state companies in the 1950s, 1980s, and 1990s, they succeeded in improving the efficiency of those firms but since they were committed to the idea of subsidizing manufacturing as little as possible, factory output and GDP continued to grow at a slow pace.

The manufacturing variable also can explain why during the 1950s the Soviet Union and several other communist countries in Eastern Europe that had nationalized the whole economy and had totally annihilated the private sector managed also to grow faster than Britain. While Britain averaged a GDP rate of less than 3 percent annually, the USSR got more than 4 percent in real terms, East Germany got more than 5 percent, and several other CMEA nations got more than 3 percent (that is, once adjusted to the growth measurements of the most efficient capitalist economies). It is probably not chance that factory output in the USSR and Eastern Germany expanded by about 12 percent per year and in several other CMEA nations by a rate about three times that of Britain.[15]

If British policy makers had been right about their free market beliefs, then CMEA nations should have experienced a constant recession the whole time because they totally distorted market forces. The fact that most of them grew even faster than Britain during the 1950s very clearly demonstrates that the basic ideas prevailing in London were simply missing the bottom line.

If it gets assumed that manufacturing is the fundamental generator of technology and therefore wealth, it is fully understandable why despite the massive inefficiencies of CMEA countries, most of them managed to outperform Britain during the 1950s. Since their promotion of manufacturing was so much stronger than Britain's, their technological development was faster and so the pace by which they created wealth was also faster. They wasted a huge amount of resources as a result of their severe distortion of market forces, but since factory output grew so fast, it managed to overcompensate and growth was therefore faster. That would also explain why Italy, with its abundance of state companies, grew much faster than Britain. There was obviously no need to nationalize or create state companies in Britain or anywhere else, but much suggests that there was a strong need to enthusiastically promote private manufacturing.

Many argued that Britain's much slower rates of growth during the 1950s relative to most OECD countries was the result of the considerable increases in social welfare expenditures and the correlative increases in taxation that prevented the private sector from putting more work effort and investment in the economy.[16]

Social welfare indeed increased considerably as a share of GDP and such a purely consumption activity was evidently not contributing anything to growth, as it was subtracting resources for possible investment. However, the fact that the bulk of OECD governments also considerably increased their social welfare expenditures during the 1950s and still managed to grow much faster than Britain clearly indicates that these increases were not the main cause of the UK's weak performance. Scandinavia and West Germany spent a larger share of their GDPs on social welfare than Britain did and they nonetheless grew at a much faster pace. Such facts suggest that the increase in social welfare did not even play an important role in growth.

Evidence suggests that what did play a determinant role was manufacturing. Scandinavian countries and Germany, for example, despite the large increase on welfare, promoted the sector much more decisively than Britain and their rates of factory output and GDP were also much faster. Japan did not increase its welfare expenditures as much as most in Western Europe but the share of GDP that it allocated to this area was much larger than during any preceding decade, and in spite of that it grew much faster than ever before. That coincided with the most decisive support of manufacturing in the history of Japan until then, and with unprecedented rates of factory output.[17]

The area that most forced London to decree support for manufacturing during the second half of the twentieth century and more particularly the 1950s was defense. As the second most important member of NATO and in a decade when the cold war heated up considerably, Britain was obliged to allocate a large share of its GDP for defense. During this decade, military expenditures accounted for almost 7 percent of the whole economy. However, contrary to the United States, that allocated the bulk of its defense expenditures for the fabrication of weapons, Britain allocated a noticeable share of its defense budget to the importation of American weapons. During World War II, Britain imported a tremendous number of American weapons, and after the war, even though Britain was no longer desperately needing them, London decided to continue to import some because of their superior technology.[18]

Not only was defense expenditure lower than the approximately 8 percent of GDP that the United States spent, but London also did not use all of it for the production of weapons. National security was the only domain where the entrenched ideas of Britain's political class about not disturbing market forces were put aside. Insuring the nation against external threats was such a life-and-death matter that it was put above all economic and political considerations. However, the particular confluence of events of the second half of the twentieth century and in particular the 1950s drove London to promote the production of weapons in a fairly unenthusiastic way.

Although World War II had made it clear that there was a need to possess large and modern arsenals because a large-scale conflict could erupt at any moment, the fact was that the major threat to Britain's security had been totally defeated. Germany was permanently occupied and partitioned by four major powers, and two of them had enough military capabilities to individually defeat her if a new aggression should erupt. Having been neutralized, Germany was therefore no longer seen as a major threat.

The USSR turned rapidly into a threat, but the likelihood of a reckless invasion or attack of any sort was very small. Stalin, Khrushchev, and the rest of the Party leadership constantly gave signs of not being interested in engaging in any war. World War II had already caused the death of almost 25 million Soviets and the Politburo gave clear signals of being fundamentally interested in the economic development of the country.[19]

Not to mention that it was the United States that absorbed the bulk of the responsibility for containing communism worldwide. In addition, there was a rapid process of decolonization that London felt politically, morally, and economically obliged to follow. As it began to grant independence to its most precious colonial possession in 1947 and in the following years to other territories under its control, British policy makers could no longer justify as large a navy and commercial fleet as before. Subsidies for shipbuilding therefore fell significantly.

In the United States, where a similar and strong noninterventionist vision of the world prevailed, the government felt forced during the 1950s and 1960s to allocate a larger share of the nation's resources to the factories that produced weapons and related goods. The intensity of the cold war left Washington without a choice, and this situation delivered an overall higher level of support for manufacturing. In Britain, however, the cold war never took on such large dimensions because the United Kingdom played a secondary role in it plus its major historical enemy had been totally neutralized. Worse still was that as it dismantled its empire, the demand for weapons and related goods fell significantly. National security concerns, therefore, on average were not very large and consequently they did not pressure the government to decisively promote the production of weapons and related factory goods.

A motivation considerably less strong than national security but which also drove London to supply some support to manufacturing was regional economic disparities. In all nations of the world certain regions are more developed than others, and Britain was no exception. Since growth had been weak almost uninterruptedly since the mid-nineteenth century, the problem of regional disparities was compounded by very high levels of underemployment and unpleasant levels of unemployment. During the 1950s, unemployment was about 5 percent and underemployment affected about a fifth of the adult population.[20]

In an effort to create employment in the regions where an insufficiency of jobs was more noticeable and in an effort to attain economic equality among all regions, London supplied support for manufacturing in depressed areas of the country. The government did not target this sector exclusively but since full-time and high-paying jobs throughout history were closely associated with this sector, much of the support supplied to these regions was used to promote factories.

The support, however, was never great because policy makers were ultimately convinced that market distortions, even when they worked to strengthen the cohesion of the country by reducing regional disparities, tended in the long term to cause more economic harm. As a result, the sums of money allocated for this purpose were much smaller than in most West European countries that pursued the same thing. Even though this money accounted for a much smaller share of budgetary allocations than in countries such as Germany, Italy, and France, a large share of the political class and most economists saw it as a tremendous amount of funds being largely wasted.

State subsidies set up to induce factories to set operations in depressed areas during these years went mostly to shipbuilding, machine tools, motorcycles, and several other fields. Since a strong antisubsidy attitude prevailed in intellectual and policy-making circles, every time one of these factories failed, most tended to see it as evidence that it was all a bad idea.[21]

In Germany, Italy, and France, countless factories subsidized by the state also went bankrupt, but because overall levels of support for manufacturing were much higher, factory output was much faster, and that coincided with much faster rates of economic growth. Much evidence suggests that what was ultimately important was not whether every manufacturing enterprise that received support from the state became a total success, but the overall rate of output by which the sector expanded. For this sector to grow rapidly on a constant basis, it was necessary that the state supply it a constant and strong level of support. In Japan, also, there were thousands of subsidized manufacturers that went bankrupt during the 1950s, but since government support for the sector was stronger than in any other OECD country, rates of factory output and rates of GDP were still faster.

The 1960s

During the 1960s, national security concerns remained largely unaltered and as a result the share of the economy that was allocated for defense and for making weapons remained largely the same. Armament and weapon-related civilian factory goods consequently grew at about the same pace.

On the one hand, the cold war remained with the same intensity as during the preceding decade, and on the other hand the continuation of the process of decolonization meant a lower number of territories that needed to be defended by the navy. However, the most important British possessions were decolonized in the 1940s and 1950s. In the 1960s, therefore, the subsidies supplied to the producers of warships and commercial sea vessels dropped by only a small amount. In consequence, overall subsidies to armaments and related factory goods during this decade were very much the same as during the 1950s.

Problems regarding regional disparities, underemployment, and unemployment also remained largely unaltered as levels of economic inequality among Britain's regions remained the same and as levels of underemployment and unemployment also remained unchanged. Consequently the government supplied about the same level of incentives to manufacturers for the purpose of combating these problems.

However, on matters relating to the effects of World War II there was a change in the level of support. Since the ravages of World War II were felt less directly than in the preceding years, the incentives that were supplied for the sake of rebuilding and rehabilitating the destroyed and paralyzed factories were reduced. Since by the late 1950s, the scarcities of the war were less pronounced than before, incentives for the production of consumer goods (of which the majority were manufactures) were also less numerous during the 1960s. Since foreign exchange reserves had grown noticeably and balance-of-payments concerns were not as pressing, export subsidies were reduced.

During the 1960s, there was a new variable that affected the measures British policy makers took concerning the level of support they supplied to the manufacturing sector. This new factor was the example of West European countries during the preceding decade. This was the first time in history that so many West European nations had grown faster than Britain, and on top of that it was the first time that the growth gap was so large. The strong historical linkages and the geographical proximity to these countries forced British politicians and economists to take notice of what these governments were doing; and the fact was that all of the nations in Western Europe that grew much faster than Britain during the 1950s promoted manufacturing in a much more decisive way.[22]

In the 1960s, therefore, London felt compelled to imitate its continental neighbors, and more fiscal, financial, and nonfinancial incentives were offered to manufacturers. The example of France was the one that especially caught the attention of London because France grew much faster during the 1960s than in the preceding decade. The fact that Paris significantly increased its factory promotion efforts during the 1960s also suggested that such a policy at least had something to do with growth. So London decreed more tax reductions, more grants, more subsidized financing, and even engaged in a few manufacturing collaborative projects with the French.

However, since laissez-faire ideas were the fulcrum of London's economic vision of the world, the support was far from being as strong as that of Paris.

As a result, therefore, of national security motivations similar to those of the 1950s, of similar regional disparities and unemployment concerns, of lower post-war reconstruction and balance-of-payments motivations, and of higher pressures from the example of West European nations, London ended up decreeing a level of support for manufacturing during the 1960s that was largely the same as that of the preceding decade. This coincided with rates of factory output that were almost identical to those of the 1950s.[23]

During the 1960s, manufacturing production averaged about 3.3 percent annually after having grown by about 3.1 percent in the 1950s, and the economy expanded by about 2.8 percent after having grown by about 2.7 percent. Paris, which promoted the sector much more decisively, attained a rate of factory output of about 7 percent per year and GDP grew by about 6 percent. France was among the few large countries in Western Europe that experienced a considerable acceleration in its GDP figures and it was also among the few where support for the sector considerably increased. In Japan, there was also a significant increase in the factory promotion efforts of the government, and there too the rate of output accelerated significantly, going from about 11 percent per year in the 1950s to more than 13 percent in the 1960s. The economy did likewise and accelerated from 9 percent to 11 percent.[24]

The 1960s, was the second postwar decade in a row when Britain performed considerably below most nations in Western Europe. Prior to World War II only a few nations, such as Germany, had managed to grow much faster than Britain, while several others, such as France, had managed to grow only slightly faster. This was the first time that such nations had grown much faster and this was also the first time that nations such as Spain, which had systematically grown more slowly than Britain in all of the preceding history, managed not just to grow faster but to grow much faster. During the 1960s, Spain grew at an average annual rate of about 7 percent while in the 1950s it grew by about 5 percent. This correlated with a very decisive factory promotion effort of Franco's military regime and with rates of manufacturing output that averaged about 7 percent in the 1950s and 9 percent in the 1960s.[25]

The Scandinavian countries, the Benelux countries, Austria, Switzerland, the United States, and even Portugal, Greece, and Ireland as well as Canada, Australia, and New Zealand during the 1960s grew at a faster pace than Britain. In each one of these countries, this situation coincided with stronger levels of government support for manufacturing and faster rates of factory output. Even though practically every OECD country managed to outperform Britain and the evidence strongly suggested that support for manufacturing at least had some positive effects on growth, the majority of British policy makers and economists continued to complain about the few subsidies that the government supplied. Most argued that the factories, the technology that emanated from them, and the jobs that they created would have been just as well created without the subsidies.[26]

Most of them even argued that if the subsidies had not been supplied, much better results would have been obtained. Some even went as far as to assert that the subsidies were the cause of the weak GDP figures relative to the rest of OECD countries. Since their basic economic premise was that all market distortions were intrinsically wrong and their adherence to such an idea took almost religious proportions, it was inevitable that only reluctantly did they supply a few subsidies to the sector. It was inevitable that only under the pressure of national security concerns and other motivations of the sort did they supply some support for manufacturing. It was also inevitable that the few fiscal, financial, and nonfinancial incentives that were offered were constantly seen more as a hindrance on growth than as a propeller.

Every subsidized factory that went bankrupt or showed losses was seen as clear evidence of the failure of a manufacturing promotion policy. The large investments that were made jointly with the French government to build the Concorde, for example, were seen as a huge waste of resources during the 1960s, simply because there was no private sector demand for supersonic commercial jets. When the airplanes made their debut in the 1970s and their excessive noise and high costs made them unattractive to most airports and airlines, most British policy makers and economists felt their ideas had been confirmed.[27]

Their fixation with noninterventionist ideas led them not to notice that Paris, which invested more in the Corcorde project and more fully subsidized the whole manufacturing sector, attained much faster rates of economic growth. It also led them not to notice that Tokyo significantly subsidized the production of commercial airplanes. The effort failed to deliver cost-effective and technology competitive large planes, but in spite of that failure, Japan attained the fastest rates of economic growth in the world during the 1960s. Numerous other factories in other fields that were subsidized by Tokyo also failed, but in spite of that, Japan grew almost four times faster than Britain.[28]

London was right in believing that the bulk of market distortions had a negative effect on the economy, but history suggests that it was totally wrong in believing that this principle applied to manufacturing. History strongly indicates that this sector is fundamentally responsible for the creation and reproduction of technology, and as such it is extremely investment-intensive. Since the development of technology requires such massive investments, this sector is by its very nature extremely risky; and because it is so risky, without government support the private sector opts for not allocating its resources into it. Under these circumstances, technology does not get generated, and in consequence wealth is not created. The economy therefore stagnates and living conditions do likewise.

London was constantly more interested in avoiding market distortions than in anything else. That is why it enthusiastically worked to liquidate the large public-sector debt resulting from World War II, and such a goal was finally achieved in 1969.[29]

Aiming to eliminate government debt was by all means a positive undertaking, but the historical evidence of Britain and of other nations had by then already demonstrated that the benefits of such efforts were only marginal. The gigantic public debt that Britain acquired during World War II actually coincided with the fastest rates of economic growth in the whole history of the country. Also worth noting is that in the 1950s, when the debt was much higher than in the 1970s, the economy grew by about 3 percent while in the 1970s it grew by just 2 percent.[30]

In the United States and Canada, growth was also outstanding during World War II while government debt reached unprecedented proportions, and decades later when it was much lower the economy grew at a much slower pace. These experiences as well as numerous others clearly demonstrated that an extremely powerful growth-generating variable had managed to supersede not just the gigantic debt and the encompassing distortion of market forces, but also the gigantic wastage of resources used for the production of weapons during the war. The fact that support for military manufacturing and related fields was very strong and that overall factory output grew at a double-digit pace in Britain, the United States, and Canada, leads to the conclusion that this impressive variable was manufacturing.[31]

Constant efforts to balance the budget in the postwar decades were also undertaken by the British government, and frequently surpluses were even attained. Here again it was clear that fiscal rectitude was a positive policy, but that it was not as important as British policy makers thought. Very large budget deficits were run in Germany during the Nazi 1930s as well as in Britain, the United States, and Canada during World War II, and extremely fast economic growth was nonetheless attained in all four countries. Such events clearly indicated that sound fiscal policies were not essential for the generation of growth.[32]

In the postwar decades, Britain's political class was so convinced of the merits of fiscal rectitude that even Labor governments often cut back the investment plans of nationalized factories in an effort to adjust government expenditure. Because policy makers were so obsessed with the idea of not distorting market forces, they could not realize that fiscal rectitude could be maintained while simultaneously giving very strong support for manufacturing. Germany, for example, ran budget surpluses during the 1950s and

1960s while the government simultaneously supplied an abundance of incentives to factory producers.[33]

As a result, inflation was very low and actually lower than in Britain, while factory output and the economy grew at a much faster pace. Japan also ran budget surpluses during these decades and it subsidized manufacturing even more than Germany, which coincided with even faster rates of factory output and GDP, and with a relatively low rate of inflation.[34]

Many argued that the reason growth in Britain during the postwar decades was way below most OECD countries was that wages rose too fast as a result of the excessive wage demands that the country's militant unions exacted. Wages indeed grew much faster than the economy, while in most OECD countries their growth was slower or as fast as that of the economy. This situation evidently hurt the competitiveness of British goods, in particular exports, but the fact is that the United Kingdom was not alone. France, for example, also experienced a much faster rate of growth of wages relative to GDP rates, and in spite of that it attained a much faster pace of economic growth than Britain.[35]

The faster growth of wages relative to GDP was evidently an error, but this was obviously not the fundamental cause of Britain's slow growth. The fact that France grew almost twice as fast while committing the same error suggests that Paris did something that had such impressive growth-generating powers that it was capable of overcoming the burden on competitiveness that an excessive growth of wages produced. Strong subsidization of manufacturing seems to have been what made the difference, because not only did factory output grow about twice as fast in France as in Britain, but exports also expanded at a much faster pace. While factory output and exports in Britain during the 1960s averaged about 3.3 percent and 4.8 percent respectively, in France the figures were 6.9 percent and 8.5 percent.[36]

Others argued that Britain's slow growth was the result of the neglect of education and the cultural dislike of the field of engineering. Many analysts noticed that in the 1950s and 1960s, the number of engineers in Britain per capita was much smaller than in Germany and many other countries in Western Europe, as well as in Japan and the United States. Since this was a situation that had also been observed since the late nineteenth century, many concluded that there was something in the culture of the country that predisposed the British to dislike the fields of engineering. Since this field of knowledge was largely associated with export goods, it was concluded that this situation was a hindrance for growth.[37]

However, it was not always like that. For centuries up to the mid-nineteenth century, Britain had the highest number of engineers in the world per capita, which coincided with policies during all those centuries in which government support for manufacturing was stronger than anywhere else in the world. It also had the highest number of inventors, scientists, technicians, and skilled workers in the world. In the mid-nineteenth century, the governments of several other nations began systematically to promote manufacturing in a more decisive way than London did. This coalesced with a proliferation of engineering universities and technical schools, as well as a rapid multiplication of scientists, inventors, technicians and skilled workers in those countries.

In the post–World War II decades, most OECD countries promoted factories more enthusiastically than ever before and that coalesced once again with an unprecedented proliferation of engineers, inventors, technicians, and skilled workers. London, which promoted the sector in a far less enthusiastic way, did not see a rapid increase in their numbers. There is much to suggest that there was no cultural proclivity in Germany, Japan, and several other OECD countries toward engineering or anything of the sort, and neither was the opposite true about Britain. Since this sector is strongly linked to fields of knowledge such as engineering, the supply of engineers and technicians had to rapidly multiply in the countries that promoted this sector considerably.

Part of the factory-promotion efforts of these governments consisted in rapidly increasing the number of schools and universities that specialized in engineering, natural sciences, and technical matters. Since London was not targeting this sector, it did not see necessary to allocate large funds for the creation and development of such schools and universities.

During the 1960s, savings as a share of GDP in Britain were much lower than in most of its West European neighbors, and many analysts saw in this situation a major cause of the much slower rates of growth. The fact that in the 1950s and in most of the preceding decades savings had also been lower than in many Western European countries led many to conclude that there was something in the British culture that conspired against parsimony. Since a low pool of savings was seen as hindering the possibilities for investment, it was also seen as a hindrance for growth.[38]

The argument seems sensible at first glance, but it ultimately does not add up consistently with the evidence. History has supplied a considerable number of situations in which nations managed to grow fast in spite of having a low level of savings. There are cases in which even spectacular rates of growth were attained notwithstanding extremely low levels of savings. Japan, Germany, and Italy, for example, during 1946 to 1949 grew almost four times faster than the rate Britain attained in the 1960s, while their level of savings was almost nonexistent. In the UK during the 1960s, savings as a share of GDP averaged about 11 percent.[39]

Also worth noting is that during the 1960s savings in the United States averaged about 8 percent of GDP, and despite that low level, the American economy grew by about 4.2 percent annually while that of Britain grew by only 2.8 percent. The case of the United States clearly demonstrated that the low pool of savings was not a hindrance for the attainment of much faster rates of growth. The cases of Japan, Germany, and Italy from 1946 to 1949 demonstrated that even if savings in Britain had been practically nonexistent, it was still possible to grow even at a double-digit pace.

None of the explanations that were presented to diagnose the cause of Britain's persistent underperformance managed to consistently match the empirical evidence. A thorough review of the numerous arguments used to explain the United Kingdom's weak performance in 1970 by a group of reputed British economists concluded that the causes remained obscure and no confident advice could be given to the government on how to set things right. Much suggests that the main factor that inhibited economists from deciphering the cause of the problem was their insistence on believing that all market distortions were intrinsically harmful for the economy. As long as they departed from that premise, it was impossible for them to appreciate the impressive growth-generating powers of manufacturing.[40]

Had they not been blinded by this idea they would have probably noticed that during 1946 to 1949 Tokyo, Bonn, and Rome supplied an abundance of subsidies to manufacturing. That is why notwithstanding the almost nonexistence of savings, factory output grew at a double-digit pace and the economy did likewise. They would have also noticed that during the 1960s, Washington decreed a much higher level of support for the sector and factory output grew by about 5.3 percent annually, while in Britain manufacturing expanded by just 3.3 percent.[41]

The historical evidence clearly indicates that economic growth is not dependent on savings. Savings seem to be more the result of growth, not its cause. That would also explain why at times extremely high levels of savings have not coincided with fast growth. In 1990 to 1997, for example, Japan had the highest level of savings among OECD countries, averaging about 17 percent of GDP, and in spite of that the economy expanded by just 2 percent per year.

Growth seems to be the result of manufacturing and for this sector to grow governments must give it support. The stronger the support, the faster its growth. Even in a situation where the level of savings is zero, governments can still supply an abundance of support by transferring a large share of the available resources to this sector. That is why growth was strong in the immediate postwar years in Japan, Germany, and Italy, and that is why growth was weak in Japan fifty years later. Since Tokyo provided few incentives to manufacturers during the 1990s, the large pool of savings ended up being of little use and the economy inevitably grew slowly.

The 1970s

The 1970s witnessed a considerable plunge in Britain's rate of economic growth. After having averaged almost 3 percent in the preceding decade, the figure fell to just 2 percent. Practically every single OECD country experienced a noticeable deceleration, but this could not console the British because once again practically all of these nations managed to grow faster. Worse still was that the rate was so slow that it considerably worsened the already high underemployment level and the already unpleasant rate of unemployment. After having averaged about 5 percent during the 1960s, the unemployment rate rose to almost 7 percent in the 1970s and underemployment affected an even larger share of the population than before.[42]

Much was debated about the causes of the deceleration and many concluded that it was basically the result of the new competition coming from East Asia. It was noticed that the rate of manufacturing output experienced a considerable deceleration and it was concluded that the rapid increase in East Asian manufacturing exports was the cause of such a situation.[43]

The fact that Britain's share of world manufacturing exports went from about 26 percent in 1950 to about 9 percent in 1975 was seen as a clear evidence of the above. The correlation, however, was weak because a large amount of related data did not add up. If East Asian competition had been the main cause of such a situation then the other European countries should have had the same experience. The fact that several did not clearly suggests that this variable was not essential. France, for example, managed to retain the exact same share of world manufacturing exports in 1975 as in 1950, and Germany actually saw a large increase, going from 7 percent to 20 percent.[44]

Also worth noting is that East Asia's capacity to export progressively increased after the end of World War II. It even experienced a major boost in the 1960s as Japan grew at an unprecedented pace, as South Korea and Singapore began to grow and export rapidly, and as Taiwan changed its import substitution strategy into one of export promotion. If Asian exports had been the cause of the deceleration of the 1970s, then the 1960s should have also witnessed a noticeable slowdown in the GDP figures of Western nations because there was actually a larger increase in exports during this decade than in the 1970s.

Many Western nations indeed experienced a deceleration in the 1960s, but several others did not. France, Spain, and the United States, for example, grew faster in this decade than in the 1950s. These nations also did not experience a deceleration of their rates of manufacturing production. Factory output actually accelerated. In other nations like Germany and Italy, factory output and GDP did decelerate but to a rate that was still much faster than that of Britain.

What all of this information ultimately reveals is that the advent of East Asian competition can not be even held marginally responsible for the deceleration of the GDP figures of Britain or of any other Western nation. Government support for manufacturing in the United States, Spain, and France increased during the 1960s, and this situation strongly suggests that the United Kingdom's problem in the 1970s as well as before was strictly of an endogenous nature. London's weak promotion of the sector seems the thesis best capable of explaining the deceleration of factory output, of GDP, and of exports during the 1970s and the constantly slower rates of the three in the preceding decades. It was like the very weak rates of growth that Britain had during the late nineteenth century and early twentieth century, even though East Asian competition then was nonexistent. However, then as in the 1970s, London's promotion of the sector was weak and rates of factory output were slow.[45]

Tied to this argument is the one stating that world demand grew more slowly during the 1970s and since the British economy was so opened to foreign trade, it was inevitable that it caused a drop in output. Such a position, however, cannot explain why Hong Kong, which was much more open to foreign trade, managed to grow by more than 9 percent annually. Since its foreign trade as a share of GDP was many times that of Britain, a drop in world demand should have affected it even more, but the fact is that it grew almost five times faster.[46]

This thesis also cannot explain why Singapore and South Korea, which were also more open to foreign trade, managed to grow not only about five times faster than Britain but also faster than in the 1960s. If the world decelerated its rate of consumption, then even if a fast output could have been maintained, it should have been less fast than formerly. The fact that it was not clearly demonstrates the fallacy of this argument.[47]

The case of the NICs very clearly demonstrated that growth is fundamentally an endogenously led phenomenon. It also suggested that the bottom line of growth resided in manufacturing because government support for this sector during the 1970s was very strong. While in Britain factory output averaged about 2 percent annually, in the NICs the rate was about 14 percent.[48]

Others argued that Britain's deceleration during the 1970s was the result of the oil shocks. Here again the evidence did not add up. The price of oil did increase considerably during the 1970s but in the 1980s it fell, and then in the 1990s it fell even more and the economy nonetheless continued to expand slowly during these last two decades. The idea that oil was not determinant nor even important for growth was further substantiated by the fact that in the late 1970s large oil deposits were discovered in the North Sea and Britain went from being a net importer to being a net exporter (since 1980). However, the economy grew only barely faster during the 1980s than in the preceding decade and during 1990 to 1997 it grew ever more slowly than in the 1970s.[49]

In the 1950s and 1960s, it had been argued that the Bretton Woods system of fixed exchange rates had instilled rigidities that inhibited the British economy from responding in a flexible way to the changes of the international economy. In the 1970s, the weak GDP figures continued to be blamed on that variable. However, in 1972 the pound was floated and allowed to respond fully to market forces. Balance-of-payments constraints began immediately to diminish, and with the advent of large oil revenues in the late 1970s the constraints diminished even more. The economy (in the 1970s), however, did not improve and during the 1980s GDP figures only barely improved.[50]

There were others who asserted that the rise in inflation had traumatized the economy and created downward pressure on output. Inflation indeed accelerated considerably during the 1970s. After having averaged about 4 percent annually in the 1960s, it averaged about 13 percent in the following decade. Much in history indicates that high inflation delivers negative effects on the economy, but also much suggests that its effects on growth are only marginal.[51]

In South Korea, for example, during the 1970s inflation grew much faster than in Britain, averaging about 18 percent per year, but the economy had no problem growing by more than 10 percent annually. In Brazil, inflation grew even faster, yet the economy also grew rapidly. Inflation averaged about 32 percent annually and the economy grew by more than 7 percent. By the mid-to-late 1990s, prices in Brazil, for the first time in more than half a century, were finally at low levels, yet economic growth was way below the rates of the 1970s. This situation goes a step further to substantiate the thesis that there is no causality linkage between inflation and growth.[52]

This also could not have been the cause of Britain's decelerated and slow rates of growth during the 1970s. Once again, an abundance of data points in the direction that it was London's reluctance to promote manufacturing that caused the weak performance. In South Korea, for example, support was very decisive and in Brazil it was less strong, but nonetheless very enthusiastic. Factory output averaged about 18 percent in South Korea and about 13 percent in Brazil during the 1970s. By the mid- to late 1990s, however, Brasilia's factory promotion efforts were weak and the sector grew by about 3 percent.[53]

In the 1950s, many economists argued that the rapid growth of wages was a main cause for the slower growth relative to most other OECD nations. In the 1960s, more economists and policy makers became convinced that this situation was severely hampering competitiveness, as wages continued to grow rapidly and the British economy continued to grow more slowly than in the other OECD countries. As the economy decelerated significantly in the 1970s, many concluded that it was the result of all of those years in which wages grew so fast. During this last decade there were numerous wage-limitation agreements between the government and labor in order to reverse the excesses of the past, but that did not bring an improvement in the economy.[54]

In the 1980s, there were even more wage-limitation measures and wages grew even at a slower pace, but the economy only barely accelerated its pace. In the 1990s, the policy of wage restraint continued in earnest but the economy once again performed poorly and actually experienced a deceleration. Not only did the case of Britain itself demonstrate the absence of a correlation between wages and growth, but numerous other nations also showed that there could not possibly be a causality linkage between the two. South Korean wages during the 1980s, for example, grew much faster than the economy and productivity, yet GDP grew by about 10 percent annually. During 1990 to 1997, wages grew at an even faster pace than the economy and productivity, yet the GDP figures were once again impressive, averaging more than 7 percent per year.[55]

The fact is that London supplied few incentives to manufacturing during the 1970s, 1980s, and 1990s and that factory output averaged just a little over 2 percent per year during these decades. The fact that Seoul continued to decisively promote the sector during the 1980s and 1990s and factory output averaged about 14 percent and 8 percent in these decades gives further credence to the thesis that the determinant variable explaining growth resides in manufacturing.

During the 1970s, Britain experienced a trade deficit and the fact that it also endured a deficit during the two preceding decades drove many to the conclusion that such imbalances were a major factor negatively affecting the economy. History, however, supplies a considerable number of examples in which fast growth was attained notwithstanding the existence of trade deficits. When fast growth was sustained for a long period of time, the deficits almost always disappeared. However, in the short term, they coexisted with the fast growth. History, therefore, suggests that Britain's slow growth was not the result of the trade deficits but that they were actually an effect of the slow growth.[56]

Something else caused the slow growth and as a result exports grew at a slower pace than imports. Several Western European countries endured trade deficits in the immediate years after the war but sometime later they overturned them and ran surpluses. The fact that their rates of GDP were much faster than those of Britain gives credence to this idea; and the fact that the NICs also experienced trade deficits in their initial years of fast growth and later ran surpluses substantiates such a position even more. Britain's continued slow growth in the 1980s and 1990s and the fact that trade deficits continued to prevail adds further weight to the idea that if there was a causality linkage then it was in the opposite direction.[57]

Once again, the evidence suggests that the slow growth and the trade deficits were fundamentally the doing of London's constant weak support of the sector. Since support for manufacturing in the other West European nations, which were the main trading partners of Britain, was much stronger, factory output grew much faster. Since manufactured goods accounted for the bulk of Britain's trade and its neighbors had more of these goods, then it was inevitable that the trade balance shifted in favor of these West European nations. Since factory output was slow in Britain, the economy also had to grow slowly.

Britain joined the European Economic Community in 1973 and many policy makers and economists argued that such a move prevented the country from responding in the most flexible way to changes in the world economy and in the domestic economy. Many, therefore, blamed the deceleration of the 1970s on this situation. However, if that argument had been valid, then the six founding members of the Coal and Steel Community in 1950 and the Treaty of Rome in 1957 should have attained slower growth than Britain in the 1950s and 1960s. The rules of these organizations restrained the capacity of each member government to dictate its own policies and yet, all of these countries grew much faster than Britain during the 1950s and 1960s.[58]

Being part of an organization inevitably limits a country's capacity to decide its policies freely, but that is not something that hinders growth. It could hinder growth if the organization's structure hampered the growth of the variable that is fundamentally responsible for increasing output, but the fact that Germany, France, Italy, and the Benelux countries grew much faster than Britain during the 1950s and 1960s suggests that that was not the case.

There was nothing in these organizations that acted to promote manufacturing. However, since the right to supply incentives to this sector rested exclusively within the jurisdiction of each country and the six founding members were strongly motivated to support it, factory output grew much faster than in Britain and the economy did likewise.

Others asserted that a major factor explaining the decelerated and slow growth of the 1970s in Britain was the significant increase in the size of the state. Government expenditure as a share of GDP went from about 32 percent in 1950 to about 45 percent in 1979. Worse still was that the area in the budget that saw the largest increases was social welfare.[59]

It was argued that government expenditure deprives the private sector of resources and that the private sector makes better use of them. Resources were not being maximized and, on top of that, by spending so much on purely consumption activities such as social welfare, London was inhibiting the possibilities for investment. The empirical evidence of numerous countries indicates that all of this was true, but the evidence also indicates that such a situation only affected growth marginally.[60]

The fact that several West European countries experienced a similar fast enlargement of government expenditure and social welfare expenditures during the same period, and nonetheless grew much faster than Britain, goes a long way to substantiate this idea. Several of them, such as the Scandinavian countries, Germany, and Holland, saw an enlargement of the state and of welfare in an even more pronounced way than Britain. The fact that they still attained much faster GDP figures solidifies the idea that even though negative, this situation played only a very small role in Britain's decelerated and slow growth of the 1970s.

The British government implemented numerous measures to accelerate the pace of the economy, but all of them failed. Cutting taxes was one of them, and increasing government expenditure was another. Increasing government expenditure for education, job training, infrastructure, housing, and other forms of construction was the favored macroeconomic stimulating instrument of this decade. The inability of these efforts to accelerate the economy and to create employment meant that by the late 1970s the political class had largely lost faith in it. Even the Labor Party, which was the one that most strongly endorsed these stimulation policies, seriously began to question its capacity to deliver positive results.[61]

The problem was tackled from numerous other perspectives and of all the efforts that were undertaken, not a single one assumed that the genesis of the problem resided in manufacturing. Such a possibility was simply incompatible with the paradigmatic vision of British economists and politicians. Their ideologically structured conceptual framework just did not allow for such an interpretation of reality. In consequence, support for manufacturing decreased during the 1970s.

By the late 1960s, the efforts to reconstruct the country from the ravages of World War II were seen by most policy makers as having largely completed their goal. All the factories that had been destroyed or paralyzed had been rebuilt and rehabilitated, and overall factory output was way above the peak level reached during the war. Incentives from this motivation were therefore largely eliminated. Balance-of-payments concerns were still high, due to the persistence of the trade deficits, but they were nonetheless lower than in the postwar years. As a result, export subsidies were reduced.

On top of that, national security concerns diminished noticeably as Washington and Moscow entered a process of détente in the early 1970s and tensions between NATO and the Warsaw Pact were reduced. As a result, London cut defense expenditures, and the share of GDP allocated for the production of weapons fell. There was also the fact that practically all Western European nations decided to considerably decrease their level of support for manufacturing during the 1970s. As a result, the pressure on London to imitate the policies of its neighbors of strong support was no longer there.

Under those circumstances, the overall level of fiscal, financial, and nonfinancial incentives supplied to the sector was considerably lower than in the 1960s, and that coincided with considerably slower rates of factory output and GDP. Factory output went from 3.3 percent to 2.0 percent and GDP went from 2.8 percent to 2.1 percent.[62]

The 1980s

In 1979, a new government intent on radically changing the status quo took office. The Thatcher government was convinced that for too long London had endorsed policies that were detrimental to the economy. The Tories, and in particular the new prime minister, were convinced that enterprise had been discouraged by too much state intervention in the economy. So they set to reduce the scope of government by abolishing restrictions on banking, international finance, and the labor market, by privatizing numerous state companies, by cutting government expenditures, and by endorsing sound monetary policies.[63]

The tax system was also reformed and rendered more homogeneous so as to reduce distortions. Subsidies were reduced for the same reason, exchange controls were lifted, labor unions were weakened, unemployment benefit (which was seen as a disincentive for work) was reduced, and all fields of the economy were deregulated.[64]

About one hundred companies were privatized during the 1980s, which accounted for almost all firms in the hands of the state. Even companies that had traditionally not been seen as capable of being privatized were handed over to entrepreneurs. Areas of the economy that were seen as a prerogative of the state because of their sensitive infrastructure nature, such as ports, airports, water companies, sewage firms, and electricity enterprises, were sold off.[65]

The unions were tamed from every perspective. Wage demands fell significantly and the workforce became considerably more flexible. Unionization as a share of the labor force decreased from 55 percent in 1980 to about 39 percent in 1990, and hours lost to strikes and other disturbances fell by about 70 percent during the same period.[66]

Partly as a result of these measures and partly as a result of the uncompromising position of the government on granting concessions to labor, as well as the government's efforts to promote a better work ethic, the average number of hours worked per week rose during the 1980s. Britain was among the few OECD countries that, instead of experiencing a decline in the number of hours worked per week, actually witnessed an increase.[67]

Convinced that the high inflation of the 1970s was the result of lax monetary policy, a tight policy on the matter was immediately instituted. The government was also convinced that its main role was to provide a sound currency by containing the growth of the money supply, and once that was attained, fast economic growth would inevitably derive from it. So the Central Bank was ordered to follow rigid monetary targets.[68]

The tight monetary policies did reduce inflation in Britain significantly, but not by as much as in several other OECD countries. After having averaged 14 percent annually, inflation fell to just 6 percent in the 1980s. In Japan, Germany, and France, however, it fell to less than 3 percent per year. The labor reforms delivered even less satisfactory results because unemployment and underemployment rose. After having averaged about 6.7 percent in the 1970s unemployment rose to about 8.3 percent in the following decade.[69]

The privatized companies did improve their efficiency, but that did little for the creation of employment and the acceleration of the economy. After the government privatized, liberalized, and deregulated, applied sound monetary policies, applied sound fiscal policies, diminished distortions, reduced inflation, and made the workforce more hardworking and flexible, the economy only barely accelerated. After having averaged about 2.1 percent during the 1970s GDP grew by about 2.4 percent in the 1980s.[70]

The economy indeed improved its efficiency as a result of the reforms, but it is evident that even their concomitant effect was not essential for the generation of growth. It is even questionable if the slight acceleration was the result of the improved efficiency. The case of other countries that embarked on similar reforms and in spite of them experienced a worsening of their GDP figures raises strong doubts about the capacity of these reforms to generate growth.

In the United States, similar reforms were undertaken and the economy actually grew more slowly than during the 1970s. After much liberalization, deregulation, sound monetary policies, a weakening of unions, moderate wage demands, and a more flexible workforce, the economy grew by just 2.4 percent in the 1980s after having grown by about 2.8 percent per year in the preceding decade.

In the United States, the reforms were not as wide-ranging as in Britain, but in New Zealand they were endorsed even more enthusiastically. In 1984, New Zealand embarked on a whole-scale liberalization program that lifted regulations on trade, finance, and services. It cut social welfare programs, institutionalized labor flexibility and restraint on wages. It reduced the budget deficit and government expenditure, it constitutionally engraved the independence of the central bank, and charged it with the sole mission of keeping inflation at very low levels. It removed most subsidies, privatized most state companies, reduced distortions on the tax system, and reduced regulations on business.[71]

However, notwithstanding this wholehearted effort to let market forces become considerably more operational, the economy decelerated. After having averaged about 2.4 percent annually during the 1970s, GDP averaged only 2.2 percent in the following decade. The decelerated rate of growth was so weak that it was of no use for the goal of creating employment. As in the United States and the UK, in New Zealand also unemployment and underemployment rose.[72]

Efficiency in the three countries improved noticeably and it is clear that the market reforms were responsible for that. However, the fact that the three attained weak GDP figures and that two of them even experienced a deceleration very clearly indicated that the reforms affected growth only in a marginal way. While countries such as these, which were doing practically everything to let market forces operate freely, attained slow rates of growth, others like South Korea which distorted market forces in a very encompassing way had the fastest rates of growth in the world. In the 1980s, the very regulated South Korean economy grew more than four times faster than these three countries, averaging almost 10 percent per year.

China, which was one of the most distorted economies in the world still operating under a centrally planned system, grew in real terms by more than 7 percent per year. It was evident that although positive, the Thatcher reforms were not stimulating the variable that was fundamentally responsible for growth. The case of South Korea and more particularly China strongly suggested that such a variable had such impressive growth-generating powers that even under the most encompassing market distortions it was still capable of superseding these barriers and delivering positive results.

The only variable that gave strong signs of possessing such characteristics was manufacturing. The efforts of London, Washington, and Auckland to promote this sector during the 1980s, for example, were weak, and factory output in the three countries averaged about 3 percent per year. In South Korea, however, the government supplied an abundance of incentives to the producers of manufactured goods and this sector's rate of output averaged about 14 percent per year. Peking's support of the sector was also very decisive and factory output averaged about 16 percent annually.

London's weak subsidization of the sector stemmed fundamentally from the entrenched noninterventionist beliefs of most policy makers and economists. History and logic suggests that eliminating all subsidies to agriculture, mining, fishing, forestry, construction, and all services is a very rational policy undertaking, but that doing the same for manufacturing is a major error.

However, notwithstanding the reluctance to promote factories and the Thatcher government's commitment to cut subsidies there too, a number of circumstances forced it to actually increase the support to this sector.

There was on the one hand the rise in cold war tensions in 1979 as the Soviets invaded Afghanistan, which drove the United States in the early 1980s to raise defense expenditures and to pressure its NATO partners to do likewise. Britain was not only the second most important member of this military alliance, it was also the one that showed the most interest in following the American line due to the staunch anticommunist position of the Thatcher government. In consequence, defense expenditures rose and armament output grew at a faster pace.

There was also the war against Argentina over the Falkland Islands in 1982 that further raised more the national security concerns of the country. Although Britain retook the islands, the dispute remained unresolved during the rest of the 1980s and London felt forced to spend a little more on weapons. Since support for civilian manufacturing remained largely the same as during the 1970s, the increased output of military manufacturing gave a small boost to the overall rate of factory output. After having grown by about 2.0 percent during the 1970s, factory output expanded by about 2.8 percent during the 1980s.

The Thatcher government wanted to eliminate the subsidies to civilian manufacturing, but it soon discovered that very strong forces impeded such a measure. The increased national security concerns, for example, reminded the government that the numerous civilian factory goods that were needed to make weapons had to be supplied in sufficient amounts and with high levels of technology. Otherwise, the capacity of the country to defend itself, project force, and intervene abroad could be jeopardized. This situation therefore forced London to maintain and even increase the subsidies to the civilian factory fields that were directly related to the production of weapons.

There was also pressure from the example of the other OECD countries. Many governments in Western Europe that exerted pressure on Britain because of their proximity also supplied larger incentives to this sector during the 1980s, making it hard for London not to imitate them at least to some extent. The fact that many of these incentives were used to fabricate goods that were urgently needed and that the private sector was not interested in making them left the Thatcher government with no alternative but to come up with the money. Environmental factory goods are the most illustrative of these goods, and for these, many countries in continental northwestern Europe decreed a considerably large number of incentives.[73]

Another factor discouraging the British government from eliminating or even cutting by much the subsidies to civilian manufacturing was the desire of the British to be at the top in world technology. Since technology is strongly and tightly linked to manufacturing, efforts to develop new technology or to maintain parity with other OECD countries on this matter translated into maintaining a considerable flow of subsidies to numerous factory fields classified as high-tech.

There was also a very long tradition of supporting the sector, dating to the late eighteenth century, and which had constantly correlated with relative strong rates of economic growth. Notwithstanding the entrenched noninterventionist beliefs, the political class could not completely ignore the fact that during the preceding two centuries the sector had been constantly supplied with much larger incentives than before and that economic growth had been constantly much faster than in all the centuries prior to the late eighteenth century. The historical evidence ultimately made most policy makers suspect that there was something about this sector that delivered positive results.

The 1990s

During the 1990s, the Thatcherite liberal reforms continued to move forward as the Tories remained in power and pushed such an agenda further. By the later part of the decade, the Labor Party took power and, in a surprising change of ideological orientation, they too enthusiastically endorsed the policies of liberalization, privatization, fiscal rectitude, and sound monetary policies. Once again, these policies gave clear signs of improving the efficiency of the economy, but the GDP figures once again failed to respond positively. The economy not only failed to accelerate, it even decelerated significantly. After having grown by about 2.4 percent annually during the 1980s, the GDP figures dropped to just 1.7 percent during 1990 to 1997.[74]

As deregulation of the labor market continued, the British economy became reputed in the 1990s for its flexible work practices and low wage demands. Its labor structure was the most liberal in Western Europe. Laws on sacking workers were the loosest, it had the least employee protection regulation, restrictions on the number of hours worked were nonexistent, and it had among the lowest labor costs. All of that, however, did little to reduce unemployment, which only barely decreased from an average of 8.3 percent in the 1980s to 7.7 percent in the 1990s, while underemployment rose significantly as low-paying part-time jobs multiplied and full-time jobs decreased.[75]

This situation inevitably ended up by affecting incomes. While in the

1980s real incomes only barely increased, in the 1990s they remained largely stagnant. On top of that, income distribution worsened. In the 1980s, income distribution had slightly worsened and during the 1990s it worsened even more. Such a situation was inevitable considering the weak rates of economic growth and the scarcity of full-time and high-paying jobs.[76]

Although the evidence strongly suggested that even the concomitant effect of all the policies endorsed by London during the 1990s was not correctly targeting the fundamental variable responsible for economic growth, the bulk of economists and policy makers refused to reach that conclusion. Most chose rather to blame the whole uncomfortable situation on the phenomenon of globalization. The end of communism in the early 1990s convinced practically all ex-centrally planned economies and developing countries that had formerly been closed or semiclosed to trade and investment to open up their economies.

As a result these countries began to attract a large number of British companies that decamped from the United Kingdom and transplanted production there. At the same time, in part as a result of the foreign direct investment (FDI) and in part as a result of the increased production from domestic companies, these countries began to export a growing amount of goods to the United Kingdom and the rest of the world. The share of manufacturing exports of developing countries from the world total went from 10 percent in 1980 to about 27 percent in 1997.[77]

It was, therefore, argued that the outflow of British firms to these countries and the avalanche of imports from these nations was undermining investment and bankrupting the companies that stayed in British soil. Since the ex-socialist and developing countries had much lower labor costs, most economists and policy makers arrived at the conclusion that little could be done to counter such a situation. Many concluded that Britain and the rest of OECD countries had to resign themselves to slower rates of growth and higher levels of unemployment or underemployment.[78]

Since a very large number of new countries in the 1990s were competing for OECD investment, most concluded that capital would be spread very thin as the bidding for a supposedly limited stock of capital rose. That supposedly left less available for investment in OECD countries and predisposed them therefore to endure decelerated rates of GDP. It was claimed that the only way to reverse such a situation was if OECD wages fell considerably and if working hours rose significantly. Since this possibility was unacceptable, some economists and policy makers therefore saw no other alternative but to raise trade barriers in order to slow down the avalanche of imports, and to tax more heavily the firms that moved operations abroad.[79]

However, since the majority of British economists and policy makers were convinced that free trade and free flows of capital were good for the economy, only a few of these market-distorting measures were undertaken. Most argued that the best way to confront this situation was by improving education, job training, and infrastructure and by liberalizing and deregulating the economy further. Since these measures were undertaken and the results were far from being the desired ones, most concluded that all that these measures could do was to slow down the exodus of companies and the inflow of imports.[80]

The bulk of OECD countries also experienced larger outflows of capital and larger inflows of imports than formerly in the 1990s, and there too the economy also suffered a deceleration. Since even Japan, which had been seen up to the 1980s as capable of growing rapidly in an indefinite way, suffered also a decelerated and slow rate of growth during 1990 to 1997, the idea about the inevitability of such a situation was reinforced.

There is much to suggest that London made a wise decision when it opted for maintaining a policy of free trade and free capital movements, but much also suggests that it was completely wrong when it concluded that slow rates of growth were inevitable. The fact that Britain attained GDP figures that were almost as slow during the 1970s, at a time when globalization was nonexistent, clearly indicates that some other variable played a much more important role. The fact that during the whole 1870 to 1910 period GDP figures were just as low as in the 1990s, even though at that time world trade and investment flows were almost exclusively limited to the future OECD countries, further reinforces the idea that globalization was not responsible for Britain's weak performance.

The fact that Ireland managed to grow very rapidly during the 1990s, despite being an OECD country, added still more force to the idea that globalization was not at fault. Ireland grew by about 6 percent annually from 1990 to 1997 after having grown by just 2 percent during the 1980s. Not only did the case of Ireland demonstrate that an OECD country could grow rapidly under a globalized world, but it also showed that rates could be massively accelerated. If Ireland, which had grown at the same pace as Britain in the 1980s, managed to impressively accelerate its economy during the 1990s, then it was clear that Britain's problems were all home-grown.[81]

Once again, the evidence strongly suggested that Britain's problems resided in the reluctance of the government to promote manufacturing in a decisive way. There was weak support of this sector in the 1980s and there was weak support again in the 1990s, while in Ireland there was weak support in the 1980s but strong support in the 1990s. In the 1990s, the state-sponsored Irish Industrial Development Agency offered abundant tax exemptions and tax reductions, as well as ample grants, soft loans, free land for the factory, and other nonfinancial incentives to domestic and foreign manufacturers. Taxes on this sector were the lowest in Western Europe and overall subsidies as a share of GDP were also the highest in the region. During the 1990s, Ireland attracted larger per capita amounts of FDI in manufacturing than any other country in the region. That coincided with the fastest rates of factory output and the fastest rates of economic growth in the whole of Western Europe.[82]

While manufacturing production in Ireland averaged only 3 percent in the 1980s, during 1990 to 1997 the figure jumped to almost 8 percent per year. In Britain, however, factory output remained weak. It was less than 3 percent in the 1980s and about 2 percent in the 1990s. The fact that London during the 1970s and the 1870 to 1910 period also supplied manufacturing with just a weak level of support and that factory output during those periods averaged about 2 percent per year, substantiates this idea even more. The fact that subsidies for the sector were few in most OECD countries during the 1990s (including Japan) and that rates of factory output among them oscillated between 1 and 3 percent, reinforces further the idea that globalization was not the cause of their troubles.

Singapore and Hong Kong in the 1990s were the most globalized economies in the world due to their absolute practice of free trade and free capital flows. The fact that they managed to grow very fast clearly shows that blaming globalization lacked scientific substance. The fact that these two economies experienced an even larger outflow of companies to low-cost countries during the 1990s as well as an avalanche of imports from formerly closed or semi-closed nations, and still managed to grow fast, cements this idea even more. It was evident that globalization could not be held even marginally responsible for Britain's weak performance. Hong Kong grew by about 5 percent annually during 1990 to 1997 and Singapore grew by about 8 percent.[83]

It is worth noting that even though these two economies were not members of the OECD club, they nonetheless had higher levels of per capita income than Britain and the bulk of OECD nations. Even though their labor costs and overall production costs were as high as the OECD average, they nonetheless succeeded in attracting massive amounts of FDI and of inducing very large domestic investments. That coincided with strong support for manufacturing and with fast rates of factory output. Of the two, the government of Singapore provided the most subsidies; and it was there that factory output grew the fastest and where the economy expanded the fastest. Manufacturing production grew by more than 5 percent per year in Hong Kong and in Singapore it by more than 9 percent.[84]

Although globalization was not even marginally responsible for Britain's weak performance, there was something about the way in which globalization came into being that did affect the level of government support for manufacturing that London decreed. With the end of the cold war, the national security concerns of numerous nations were significantly reduced. The dissolution of the Warsaw Pact, and the friendly attitude toward NATO that its former members took, drove London and its allies to reduce defense expenditures. After having averaged about 5 percent as a share of GDP during the 1980s, Britain's defense budget fell to a little more than 3 percent during 1990 to 1997.[85]

That delivered a considerable deceleration in the rate of armament output. Since most of the resources that were no longer allocated for the production of weapons were not utilized to promote civilian manufacturing, the overall rate of factory output had to decelerate, and along with it the rest of the economy. Not all of the cuts in defense ended in nonmanufacturing sectors, but the majority did, because there was practically nothing within London's vision of the world that stated that support for this sector was determinant or at least important for the generation of growth.

A similar situation occurred in most OECD countries. In consequence, none of them transferred much of the cuts in defense to civilian factories. During the 1990s, most OECD governments were fundamentally interested in liberalizing and deregulating their economies. The collapse of communism drove them to the conclusion that it was in that direction where policy efforts were most needed. The end of central planning was indeed a very positive event for the whole world because the countries under that system and the numerous others that because of their influence heavily regulated their economies no longer had to waste so many resources. It is just a tragedy that it took so long for so many governments to realize that those policies generated countless inefficiencies.

However, much suggests that liberalization is not the bottom line for growth. Without a policy of strong support for manufacturing, the benefits of liberalization are only barely felt, and at times these policies even have negative side effects. The financial system was perhaps the area where the efforts to let market forces operate freely was most enthusiastically pursued in Britain, most OECD countries and numerous other nations. Deregulation in this area, however, produced untold problems, ranging from excessive financial speculation to massive gyration in exchange rates, recessions, depressions, and even food riots.

Deregulation in Britain, for example, led to a steep recession in the early 1990s as speculation led to excessive flows of funds into real state, the stock exchange, and other quick-return activities. In Canada, Australia, and New Zealand, financial deregulation also led to the same excesses in the early 1990s and to a contraction of the economy.[86]

The United States also fell into recession in these years, but the speculative excesses occurred mostly during the 1980s when hundreds of banks went bankrupt as a result of irrational lending to real state projects. That was the result of the deregulation of the financial system that began in the late 1970s. More than a thousand banks had to be closed by the Federal Deposit Insurance Corporation during the 1980s and the government had to absorb many of the losses. Also worth noting is that economic growth in the United States during the 1980s was slower than before, clearly giving signs that deregulation of banks contributed, at best, only marginally to growth. In the 1990s, the economy decelerated further, giving more evidence that such a policy had very little influence on the generation of growth.[87]

Scandinavian countries such as Sweden, Norway, and Finland also deregulated their financial systems in the early 1990s and that coincided with a severe recession in these three countries during those years. Here also, once bankers were left to do as they pleased they began to allocate a large share of funds into quick-return activities such as the stock exchange, real estate, and currency dealings. Eventually, the bubble erupted and prices of stocks, real estate, and so on plunged, along with the rest of the economy. To make matters worse, the government ended up budgeting large amounts of funds in order to rescue the failed banks.[88]

Sweden was indeed the most enthusiastic deregulator of the three, and not just on financial matters. Stockholm privatized numerous enterprises, significantly downsized the welfare state, made the labor market much more flexible, and even became a member of the European Union in 1995. What it got in exchange for its efforts was not only the worst banking crisis since the 1920s, but also the worst GDP figures of the twentieth century. During 1990 to 1997, GDP in Sweden averaged only 0.6 percent annually. Unemployment, which until 1989 had regularly been at about 2 percent, jumped to 12 percent by 1994.[89]

Latin America also deregulated, privatized, liberalized, and made numerous reforms of the sort during the 1990s, and like OECD countries, it paid particular interest to letting market forces operate as freely as possible in the financial system. Mexico was among the most enthusiastic deregulators of banking and in 1995 a major crisis erupted as financial speculation resulted in a massive plunge in the exchange rate, which coincided with a massive economic contraction. Despite all the reforms that were undertaken during those years and despite being a developing country (which supposedly allows for the possibility of faster growth), the economy expanded by less than 3 percent annually during 1990 to 1997.

In Japan, financial deregulation during the 1980s and 1990s led to rampant speculation and reckless lending, in particular to real estate developers but also to currency dealers and stock traders. In the 1990s, property prices crashed, as did the stock exchange, and that coincided with stagnation. Worst of all was that the government had to allocate a large amount of funds to the rescue numerous banks and more resources were therefore diverted to activities that evidently contributed little to growth. During 1990 to 1997, GDP figures in Japan were on average slower than in any other decade of the twentieth century.[90]

Other countries in East Asia also endorsed financial deregulation during the 1990s and at first nothing happened, but in 1997 a major banking crisis erupted in several countries, including Thailand, Indonesia, South Korea, and the Philippines. The liberty that was given to bankers was once again utilized to try to make money at an extremely fast pace, and a abundance of loans were made to real estate developers, currency dealers, and stock exchange operators. Eventually, the currencies of these countries collapsed along with the stock exchange and the economy. The economy, which had been growing very fast, suddenly decelerated or even contracted.[91]

Dani Rodrik, a Harvard economist, analyzed almost one hundred countries from 1975 to 1989 in order to measure the impact of liberalization of capital flows on growth, investment, and inflation. The study revealed that the free mobility of capital has no significant influence on any of the three factors. There was no statistically significant evidence that capital controls reduce investment and growth or that they generate higher inflation.[92]

Aside from the recent cases of the 1990s, history also has abundant examples in which an absence of controls on capital markets led to speculative excesses that coincided with economic contractions. The crash of the New York Stock Exchange in 1929 and the massive depression that followed is just the most illustrative of these examples.

However, despite the numerous examples that seem to indicate that financial deregulation has serious negative effects on the economy, the compounded analysis of the bulk of world history does not point in that direction. History, in particular of the last two centuries in the West and East Asia, suggests that free movements of capital have a marginally positive effect on the economy as it allows for a more efficient use of this resource. By increasing competition as much as possible in banking, a nation can squeeze more output from the available capital.

Although history suggests that regulations on banks are generally harmful to the economy, it nonetheless indicates that there is one particular area of banking that needs to be decisively guided, not just to avoid speculative excesses but also so that they make a large contribution to the generation of economic growth. This area deals with the capacity of bankers to allocate the resources under their control. If bankers are left to do what they want, they will allocate most, if not all, of their assets to nonmanufacturing sectors, in particular construction and services, because of their low-investment and quick-return nature. Within these two sectors, there are several areas that are highly speculative in nature and which tend to create chaos in an economy when too much money flows into them.

If the government decrees that only a very small share of bank assets may flow into speculative areas, then it will succeed in avoiding financial crises. However, it will not generate fast economic growth because bankers will continue to allocate the bulk of their capital into nonspeculative services, construction, and primary activities. As long as they are free to do as they please, they will constantly abstain from making even a small share of their loans to manufacturing because of the inherent high-risk nature of this sector. Financing this sector requires gigantic investments relative to the other three, and in consequence carries a higher level of risk. Recouping an investment made in manufacturing also requires a much longer period of time than in the other sectors, and bankers are instinctively driven to want to see the fastest possible rate of return.

Throughout history, bankers in practically every corner of the world behaved in this way, and that coincided with perennially stagnant rates of manufacturing output and GDP. Only when governments supplied a strong support for manufacturing did bankers decide to allocate a small share of their resources to this sector. Because the government was significantly reducing the inherently high risks of this sector by supplying ample tax exemptions, grants, and nonfinancial incentives, bankers thought it worthwhile to lend a small share of their assets to factory producers. However, since they continued to be free in their decisions, they continued to allocate the large majority of their money to services, construction, and primary activities.

As a result much continued to flow into speculative areas, which is what took place during the East Asian crisis of the late 1990s. Strong state support for the sector in these nations delivered fast factory output and fast GDP, but since bankers were largely free to lend to real estate and other speculative activities, a very large share of banking assets flowed into those areas. It was like the situation of the 1870s in the United States and Germany when a strong factory promotion-effort by Washington and Berlin delivered fast factory output and fast GDP figures. At a given moment, however, the freedom of bankers drove too much money into speculative activities and a bubble erupted, causing the stock market to crash and a recession.

History, therefore, suggests that the best way to lead banks to be highly useful to society is by guiding them to lend fundamentally to manufacturing. This way, very fast rates of economic growth are attained and at the same time speculative crises are avoided. A system of incentives must be instituted so that at least the majority of their assets are lent to the manufacturing sector. The system should very heavily tax those banks that lend little to the sector, and banks that lend the bulk of their assets to manufacturing should be almost tax-exempt. Banks that under this system of tax incentives still refuse to allocate at least a large share of their assets to this sector should risk having their banking licenses revoked. With such a system, financial crises are highly unlikely to occur because too little capital is left for speculative activities.

The rest of the regulations that have been applied to banks throughout history are unnecessary and should be eliminated. With the exception of the criteria set by the Bank of International Settlements, the numerous existing regulations in most countries of the world should be totally phased out.

Manufacturing and Technology

Banks need to be guided to allocate the large majority of their assets to this sector because of the particular investment-intensive nature of it, which is fundamentally the result of the particular capacity of this sector to create and reproduce technology.

The evidence in support of such a vision of things is considerably strong and that would consistently explain why this sector throughout history has been so directly linked to growth. Since technology is the fundamental creator of wealth, it follows that whatever is responsible for the creation of technology is also responsible for growth.

The second half of the twentieth century in Britain as well as the preceding history of the country systematically showed a tight correlation between levels of support for manufacturing and rates of technology. They constantly fluctuated in unison. Innovation, for example, experienced a considerable deceleration in the 1950s compared with the preceding decade, coinciding with a considerable decrease in the factory promotion efforts of London. World War II forced the government to allocate a very large share of the nation's resources to factories for the production of weapons and related goods, and factory output averaged about 6 percent in the 1940s. Inventions, in particular during the first half of this decade, multiplied very rapidly. In the 1950s, support decreased significantly, factory output averaged about 3 percent per year, and inventions grew at a much slower pace.[93]

During the 1960s, support for the sector was almost at an identical level as in the 1950s, which coalesced with an almost identical rate of factory output and an almost identical rate of technological development. The compounded rate of inventions and the rate by which technology was imported remained largely the same. In the 1970s, patents were registered at a considerably slower pace and technology imports also grew less fast, coinciding with a considerable reduction in the subsidies that London supplied to the sector and a decelerated rate of factory output. In the 1980s, there was a slight increase in the pace of technology and there was also a higher level of incentives offered to factory producers. In the 1990s, the factory-promotion efforts of the state decreased and the pace of technology did likewise. Such a striking correlation almost certainly implies causation.

Giving further credence to the idea that such a correlation is not the result of chance is the way this sector constantly absorbed research and development (R&D) funds. During the second half of the twentieth century, the factories producing airplanes, motor vehicles, chemicals, mechanical engineering goods, electrical engineering goods, and other engineering goods absorbed about three-fourths of overall R&D spending. If nonengineering weapons, pharmaceuticals, and the remaining manufacturing fields are added, then the sector absorbed almost all of the R&D expenditures made by the state and the private sector.[94]

The fact that technology constantly found itself embodied in manufactured goods goes a step further to substantiate the belief that the correlation reflects causation. Practically all registered patents during the second half of the twentieth century found themselves embodied in a manufactured good. Technology imports during this period also came fundamentally in the form of factory goods such as machinery and equipment. Electric trains during the 1950s and gas-cooled nuclear reactors in the 1960s were among the most outstanding technologies of this period and they were also among the most outstanding factory goods.

Another situation that reflected the extremely investment-intensive nature of this sector and at the same time its technology-creating profile was the fact that during this whole half century the British government supplied a very large share of the overall R&D expenditures the country made. London paid for about half of all R&D expenditures.[95]

Since creating technology carries such high costs and the private sector constantly proved to be unwilling to absorb them, the government felt forced to pay for about half of the R&D costs in order to persuade private firms to pay for the rest. Also worth noting is that the bulk of the R&D funds provided by the government and the private sector flowed into the laboratories of manufacturing companies. A similar phenomenon occurred in practically all other OECD countries.

Also, over time in Britain, the government decreased its share in the total of R&D expenditures that the country made. In the 1950s and 1960s, London paid for about 60 percent of overall R&D costs and in the following three decades it paid for just 45 percent, which coincided with a faster rate of technological development in the first two decades of this period. In North

America, Western Europe, and Japan, the state also paid for a much larger share of overall R&D costs during the 1950s and 1960s than in the following decades. That situation was also accompanied by faster technological development and faster economic growth in the first decades.[96]

Since the creation of technology is such an extremely expensive effort, it was only natural that only when the government reduced its costs by absorbing part of them would capitalists consider that the risk was not too high and therefore decide to invest in the sector. It was, therefore, also natural that as the state absorbed a larger share of the R&D costs, the interest of the private sector in creating new technology rose proportionately, because their risks were diminished and their possibilities for profits were increased.

From the 1950s until the 1990s, Britain endured a trade deficit with the rest of the world in high technology goods, and it is also true that support for manufacturing in that period was much weaker than in numerous countries of the world, which were Britain's main trading partners. Since factory output grew much more slowly in Britain, it was inevitable that technology progressed more slowly than in these countries and that these countries had more high technology goods capable of being traded.

By the 1980s, however, it was continental Northwestern Europe's turn to begin a period of constant trade deficits with the rest of the world in high technology goods. That coincided with a much stronger factory promotion effort of several governments in the world (in particular those of East Asia) during practically the whole second half of the twentieth century. Since these countries were considerably lagging technologically behind Western Europe by the middle of the century, they took some time to catch up despite their impressive rates of factory output, but by the 1980s they were exporting massive amounts of high-tech goods.[97]

Further evidence substantiating the thesis that manufacturing is responsible for the creation and reproduction of technology is found in the way productivity behaved. During the second half of the twentieth century, levels of support for the sector correlated with rates of productivity. During the 1950s and 1960s, productivity in Britain grew much faster than in the following three decades, coinciding with a much stronger subsidization during the first two decades. Productivity averaged almost 3 percent per year and factory output grew by more than 3 percent. In the following decades (1970s, 1980s, and 1990s), manufacturing grew by more than 2 percent per year and productivity averaged a little more than 1 percent.[98]

In most OECD countries the same phenomenon was observed. In the 1950s and 1960s, at a time when abundant subsidies were supplied to the sector and factory output grew quickly, productivity grew also quickly. From then onward support decreased, rates of factory output decelerated, and rates of productivity did likewise.

Also worth noting is that during the whole period, rates of productivity among OECD countries were proportionate to the rates of factory output of each country. For example, during 1950 to 1997 Britain promoted the sector the least, and it also attained the slowest productivity figures. On the other hand, Japan attained the fastest rates of factory output and also had the fastest rates of productivity. Factory output averaged about 2.6 percent in Britain and productivity grew on average by about 1.7 percent annually. In Japan, factory output averaged about 7.5 percent and productivity grew by about 5.8 percent. The bulk of Western European governments which promoted the sector more decisively than London also attained faster average rates of productivity.[99]

Further evidence in support of this thesis was found in the fact that this sector attained the fastest rates of productivity. Services was the only other sector large enough to have had the capacity to propel the economy, and most economists increasingly argued that this sector was fundamental for the generation of growth in a developed economy. However, the fact is that the rates of productivity of services were systematically the lowest. While manufacturing productivity in Britain averaged about 2.3 percent during the second half of the twentieth century, that of services averaged only 0.7 percent per year. In the other OECD countries, the same phenomenon was observed. Manufacturing productivity constantly grew much faster than that of services.[100]

Technology is the fundamental factor responsible for the growth of productivity, and it was inevitable that the sector that was mainly responsible for the creation of technology was also the one that attained the fastest productivity rates. Since the pace at which technology develops is fundamentally determined by the level of support the government supplies to the manufacturing sector, it was also inevitable that the countries with the fastest rates of factory output attained as well the fastest rates of productivity.

The way in which investment behaved also gave much credence to the idea that this sector is fundamentally responsible for the development of technology. Levels of support for the sector constantly correlated with levels of investment. During the 1950s and 1960s, London's factory-promotion efforts were stronger than in the following three decades, which coalesced with higher levels of investment. While between 1950 to 1969, investment as a share of GDP averaged about 15 percent, in the following three decades the figure dropped to about 7 percent.[101]

In practically all other OECD countries a similar phenomenon took place. Support was stronger and rates of factory output were faster from 1950 to 1969 than in the following decades, and that was paralleled by higher levels of investment. For example, in Japan, which was the most decisive investor among OECD nations, investment as a share of GDP during the first two decades averaged about 30 percent while in the following three decades it averaged about 17 percent. Since manufacturing accounts for a disproportionately high share of overall investment, its movements were determinant for the whole investment scenario.

The behavior of savings was another piece of evidence suggesting that manufacturing was the bottom line for the development of technology, because it tended to correlate with the differing levels of support for this sector. During the 1950s and 1960s, Britain's level of savings from GDP averaged about 10 percent and during the following three decades it averaged about 6 percent, coinciding with rates of factory output that averaged about 3 percent during the first period and 2 percent during the second.[102]

In practically all other OECD countries, savings were much higher between 1950 to 1969 than in the following three decades, coalescing with much stronger levels of government support for the sector and much faster rates of factory output in the first period.

Since the possibility of increasing savings is larger when wealth is created at a fast pace, and since manufacturing is the prime creator of technology and therefore wealth, it was inevitable that when the sector was decisively promoted and wealth was created at a fast pace, savings rose. Since the bulk of OECD countries, including Britain, subsidized the sector more during the 1950s and 1960s than in the following decades, it was inevitable that wealth grew faster and savings were higher during these years.

The development of the nonmanufacturing sectors gave also much data substantiating the idea that factories were at the epicenter of technology. Agriculture and the other primary activities, for example, grew much faster during the decades when London promoted factory production in the most committed way (the 1950s and 1960s). During these decades, agriculture grew on average by about 2.6 percent per year while during the last three decades of the century it averaged only 0.5 percent.[103]

In the majority of the other OECD countries, the same phenomenon took place. Primary activities grew the fastest when subsidies were more abundant for factory production (1950s and 1960s). Precisely when resources were subtracted from the nonfactory sectors in order to transfer them to manufacturing, did agriculture, mining, forestry, and fishing grow the fastest. The average farm output of OECD countries during 1950 to 1969, for example, was about 1.5 percent annually while the average of the following three decades was about 0.9 percent.

The same occurred with construction and services. It was when resources were transferred from all the other sectors to manufacturing that construction and services grew the fastest. This phenomenon was observed in Britain and in practically all other OECD countries. Services, for example, grew on average by about 4.8 percent for all OECD countries between 1950 and 1969 and by just 2.9 percent during the following three decades.

What seems at first glance a paradox becomes clear the moment it no longer is assumed that all sectors are equal. If instead of assuming what mainstream economies believe in, it is assumed that manufacturing has a unique capacity to generate technology, then it becomes understandable why, when resources were transferred to the factories, agriculture, construction, and services grew faster than when these sectors were given a larger share of overall investment. A larger share of resources to manufacturing translated immediately into an accelerated pace of technological development, and as larger amounts of technology flowed into all of the other sectors, their output immediately increased.

That would explain why the productivity of agriculture, construction, and services was higher during the 1950s and 1960s than in the following decades in most OECD countries. Since the technology created by manufacturing grew at a faster pace than afterward, it flowed at a faster pace into primary activities, construction, and services. Since it is technology that is fundamentally responsible for the growth of productivity, productivity in these sectors had to grow faster.[104]

The large increase in the share of services in the GDP of developed countries during the second half of the twentieth century led many analysts to the conclusion that this was the most important sector for the generation of growth in a mature economy. In Britain, for example, the share of services went from about 35 percent in 1950 to about 66 percent by the late 1990s. In the other OECD countries, a similar shift took place.[105]

Such a vision of the world, however, is simply not compatible with the empirical evidence. It is just not logically credible to assert that services are determinant for growth considering the perennially low level of productivity that this sector has attained, not just in OECD countries but in practically all other nations of the world. Service productivity has been so low that it has almost constantly performed way below that of manufacturing as well as below that of primary activities and construction. By the late twentieth century, at the time when services in OECD countries had increased to an unprecedented share of GDP, their rates of productivity were almost approaching zero.[106]

In addition, as the share of services in GDP grew during 1950 to 1997, economic growth progressively decelerated. By the 1990s, services had attained an unprecedented size in OECD countries and the GDP figures of the bulk of these countries were at their lowest levels during the whole half century. The correlation actually went in the opposite direction and suggested that a growth in the share of services was detrimental to the economy.

Overall levels of productivity were also at their lowest in the 1990s, at the time when the share of services had reached a size larger than that experienced at any other moment in history. The evidence once again suggested that a larger share of services in GDP delivered lower overall levels of productivity.

In reality, such a causality linkage was absent, for the cases of Singapore and Hong Kong demonstrated that while having a share of services as large as that of OECD countries, it was still possible to attain very fast rates of economic growth and productivity. In the 1990s, these two economies attained very fast GDP figures and very fast rates of productivity while services accounted for more than two-thirds of the whole economy. The difference, however, between these two and the OECD nations was that the governments of Singapore and Hong Kong during the 1990s promoted manufacturing in a much more decisive way and attained much faster rates of factory output.[107]

The Singaporean government, which was the most enthusiastic promoter of factories of the two, succeeded in making the sector grow by about 9 percent annually, GDP by about 8 percent and overall productivity by 5 percent. The factory output of OECD countries during 1990 to 1997, on the other hand, averaged about 2.5 percent, GDP averaged about 2 percent, and productivity grew by about 1 percent annually. The evidence strongly suggested that the large size of the service sector was not the cause of the slow economic and productivity growth of OECD countries. The epicenter of the problem resided in the low level of support that the manufacturing sector received.[108]

During the second half of the twentieth century, Britain attained a rate of economic growth that was considerably below that of numerous other nations and much suggests that it could have grown at a much faster pace. History strongly indicates that very fast and sustained economic growth for a mature economy is easy to attain, and that the bottom line for such a feat is a government that endorses policies that are backed by rationality and not by ideology.

History and logic demonstrate that if London would supply as many incentives to private civilian manufacturing as it did during the early 1940s to military manufacturing, factory output would grow at a double-digit pace and the economy would do likewise, just as it did during those war years.

Policy Advice

The events that took place during the second half of the twentieth century in East Asia, the West, and Russia lead strongly to the conclusion that economic growth is a phenomenon that can be easily manipulated by government policy. The empirical evidence shows that growth is an endogenous phenomenon and that it is not dependent on variables such as culture, geography, climate, or population size. Nor is it dependent on economic and political events occurring outside a given nation. The evidence shows that these other variables have an effect, but only marginally. Growth is fundamentally determined by the policies that a government endorses.

The history of these nations demonstrates that growth is intimately tied to the manufacturing sector because of the almost unique capacity of this sector to create and reproduce technology. Since the generation of technology is the most resource-absorbing of efforts, this sector is inevitably predisposed to be extremely investment-intensive. As such, it predisposes the private sector not to invest in it. The inherent risk-averse nature of capitalists makes them prefer the other sectors because they demand much lower investments and because they offer a much faster rate of return. History shows that when things are left to take their natural course, investment in the sector will not materialize, and if it does not materialize, technology will not be generated. Under those circumstances, wealth is not created and economic stagnation prevails.

That is why the state plays a determinant role in the generation of growth because it alone has the capacity to reverse this natural structure of things. The state alone has the capacity to reduce the inherently high production costs of manufacturing because it has the power to offer incentives to the private sector and so overturn the high-risk nature of this sector. The more incentives it offers, the more it reduces the investment risk, and in consequence the more attractive manufacturing becomes for capitalists and entrepreneurs.

That is why the fundamental variable determining growth is the level of government support for the manufacturing sector. The stronger the level of government support, the faster the rate of factory output. The rate of this sector is the bottom line for determining the rate of economic growth. What matters is not the size of this sector as a share of GDP, but the average annual rate of growth.

However, the degree to which market forces are free to operate within an economy determines how much growth will be extracted per unit of manufacturing output. This is the second most important variable affecting growth. The higher the level of competition, the higher the levels of efficiency and, in consequence the more the extraction of wealth per unit of factory output. Governments must therefore promote the growth of this sector as much as possible while simultaneously liberalizing the economy as much as possible.

The empirical evidence indicates that the only rational way to promote manufacturing is by means of fiscal, financial, and nonfinancial incentives. The promotion efforts that history has witnessed by means of trade protection, state-owned factories, cartels, other price-fixing measures, and regulations of numerous sorts proved to deliver negative effects on the economy. These practices must be avoided. In order to attain the fastest possible rates of growth, fiscal, financial, and nonfinancial incentives must be provided simultaneously and at the highest possible level.

Fiscal Incentives

Included under the rubric of fiscal incentives are tax exemptions and tax reductions. Governments should therefore attempt to reduce the overall tax load on manufacturing companies as much as possible. There should not be an across-the-board reduction of taxes to all sectors. It must be limited to manufacturing and it must be across-the-board without making distinctions among the numerous factory fields. Policy makers must make sure that there is a large tax-difference between the other sectors and manufacturing. The larger the difference, the larger will be the shift of resources from the other sectors on to manufacturing. The ideal tax load that factories must be subjected to is zero, and not just for a limited period of time, as with the present tax concessions found in numerous industrial parks, special economic zones, and similar areas. The tax exemptions should be supplied for an indefinite amount of time.

Personal income taxes for the people who labor in manufacturing should also be much lower than in all the other sectors. Since this sector is so technology-intensive and the creation of technology demands such a vast mental effort, there is a need to attract the best brains to it. It is not just the scientists and technicians who work in the laboratories of manufacturing companies who deliver progress, but also the blue-collar workers. With the help of a very favorable taxation for people working in it, the wage gap with the other sectors can be considerably widened. In this way, the attractiveness of working in factories is enhanced.

Financial Incentives

Included under the rubric of financial incentives are grants and subsidized loans. In this category too, they must be supplied in the largest possible amounts. In order to supply grants, however, governments do not need to run budget deficits or increase overall government expenditure. What they must do is transfer as many resources as possible from the budget to the manufacturing sector. Resources regularly utilized to subsidize agriculture, the other primary activities, construction, and services must be reallocated and used to promote the growth of factories.

For example, resources regularly utilized to provide unemployment benefits, job training, and similar programs should be used to provide grants to manufacturers. These grants should cover part of the cost of constructing the factory, of acquiring the machinery and equipment, of training the workforce, and for research and development (R&D). By reallocating budgetary resources in such a way, a faster rate of economic growth will be attained, and that is the best mechanism for reducing unemployment and income inequalities. Throughout history, poverty, unemployment, underemployment, and income disparities always diminished the fastest when economic growth was strong, and fast growth is ultimately dependent on how much of a nation's total resources are allocated into factories.

History also shows that transferring resources from agriculture and the other primary activities onto manufacturing is actually the best mechanism to accelerate the rate of output of primary sector activities. Over the centuries, in East Asia, the West and Russia, primary production always grew the fastest when support for manufacturing was strong. The same is true of all forms of construction. When governments re-channeled resources from construction and the other sectors into the factories, housing, infrastructure, and other forms of construction made the most progress.

Social welfare services such as health and education account for a large share of the budget in numerous countries of the world. These resources need to be also transferred and be used for the promotion of factories, because it is precisely in this way that the health and the education of a nation is most improved. Over and over again, it was observed that the fastest progress on health and education took place when factory output grew at the fastest pace.

History unequivocally demonstrates that the fastest improvements in living conditions in all forms always occurred when the economy expanded at the fastest pace. Since the evidence also demonstrates that economic growth is fundamentally dependent on how much manufacturing is promoted, it follows that the best means to improve living conditions is by endorsing an enthusiastic factory promotion policy.

The government should supply grants across-the-board to all fields of manufacturing without being preferential to any particular field. It should also use grants to cover the largest possible share of the overall production costs of manufacturers.

The financial system shall provide the remaining and majority share of the financing. The financial system shall be entirely in private hands. History has witnessed numerous experiences in which governments created banks particularly for the purpose of channeling funds into manufacturing, and on other occasions banks were nationalized for that same purpose. The evidence conclusively demonstrates that these were policy errors.

The government shall guide private banks into allocating the largest possible share of their financial assets into manufacturing. A system of incentives must be instituted so that at least the majority of their assets is lent to the manufacturing sector. The system should heavily tax those banks that lend only a very small share of assets to the sector, and banks that lend the bulk of their assets to manufacturing should be almost tax exempt. Banks that under this system still refuse to allocate at least a large share of their assets to this sector should risk having their banking licenses revoked. Banks should also be encouraged to lend to manufacturing at lower interest rates than to the other sectors and with longer periods of maturity.

With such a system, financial crises are also highly unlikely to occur because too little capital is left for speculative activities. Since the majority of capital within an economy would be utilized for the most long-term and stable of investments, the amount of capital left for real estate speculation, stock market dealings and currency transactions is massively reduced. That is why governments should guide banks into lending the largest possible share of their assets to manufacturing. The more they lend to it, the less there is capital for speculation.

The more they lend to factories also, the faster the economic growth, and that immediately transforms into higher profit margins for the banks. History demonstrates that bank failures were most common when policy makers endorsed policies that did not support manufacturing or that supported it in a weak way.

Throughout history, banks always performed better when the economy grew rapidly, and how fast the economy grows depends fundamentally on how many resources are allocated into manufacturing. By guiding banks in such a direction, the government is guaranteeing the profitability of banks and using them simultaneously to increase the overall wealth of the nation.

Numerous bank regulations that at present exist in numerous nations of the world should be phased out with the exception of those dictated by the Bank of International Settlements (BIS). The evidence indicates that the bulk of regulations decrease financial efficiency.

Nonfinancial Incentives

Under the rubric of nonfinancial incentives are several mechanisms for reducing costs of production even more. One is the supply of free or subsidized land for the factory. Another is the construction of the adjacent infrastructure of a production facility. Improving the overall infrastructure of the country in the most manufacturing-oriented way is another of this type of incentive, as is emphasizing the educational fields that are directly linked to manufacturing. Providing electricity, water, and other utilities at below-market prices is also a nonfinancial means to assist manufacturers.

In most countries, the state is the largest owner of land. In most nations, the combined ownership of federal, state, and local governments is about half of the total territory. Governments should supply free of any cost the land needed for the creation of any factory. There is no logic in even demanding a fraction of the value of the land. Policy makers only must make sure that the land is effectively used for the direct production of factory goods.

The state should also cover all the costs of all the infrastructure linkages that a factory needs. That would particularly apply when the factory is located outside an urban zone or, even within a city, its precise location is not fully linked with the rest of the city.

Policy makers should also concentrate on developing the type of infrastructure that is manufacturing-intensive. For example, instead of making highways and roads, which are construction-intensive, they should concentrate on railroads. If the trains are fabricated domestically, then the infrastructure effort becomes highly manufacturing-intensive. In such a way, infrastructure improves and the overall rate of manufacturing output accelerates, and thus contributes to making the economy grow faster.

Governments should concentrate on promoting as much as possible the fields of education that are tightly bonded to the sector. These fields are basically engineering, natural sciences, and mathematics. Resources within the educational budget must be radically restructured. A vast transfer of resources must take place so that these manufacturing-related fields get the bulk of the funds. Private schools and universities must be stimulated with taxes so that they give priority to these fields.

Utilities such as electricity, water, and the like should be provided to factories at below-market prices. The private-sector providers of utilities should be supplied with tax incentives so that they provide these services to factories at subsidized rates.

Governments should also offer similar incentives to foreign manufacturing capitalists. Foreign direct investment (FDI) should be promoted as much as possible. FDI should be allowed in all economic sectors, but it should be treated favorably only in manufacturing. Foreign firms in primary activities, construction, and services should be treated in exactly the same way as domestic firms. In all of these sectors, however, nationals and foreigners should be equally given less incentives than to those that invest in manufacturing. There should be no restrictions on foreign banks, but just like the national banks, they should also be guided into lending the large majority of their capital to manufacturing.

Another tool that governments regularly utilized throughout history consisted of guaranteeing private manufacturers the purchase of their output, and at prices that ensured a profit. This, however, proved to have more negative effects than positive ones. It has been frequently utilized for military procurement, and abuses were frequent. It was mostly used when a new technology appeared and production costs were so high that private-sector consumption, and more particularly mass consumption, was impossible. The evidence from the West and East Asia suggests that much better results can be attained when the government concentrates on supplying as many grants as possible for R&D so that a cost-effective technology is developed sooner.

Although the most rational approach is to promote all manufacturing fields with the exact same level of incentives, there is one field that seems to justify an additional level of support. This field is environmental manufacturing. For the factory producers of goods that reduce pollution, improve the survivability of plant and animal species, restore ecosystems, reduce climate change, or redress the damage of the ozone layer, governments should supply even more fiscal, financial, and nonfinancial incentives.

The empirical evidence from the second half of the twentieth century demonstrates that very fast and sustained rates of factory output do not lead, as many people presume, to environmental degradation. If environmental manufacturing is heavily promoted, then the technology to even redress the worst ecological problems will be uncovered and made available in a cost-effective way. Rationally managed, manufacturing can be rapidly promoted in such a way that from the start it presents no threat to the environment. The evidence suggests that with the right technology it is even possible to improve the living conditions of plants and animals beyond any present or past situation. How fast technology grows depends fundamentally on how much support the manufacturing sector receives.

History also shows that impressive economic growth can easily coexist with fiscal rectitude, low inflation, and income equality. It carries no additional cost for society to grow at a very fast pace. Quite the contrary, the evidence demonstrates that when governments promote manufacturing decisively, the chances of reducing or even eliminating public-sector debt are higher. Under such a policy, balanced budgets and even surpluses are more probable.

The impressive growth of productivity that fast factory growth delivers curtails inflation, and thus double-digit GDP growth has no problem coexisting with extremely low levels of inflation.

The data from numerous nations during the past decades strongly suggests that a decisive factory promotion effort is the most capable of reducing income disparities. This sector has systematically been shown to pay higher wages than any other sector. As a result, the increased share of the population that ends up working in factories when a policy of this sort gets endorsed inevitably enlarges the share of the workforce earning the highest possible wages. Under those circumstances, the income of the working class compared with the rest of society rises, and income inequalities decrease.

A rational manufacturist policy can only operate adequately when governments also endorse the policies that mainstream economics has elaborated over the years. The economy must be exclusively in the hands of the private sector. Privatization programs are very much needed and they should go beyond the traditional domains. Areas denominated as natural monopolies should be transferred as fast as possible to capitalists and they should be simultaneously broken up so that there is ample competition. Free trade in its most absolute way should be practiced, with no tariff or nontariff barriers. All fields and areas of the economy should also be fully liberalized and deregulated. All collusive and price-distorting practices must be phased out.

Even though a manufacturist policy considerably assists in the maintenance of fiscal rectitude and low inflation, governments must make additional efforts on this matter. They should make decisive efforts to always have a positive balance between their revenue and their expenses. They should attempt to constantly run at least balanced budgets and not just at the federal level, but also at the state and local level.

They should also constitutionally engrave the independence of the central bank and its sole role of keeping inflation at constantly low levels.

Manufacturism is a policy that has the capacity to provide to least-developed countries, middle-income nations, transition economies, and developed countries the means to vastly accelerate their rates of economic growth. All nations can attain simultaneously fast and sustained economic growth. History shows that economic growth is not a zero-sum game in which the wealth of a nation is achieved at the expense of another.

That does not mean that nations will inevitably reach similar levels of income. However, this policy can largely guarantee that least developing countries will rapidly overcome poverty, that middle income nations will attain much higher levels of development, that transition economies will rapidly see the light at the end of the tunnel, and that developed countries will eliminate unemployment and underemployment.

Appendix

Data Tables: Description

The data used to prepare the historical macroeconomic charts that follow came from numerous sources. The tables concentrate fundamentally on showing the long- medium-, and short-term development of the main macroeconomic indicators of several nations. These nations were selected because of their fast rates of economic growth through time and their relevance in world affairs. These nations are: Great Britain, the United States of America, Germany, Russia, Japan, China, South Korea, Taiwan, Hong Kong, and Singapore.

The main macroeconomic statistics for the purposes of the thesis here presented involve gross domestic product, manufacturing, and agriculture. However, data on population, inflation, unemployment, and exports is also compiled, although in a less time-consistent way. Tables are structured chronologically and there is one for every nation. Each nation's table was constructed by stitching together data extracted from numerous sources. At times, the data was elaborated from nonstatistical information. Educated inferences from historical accounts were made to statistically explain a few periods in which no precise numerical figures were found.

The rationale for concentrating on manufacturing, GDP, and agriculture stems from the central thesis presented in this book, which contends that government support for manufacturing is the fundamental determinant of economic growth. Only a long-term cross-country correlation between manufacturing and GDP can give that theory a chance of demonstrating a causal relationship. Throughout history agriculture has been the variable most frequently cited as being fundamentally responsible for growth, and presenting the long-term development of this domain is therefore necessary in order to demonstrate that a correlation is absent and therefore a causality linkage.

The tables are an effort to display the variables that have most frequently been presented as determinant for the attainment of growth. All figures are presented on an average-annual-rate basis in order to make them comparable. For simplification, manufacturing is abbreviated as *Man,* gross domestic product as *GDP,* agriculture as *Agri,* population as *Pop,* inflation as *Inf,* unemployment as *Unem,* and exports as Ex.

For countries such as Russia and China there was a need not just to stitch and infer in order to structure chronologically consistent statistics. Because during a large part of the twentieth century, these countries distorted market forces in a very extensive way, the official figures were not representative of reality. They were not comparable with the nations where the highest levels of efficiency were attained. That is why there was a need to significantly reconstruct the official figures so that they could express real economic growth.

Tables were reconstructed in proportion to the differing levels of market distortion that were experienced. During the Stalinist and Maoist years, for example, the statistics were shrunken the most. The discounting was so large that it was decided to present the discounted and the official GDP figures separately. The official statistics are presented in brackets. Only the GDP figures were reconstructed. Rates of output of manufacturing, agriculture, and all the other variables were left as the official figures presented them.

Table 1

Great Britain

	Man	GDP	Agri	Pop	Inf	Unem	Ex
18th century	1.1	0.8	0.5	0.6	0.9		
19th century	2.7	2.3	1.2	1.4	0.1		
20th century	3.0	2.5	1.4	0.5	4.0		
1700-59	0.7	0.7	0.6				
1760-79	1.3	0.6	0.1				
1780-99	2.0	1.4	0.8				
1800-49	2.9	2.5			0.2		
1850-99	2.3	2.0			0.0		
1900-49	3.3	2.7			1.8		
1950-97	2.6	2.3			6.3	6.5	
1800s	2.2	1.9					
1810s	2.8	2.3					
1820s	3.4	2.9					
1830s	3.3	2.8					
1840s	3.1	2.6					
1850s	2.8	2.3					
1860s	2.5	2.2					
1870s	2.2	1.9					
1880s	2.0	1.8					
1890s	2.1	1.8					
1900s	1.7	1.3			0.0		
1910s	3.3	2.6			0.1		
1920s	2.5	2.0			1.9	12.1	
1930s	3.4	2.8			2.1	15.4	
1940s	5.8	4.6	5.0		5.0	3.6	
1950s	3.1	2.7	2.7		3.0	4.7	
1960s	3.3	2.8	2.4		4.1	5.1	4.8
1970s	2.0	2.1	1.0		14.2	6.7	8.2
1980s	2.9	2.4			6.1	8.3	3.0
1990–97	1.9	1.7			4.2	7.7	2.1

Note: All figures are presented on an average annual percentage basis.

Table 2

United States

	Man	GDP	Agri	Pop	Inf	Unem	Ex
18 century	0.8	0.6	0.5	5.5			
19th century	4.5	3.7	1.8	5.1			
20th century	3.6	3.1	1.5	1.3	6.5		
1800–49	2.8	2.3	1.1	5.9			
1850–99	6.1	5.1	2.4	4.2			
1900–49	3.7	3.1	1.3	1.4		7.6	
1950–97	3.5	3.0	1.5	1.2	3.9	5.4	
1800–29	2.4	1.9	0.9				
1830s	3.2	2.7	1.4				
1840s	3.5	3.1	1.5				
1850s	4.0	3.6	1.8				
1860s	4.8	3.8	2.4				
1870s	6.1	5.0	2.7				
1880s	8.3	7.1	2.8				
1890s	7.5	6.2	2.4				
1900s	5.3	4.6	2.1			4.0	
1910	2.8	2.3	1.4			5.2	
1920s	4.0	3.5	0.8			5.1	
1930s	0.8	0.6	0.1			18.0	
1940s	5.4	4.4	2.3			6.1	
1950s	4.2	3.8	2.0	1.8	2.3	4.3	
1960s	5.3	4.2	0.3	1.3	2.8	4.0	6.1
1970s	3.1	2.8	0.9	1.0	6.9	6.3	6.9
1980s	2.8	2.4	3.2	1.0	4.0	6.5	3.4
1990-97	2.6	2.3	1.1	1.0	3.3	6.0	5.8

Table 3

Germany

	Man	GDP	Agri	Pop	Inf	Unem	Ex
18th century	0.7	0.5	0.4	0.2			
19th century	2.8	2.3	1.2	1.2			
20th century	4.6	3.6	2.0	0.4			
1800-49	1.2	0.9	0.6	0.7			
1850-99	4.4	3.7	2.1	1.7			
1900-49	4.7	3.5	1.9	0.3			
1950-97	4.6	3.8	2.1	0.5	3.3	5.5	7.5 (western Germany)
1800-39	1.1	0.8	0.5				
1840s	1.7	1.3	0.8				
1850s	2.4	2.0	1.2				
1860s	2.8	2.3	1.4				
1870s	4.2	3.6	2.0				
1880s	6.5	5.4	3.0				
1890s	6.2	5.1	2.8				
1900s	4.5	3.8	2.2				
1910s	1.9	1.0	0.3				
1920s	5.0	4.1	2.7				
1930s	6.8	5.0	2.4				
1940s	5.1	3.5	2.3				
1950s	9.3	8.0	4.0	1.2	2.2	8.1	15.0 (western Germany)
1960s	5.7	4.6	1.7	0.9	2.7	1.2	10.1 (western Germany)
1970s	3.2	2.6	1.5	0.1	5.3	3.0	6.0 (western Germany)
1980s	2.4	2.0	1.6	0.0	2.7	6.1	4.4 (western Germany)
1990-97	2.3	2.0	1.2	0.4	3.1	9.5	2.2 (western Germany)

German Democratic Republic (Eastern Germany)

	Man	GDP	Agri
1950s	12.2	5.1(10)	3.6
1960s	5.1	2.0(4)	2.0
1970s	6.3	2.4(5)	2.1
1980s	3.9	1.7(3)	1.8
1990-97	1.4	2.6	

Note: Two sets of GDP figures appear for the GDR (Eastern Germany). The official statistics supplied by the communist authorities for the years 1950-1989 are the ones in parenthesis. The ones to the left are the ones resulting from an adjustment effort intended to make them compatible with those of efficient capitalist economies. The ones to the left express real economic growth.

Table 4

Russia

	Man	GDP	Agri	Pop	Inf	Unem	Ex
18th century	0.8	0.5	0.3	1.1			
19th century	2.1	1.6	1.0	1.2			
20th century	6.7	2.4	1.9	0.9			
1800-49	0.4	0.3	0.2	0.2			
1850-99	3.9	3.0	1.6	2.2			
1900-49	9.0	3.1	2.0	0.8			
1950-97	4.5	1.6	1.8	1.0			
1800-39	0.3	0.2	0.2				
1840s	0.5	0.4	0.3				
1850s	0.7	0.5	0.4				
1860s	2.8	2.0	1.2				
1870s	3.2	2.5	1.4				
1880s	5.0	3.8	2.3				
1890s	8.0	6.1	2.8				
1900s	3.0	2.2	2.0				
1910s	–3.5	–2.6	–3.0				
1920s	21.1	8.0(16)	5.9				
1930s	17.2	5.8(14)	0.1				
1940s	7.1	2.2(5)	5.1				
1950s	12.2	4.1(9)	4.8				
1960s	9.1	3.0(7)	2.9				
1970s	6.2	2.3(4)	2.0				
1980s	4.3	1.4(2)	1.1				
1990-97	–9.3	–3.0(–6)	–1.2		502.1		

Notes: Two sets of GDP figures for the period 1920-1997 are presented. The ones in parenthesis are the official ones. The ones to their left are the result of a reconstruction effort intended on making them consistent with those attained by the market economies with the highest levels of efficiency.

Table 5

Japan

	Man	GDP	Agri	Pop	Inf	Unem	Ex
18th century	0.3	0.2	0.1	0.0			
19th century	1.8	1.3	0.5	0.3			
20th century	6.9	5.4	1.4	1.1			
1800-49	0.6	0.4	0.3	0.0			
1850-99	3.0	2.4	1.3	0.7			
1900-49	6.2	4.5	1.5	1.3			
1950-97	7.5	6.3	1.9	0.9	3.8	2.2	10.4
1850s	1.2	0.9	0.6				
1860s	1.7	1.4	0.8				
1870s	2.5	2.0	1.1				
1880s	3.5	3.0	1.7				
1890s	6.0	4.7	2.1				
1900s	6.8	5.2	2.6				
1910s	8.1	6.3	2.9				
1920s	3.1	2.4	1.8				
1930s	8.5	5.4	2.5				
1940s	4.3	3.0	–1.7				
1950s	11.1	9.1	3.8	1.3	3.1	3.6	14.2
1960s	13.4	11.3	4.0	1.0	4.9	0.9	17.0
1970s	6.2	5.2	1.1	1.1	8.2	1.7	9.1
1980s	5.3	4.0	0.4	0.6	1.3	2.2	7.4
1990-97	2.4	2.1	0.4	0.5	1.5	2.8	4.2

Table 6

China

	Man	GDP	Agri	Pop	Inf	Unem	Ex
18th century	0.7	0.5	0.4	0.8			
19th century	0.2	0.1	0.1	0.3			
20th century	7.3	2.7	2.2	1.0			
1800–49	0.3	0.2	0.2	0.6			
1850–99	0.0	0.0	0.0	0.0			
1900–49	0.6	0.5	0.4	0.3			
1950–97	14.1	4.9(9)	4.0	1.7			9.1
1900s	0.4	0.3	0.2	0.2			
1910s	0.7	0.5	0.3	0.3			
1920s	1.2	0.9	0.6	0.4			
1930s	2.9	2.1	1.4	0.6			
1940s	–2.3	–1.4	–0.5	0.0			
1950s	17.0	4.6(13)	5.0	2.3			7.9
1960s	10.1	2.2(5)	1.5	1.9			3.1
1970s	10.8	3.0(6)	3.2	1.9			7.0
1980s	16.2	7.1(10)	6.3	1.4	5.8		12.2
1990–97	17.9	8.0(11)	4.2	1.2	8.9		16.1

Note: Two sets of GDP figures for the period 1950–1997 are presented. The ones in parenthesis are the official statistics delivered by the Chinese government. The ones to their left are an attempt to express the rate by which "real wealth" was created during those years, which is why they are much lower.

Table 7

Hong Kong

	Man	GDP	Agri	Pop	Inf	Unem	Ex
19th century	0.9	0.7					
20th century	7.7	6.2					
1800–49	0.4	0.3					
1850–99	1.4	1.1					
1900–49	4.0	3.5					
1950–97	10.8	8.8		2.2			13.3
1900s	2.8	2.1					
1910s	4.3	3.6					
1920s	3.3	2.8					
1930s	4.9	4.2					
1940s	5.7	4.8					
1950s	15.8	12.1		2.8			14.0
1960s	14.4	10.2		2.5	2.4		12.7
1970s	10.6	9.4	–11.0	2.6	7.9		8.3
1980s	8.1	7.5		1.5	7.1	2.7	16.1
1990–97	5.2	5.0		1.5	8.2	2.4	15.3

Table 8

Singapore

	Man	GDP	Agri	Pop	Inf	Unem	Ex
19th century	1.9	1.3					
20th century	7.1	5.5					
1800–49	1.2	0.8					
1850–99	2.5	1.8					
1900–49	4.6	3.4					3.1
1950–97	9.5	7.6			2.4	7.6	9.3
1900s	4.6	3.5					
1910s	4.9	3.6					
1920s	5.1	3.9					
1930s	3.7	2.5					
1940s	4.8	3.5					
1950s	6.1	4.3				16.2	3.1
1960s	12.0	8.8	5.0	2.4	1.1	10.1	4.2
1970s	11.8	9.2	1.7	1.4	5.7	5.8	11.0
1980s	8.5	7.2	–5.7	1.2	1.7	4.0	11.9
1990–97	9.1	8.2	–1.3	2.0	2.2	1.9	16.3

Table 9

South Korea

	Man	GDP	Agri	Pop	Inf	Unem	Ex
19th century	0.5	0.3	0.2				
20th century	6.6	4.4	1.9				
1900–49	2.1	1.4	0.9				
1950–97	11.1	7.4	2.9		16.4		
1900s	0.5	0.3					
1910s	2.5	1.6					
1920s	3.1	2.0					
1930s	4.3	3.1					
1940s	0.2	0.1					
1950s	1.1	0.7	0.5		36.3		
1960s	15.0	8.4	4.2	2.3	15.1		34.1
1970s	17.8	10.3	4.8	1.9	18.2		25.7
1980s	13.6	9.7	3.3	1.2	5.0		14.2
1990–97	8.2	7.4	1.5	0.9	7.7	2.6	7.4

Note: All figures from 1800 to 1945 are for the whole Korean peninsula. From 1950 to 1997, the figures express only the economic events occurring in South Korea.

Table 10

Taiwan

	Man	GDP	Agri	Pop	Inf	Unem	Ex
19th century	0.2	0.1	0.1				
20th century	6.9	4.9	2.3				
1900–49	2.0	1.4	1.0				
1950–97	11.8	8.4	3.6	2.3	8.4		16.4
1900s	1.0	0.7					
1910s	2.6	1.8					
1920s	3.7	2.3					
1930s	4.4	3.4					
1940s	–1.8	–1.0					
1950s	12.2	7.8		3.5	20.3		6.1
1960s	14.0	10.1		3.0	2.4		25.3
1970s	13.9	9.6		2.1	10.2		30.2
1980s	11.1	8.1		1.6	5.1		13.4
1990–97	8.0	6.4		1.1	3.6		7.1

Notes

Chapter 1

1. "The New Power in Asia," *Fortune,* 31 October 1994, 46.
2. "The Magnificent Eight," *Far Eastern Economic Review,* 22 July 1993, 79.
3. "Free Markets," *Economist,* 11 February 1995, 124.
4. *The Cambridge History of Japan,* vol. 6 (Cambridge: Cambridge University Press, 1988), 19.
5. G.C. Allen, *A Short Economic History of Modern Japan* (London: George Allen and Unwin, 1972), 173.
6. *The Cambridge History of Japan,* vol. 6, 29, 15.
7. R. Dore, *Structural Adjustment in Japan, 1970-82* (Geneva: International Labor Organization, 1986), 16.
8. "The Japanese Economy," *Economist,* 6 March 1993.
9. Shigeto Tsuru, *Japan's Capitalism* (Cambridge: Cambridge University Press, 1993), 108-110.
10. Ibid., 104, 105, 115.
11. Dore, *Structural Adjustment,* 11-16.
12. *Le Japon d'Aujourd'hui* (Tokyo: Minisètre des Affaires Etrangères, 1971), 37.
13. Michio Morishima, *Why Has Japan Succeeded?* (Cambridge: Cambridge University Press, 1982), 158.
14. Erza Vogel, *Japan as No. 1* (Cambridge: Harvard University Press, 1979), 3-11.
15. J. Bremond, C. Chalaye, and M. Loeb, L*'économie du Japon* (Paris: Hatier, 1987), 12.
16. Salomon Levine and Hisashi Kawada, *Human Resources in Japanese Industrial Development* (Princeton: Princeton University Press, 1980), 35.
17. Morishima, *Why Has Japan Succeeded?* 176-178.
18. *Le Japon d'Aujourd'hui*, 62.
19. Morishima, *Why Has Japan Succeeded?* 189.
20. Allen, *A Short Economic History,* 174-176, 245.
21. *World Development Report 1981* (Washington, DC: World Bank), 137.
22. "Japan's Economy and Japan-U.S. Trade," *The Japan Times,* 1982, 36-38.

23. *Le Japon d'Aujourd'hui*, 66-69.
24. Shigeto Tsuru, *Japan's Capitalism,* 129.
25. "America's Edge in Wages," *BusinessWeek International,* 30 June 1997, 11.
26. "Inflation," *Economist,* 22 March 1997, 142.
27. Tsuru, *Japan's Capitalism,* 45.
28. Heizo Takenaka, *Contemporary Japanese Economy and Economic Policy* (Ann Arbor: University of Michigan Press, 1991), 5.
29. Dore, *Structural Adjustment,* 2.
30. Multinationals, *Economist,* 27 March 1993.
31. *The Cambridge History of Japan,* vol. 6, 17.
32. Ibid., 29.
33. Dore, *Structural Adjustment,* 97.
34. Tsuru, *Japan's Capitalism,* 109.
35. "Bank Alarms Are Blaring," *BusinessWeek International,* 26 June 1995, 16, 17.
36. M. Bruno and J. Sachs, *Economics of World Stagflation* (Cambridge: Harvard University Press, 1985).
37. "The Asian Miracle," *Economist,* 1 March 1997, 23.
38. *Trade Policy Review: Republic of Korea,* vol. I (Geneva: GATT, 1992), 15-22.
39. *The Cambridge History of Japan,* vol. 6, 30, 521-523.
40. *World Development Report 1981,* 149.
41. "More of the Same," *Far Eastern Economic Review,* 5 October 1995, 88; "Fiscal Paralysis," *Economist,* 28 February 1998, 19.
42. Dore, *Structural Adjustment,* 62.
43. Ibid., 91-93.
44. "Neither Borrower nor Lender," *Economist,* 29 April 1995, 101.
45. *Rapport sur le développement dans le monde 1991* (Washington, DC: Banque Mondiale), 229.
46. *OECD Economic Surveys: Japan* (Paris: OECD, 1992), 84-103.
47. *Trade Policy Review: Japan* (Geneva: GATT, 1992), 1.
48. "The Changing Face of Japan," *Newsweek,* 3 July 1995, 21.
49. "Restoring Growth," *Far Eastern Economic Review,* 3 June 1993, 42.
50. "No Going Back," *Far Eastern Economic Review,* 15 September 1994, 75.
51. "Japan," *BusinessWeek International,* 29 August 1994, 23.
52. "Forcing the Nikkei Out of Its Skids," *BusinessWeek International,* 5 April 1993, 48.
53. "Interest Rates," *BusinessWeek International,* 22 March 1993, 42.
54. "Big Deal," *Far Eastern Economic Review,* 27 April 1995, 70; *A Short Economic History of Modern Japan,* 102; "Fiscal Paralysis," ibid. 19.
55. "Bank Alarms Are Blaring," 16, 17.
56. "Japan Hits the Wall," *Fortune,* 1 November 1993, 60.
57. "Why Japan Can Still Say No," *BusinessWeek International,* 5 July 1993, 16.
58. "Japan: How Bad?" *BusinessWeek International,* 13 December 1993, 16.
59. "That Sinking Feeling," *Time,* 20 March 1995, 22.
60. *Rapport sur le développement dans le monde 1996,* 241, 221.
61. *The Asian Miracle,* ibid. 23.
62. "The Island Is Crazy for Chips," *BusinessWeek International,* 16 September 1996, 28.
63. "On the Chin," *Far Eastern Economic Review,* 8 June 1995, 40.
64. Tibor Mende, *Soleils Levants* (Paris: Editions du Soleil, 1975), 67, 69.
65. Dore, *Structural Adjustment,* 43.
66. Takenaka, *Contemporary Japanese Economy,* 78.

67. "Jobs for Life," *Fortune,* 20 March 1995, 74.

68. Rajendra Sisodia, "Singapore Invests in the Nation Corporation," *Harvard Business Review,* May–June 1992, 40.

69. "Japan's Aerospace Giants Are Flying Low," *Newsweek,* 5 December 1994, 31.

70. Tsuru, *Japan's Capitalism,* 129.

71. Andrew Gamble, *Britain in Decline* (London: Macmillan Education), 1985, 16.

72. "Yin and Yang in Asia's Science Cities," *Economist,* 21 May 1994, 89.

73. Allen, *A Short Economic History,* 171-176.

74. *World Development Report 1981,* 137.

75. *Rapport sur le dėveloppement dans le monde, 1996,* 241.

76. Ibid., 243.

77. "Its Wise to Deindustrialise," *Economist,* 26 April 1997, 88.

78. "World Education Rankings," *Economist,* 29 March 1997, 21.

79. "The Economics of Aging," *Economist,* 27 January 1996, 11.

80. "Long Term Growth," *Economist,* 28 August 1993, 95.

81. "Japan," *Economist,* 14 January 1995, 18; William Tabb, *The Postwar Japanese System* (New York: Oxford University Press, 1995, 140.

82. Dore, *Structural Adjustment,* 3, 64.

83. Morishima, *Why Has Japan Succeeded?* 52.

84. Dore, *Structural Adjustment,* 12, 16.

85. Tsuru, *Japan's Capitalism,* 104, 105.

86. "Soul of Inefficiency," *Far Eastern Economic Review,* 22 December 1994, 53.

87. *The Cambridge History of Japan,* vol. 6, 507-509.

88. *World Development Reports 1981, 1991 and 1997—Basic Indicators* (Washington, DC: World Bank).

Chapter 2

1. "Jeffrey Sachs and Wing Thye Woo, Structural Factors in the Economic Reforms of China, Eastern Europe and the Former Soviet Union," *Economic Policy* (April 1994), 108.

2. *China at the Threshold of a Market Economy* (Washington, DC: IMF, September 1993), 2.

3. Christopher Howe, *China's Economy* (London: Granada, 1978), 194, 195.

4. Walt Rostow, *The World Economy* (Austin: University of Texas Press, 1978), 532.

5. D. Goodman and G. Segal, *China in the Nineties* (New York: Oxford University Press, 1991), 140.

6. Thomas Rawski, *Economic Growth and Employment in China* (New York: Oxford University Press, 1979), 51.

7. Dwight Perkins, ed., *China's Modern Economy in Historical Perspective* (Stanford: Stanford University Press, 1975), 204, 222, 132.

8. Christopher Howe, *China's Economy,* xxiii, 170.

9. Ibid., 96.

10. Jack Gray, *Rebellions and Revolutions* (New York: Oxford University Press, 1990), 296.

11. John Fairbank, *China: A New History* (Cambridge: Harvard University Press, 1992), 413.

12. Perkins, ed., *China's Modern Economy,* 7, 46.

13. Ibid., 132.

14. Howe, *China's Economy,* xxii.
15. Gray, *Rebellions and Revolutions,* 302, 309.
16. C.D. Cowan, ed., *The Economic Development of China and Japan* (London: George Allen and Unwin, 1964), 133.
17. Gray, *Rebellions and Revolutions,* 315, 316.
18. Howe, *China's Economy,* xxiii.
19. "*A Survey of China,*" *Economist,* 18 March 1995, 23.
20. Ibid., 24.
21. Rostow, *The World Economy,* 532.
22. *World Development Report 1981* (Washington, DC: World Bank), 136.
23. Howe, *China's Economy,* xxiii, 196.
24. Harry Harding, *China's Second Revolution* (Washington, DC: Brookings Institution, 1987), 16-18.
25. J. Gray and G. White, eds., *China's New Development Strategy* (London: Academic Press, 1982), 119, 131.
26. Ibid., 93, 120.
27. Harding, *China's Second Revolution,* 30, 91.
28. *World Development Report 1981,* 136.
29. Rawski, *Economic Growth and Employment,* 6.
30. Gray and White, eds., *China's New Development Strategy,* 101.
31. *Trade Policy Review: Hong Kong* (Geneva: GATT, 1990), 41.
32. Goodman and Segal, *China in the Nineties,* 140.
33. *Rapport sur le développement dans le monde, 1991* (Washington, DC: Banque Mondiale), 228.
34. *China at the Threshold,* 13, 17, 38, 59.
35. "The End Is Near," *Far Eastern Economic Review,* 23 February 1995, 48-52.
36. "China and GATT," *Economist,* 6 March 1993, 61.
37. Gray, *Rebellions and Revolutions,* 386-388.
38. *China at the Threshold,* 15, 66.
39. "Hoping and Praying," *Newsweek,* 19 December 1994, 29.
40. "China's Banks," *Economist,* 27 March 1993, 102.
41. *China at the Threshold,* 21.
42. "Investing Abroad," *Economist,* 26 October 1996, 140.
43. *Rapport sur le développement dans le monde, 1991,* 229.
44. *China at the Threshold,* 34-58.
45. "Tax Trouble," *Far Eastern Economic Review,* 26 January 1995, 53.
46. Wolfram Eberhard, *A History of China* (Berkeley: University of California Press, 1977), 348, 354.
47. Gray, *Rebellions and Revolutions,* 217.
48. "New Fashion for Old Wisdom," *Economist,* 21 January 1995, 67.
49. *China at the Threshold,* 61, 62.
50. "Out of Joint," *Far Eastern Economic Review,* 10 March 1994, 42.
51. "Grease that Sticks," *Far Eastern Economic Review,* 23 March 1995, 54.
52. "All the Tea in China," *BusinessWeek International,* 17 January 1994, 14.
53. "The Rush to Modernize," *Far Eastern Economic Review,* 6 April 1995, 38.
54. "The Long March to Capitalism," *Economist,* 13 September 1997, 22.
55. *World Development Report 1997—Basic Indicators* (Washington, DC: World Bank); "Steel," *Economist,* 14 February 1998, 188.
56. "Can China Reform Its Economy?" *BusinessWeek,* European ed., 29 September 1997, 38-44.
57. *China at the Threshold,* 34, 38, 54, 58.

58. "China After Deng," *BusinessWeek International,* 6 February 1994, 41.
59. "And You Thought Japan's Trade Hurdles Were High," *BusinessWeek International,* 12 December 1994, 9.
60. "China's Yuan," *Economist,* 20 March 1993, 86; "Sleight of Hand," *Economist,* 7 March 1998, 97.
61. *A Survey of China,* 6, 10.
62. "Hard Graft in Asia," *Economist,* 27 May 1995, 71.
63. "Out of Africa—A Smoother Ride," *Economist,* 10 June 1995, 84.
64. "Shrinking the Chinese State," *Economist,* 10 June 1995, 59.
65. "A Trickle or a Flood?" *Economist,* 6 August 1994, 65; "*China's Economy,*" *Economist,* 14 February 1998, 63.
66. "Time and a Lot," *Far Eastern Economic Review,* 2 February 1995, 53.
67. *Rapport sur le développement dans le monde, 1997* (Washington, DC: Banque Mondiale), 256, 258.
68. Howe, *China's Economy,* 194-202.
69. J. Major and A. Kane, *China Briefing, 1987* (Boulder, CO: Westview Press, 1987), 27-29.
70. Howe, *China's Economy,* xxiii.
71. "Investment Ratios," *Economist,* 6 May 1995, 114; "China's Claws of Clay," *Economist,* 14 February 1998, 18.
72. Rawski, *Economic Growth,* 5.
73. "Market in the Making," *Far Eastern Economic Review,* 22 July 1992, 38.
74. *World Development Reports 1981, 1991, and 1997—Basic Indicators* (Washington, DC: World Bank).
75. Rawski, *Economic Growth,* 4.
76. Harding, *China's Second Revolution,* 142.
77. Goodman and Segal, *China in the Nineties,* 202.
78. "A Tidal Wave of Chinese Goods," *BusinessWeek International,* 12 December 1994, 8.
79. "Indicators," *Far Eastern Economic Review,* 23 June 1994, 63.
80. Rawski, *Economic Growth;* Ibid. 5; *World Development Reports 1981, 1991, and 1997—Basic Indicators.*
81. China, *Far Eastern Economic Review,* 7 September 1995, 67.
82. "Duplicitous Liaisons," *Far Eastern Economic Review,* 20 April 1995, 64.
83. Gray, *Rebellions and Revolutions,* 403.

Chapter 3

1. "The Lessons of Hong Kong," *Newsweek,* 9 May 1994, 40.
2. *Trade Policy Review: Hong Kong* (Geneva: GATT, 1990), 13-16.
3. Keith Hopkins, ed., *Hong Kong: The Industrial Colony* (London: Oxford University-Press, 1971), 1-8.
4. Ibid., 9.
5. Ibid., 2, 13, 25, 147, 148.
6. Jao Hungdah Chiu and Wu Yuan-li, eds., *The Future of Hong Kong* (New York: Quorum Books, 1987), 43-45.
7. Jon Woronoff, *Hong Kong: Capitalist Paradise* (Hong Kong: Heinemann Educational Books, 1980), 268.
8. *World Development Report, 1981* (Washington, DC: World Bank), 135-137.
9. *Rapport sur le développement dans le monde, 1991* (Washington, DC: Banque Mondiale), 229.

10. *Trade Policy Review: Hong Kong,* vol. I (Geneva: GATT, 1994), 1-3.
11. Ibid., 54-62.
12. *Rapport sur le développement dans le monde, 1997* (Washington, DC: Banque Mondiale), 257.
13. "Asia's High Tech Quest," *BusinessWeek International,* 30 November 1992.
14. "Business in Asia," *Economist,* 9 March 1996, 24.
15. "The Asian Miracle," *Economist,* 1 March 1997, 23.
16. "Drawing a New Map," *BusinessWeek,* European ed., 9 June 1997, 43.
17. Joseph E. Stiglitz, "Some Lessons from the East Asian Miracle," *World Bank Research Observer* 2, no. 2 (August 1996): 151-174.
18. "How they Stack Up," *Fortune,* 25 July 1994, 40.
19. "Schools—No Tiger," *BusinessWeek,* European ed., 9 June 1997, 48.
20. "That Foreign Touch," *Far Eastern Economic Review,* 31 March 1994, 40.
21. Hopkins, ed., *Hong Kong,* 44, 143.
22. Ibid., 141, 179, 199.
23. Chiu and Yuan-li, eds., *The Future of Hong Kong,* 44.
24. "The Visible Hand," *Economist,* 20 September 1997, 17.
25. "Accidental Surplus," *Far Eastern Economic Review,* 17 March 1994, 46.
26. "Indicators," *Far Eastern Economic Review,* 11 April 1996, 79.
27. Hopkins, ed., *Hong Kong,* 178.
28. *Rapport sur le développement dans le monde, 1997,* 257, 259.
29. "Instant Islands," *Far Eastern Economic Review,* 9 March 1995, 35.
30. *World Development Report, 1981 and 1997—Basic Indicators* (Washington, DC: World Bank).
31. Jon Woronoff, *Hong Kong,* 121.
32. "Still Singing Let It Be," *BusinessWeek International,* 7 June 1993, 29.
33. Robert Wade, *Governing the Market* (Princeton: Princeton University Press, 1990), 53, 74, 77, 96, 97.
34. "Sincere Regrets," *Newsweek,* 13 March 1995, 32.
35. Arnold Harberger, ed., *World Economic Growth* (Oakland: ICS Press, 1984), 303.
36. Wade, *Governing the Market,* 83.
37. Harberger, ed., *World Economic Growth,* 302.
38. Joel Aberbach, David Dollar, and Kenneth Sokoloff, eds., *The Role of the State in Taiwan's Development* (Armonk, NY: M.E. Sharpe, 1994), 5-120.
39. Wade, *Governing the Market,* 79, 81.
40. Aberbach, Dollar, and Sokoloff, eds., *The Role of the State in Taiwan's Development,* 193-228.
41. Wade, *Governing the Market,* 78.
42. "Borrowed Time," *Far Eastern Economic Review,* 26 May 1994, 60, 61.
43. Wade, *Governing the Market,* 53, 77, 81.
44. Harberger, ed., *World Economic Growth,* 321.
45. Wade, *Governing the Market,* 82.
46. "Into the World," *Time,* 19 June 1995, 45.
47. Wade, *Governing the Market,* 87.
48. Ibid., 83, 95.
49. Ibid., 96.
50. "The Price of Success," *Far Eastern Economic Review,* 6 July 1995, 56.
51. "The Sleeping Giant Awakes," *Financial Times,* 28 June 1993, 13.
52. "The Dream Postponed," *Economist,* 7 August 1993, 60.
53. Wade, *Governing the Market,* 97.

54. "Latin American and Asian Inflation," *Economist,* 8 April 1995, 128.
55. "Do Something," *Far Eastern Economic Review,* 10 November 1994, 50.
56. "Taiwan–Trade and Investment," *Far Eastern Economic Review,* 13 October 1994, 44.
57. "Anyone Wants to Buy a Piece of Taiwan," *BusinessWeek International,* 22 March 1993, 57.
58. "China: The Titan Stirs," *Economist,* 28 November 1992.
59. "Goodbye to All that," *Far Eastern Economic Review,* 15 July 1993, 66.
60. "A Difficult Age," *Far Eastern Economic Review,* 30 June 1994, 56.
61. "Tough as Steel," *Far Eastern Economic Review,* 20 October 1994, 85.
62. "Income per Head," *Economist,* 5 February 1994, 120.
63. Wade, *Governing the Market,* 98.
64. "Home, Sweet Home," *Far Eastern Economic Review,* 29 September 1994, 66.
65. "Class Politics," *Far Eastern Economic Review,* 1 December 1994, 58.
66. Wade, *Governing the Market,* 48.
67. Harberger, ed., *World Economic Growth,* 319.
68. "Indicators," *Far Eastern Economic Review,* 27 April 1995, 71.
69. Harberger, ed., *World Economic Growth,* 308, 311.
70. Paul Bairoch, *Economics and World History* (London: Harvester Wheatsheaf, 1993), 92.
71. "Made in Taiwan," *Fortune,* 8 August 1994, 42.
72. "How Asia's Tigers Got Their Stripes," *BusinessWeek International,* 25 December 1995, 6.
73. "Measuring the Price of Politics," *Economist,* 27 January 1996, 78.
74. "A Matter of Time," *Far Eastern Economic Review,* 7 December 1995, 71.
75. "Indicators," Ibid. 71.
76. Harberger, ed., *World Economic Growth,* 312, 320.
77. Indicators, *Far Eastern Economic Review,* 16 May 1996, 83.
78. "The New Hong Kong?" *BusinessWeek International,* 3 April 1995, 24.
79. "No Pain—No Grain," *Far Eastern Economic Review,* 16 November 1995, 89.

Chapter 4

1. "The Boom," *Far Eastern Economic Review,* 24 November 1994, 43.
2. *Trade Policy Review: Republic of Korea,* vol. I (Geneva: GATT, Geneva, 1992), 165.
3. *New Encyclopaedia Britannica,* vol. 6—Micropaedia (Chicago: Encyclopaedia Britannica, 1988), 958, 959.
4. Ibid., 960.
5. Robert Wade, *Governing the Market* (Princeton: Princeton University Press, 1990), p. 43.
6. "Shattering Our Notions of Korea," *BusinessWeek,* European ed., 10 March 1997, 7.
7. *Trade Policy Review: Republic of Korea,* vol. I, 12-16.
8. "South Korea," *Economist,* 3 June 1995, 17.
9. *World Development Report 1981* (Washington, DC: World Bank), 137.
10. "The Money Machine," *Far Eastern Economic Review,* 11 August 1994, 62.
11. "Passing Grade," *Far Eastern Economic Review,* 16 March 1995, 58.
12. *Trade Policy Review: Republic of Korea,* vol. I, 165, 144.
13. Ibid., 22.

14. *A Case of Successful Adjustment: Korea's Experience During 1980-84* (Washington, DC: IMF, August 1985), 1, 2.
15. "South Korea," *Economist,* 14-20.
16. Wade, *Governing the Market,* 43.
17. "The Cost of Inflation," *Economist,* 13 May 1995, 90.
18. *World Development Report 1981,* 135.
19. "South Korea," *Time,* 26 June 1995, 53, 54.
20. *Trade Policy Review: Republic of Korea,* vol. I, 23.
21. *Rapport sur le dévelopment dans le monde, 1991* (Washington, DC: Banque Mondiale), 229.
22. "High-Tech Jobs All over the Map," *BusinessWeek International,* 19 December 1994, 42.
23. *Trade Policy Review: Republic of Korea,* vol. I, 25, vol. II, 21, 24, 66.
24. "South Korea," *Economist,* 14-18.
25. Joseph Stiglitz and Marilou Uy, "Financial Markets, Public Policy, and the East Asian Miracle," *World Bank Research Observer* 2, no. 2 (August 1996): 249-273.
26. *Rapport sur le dévelopment dans le monde, 1995,* 257.
27. "Korea," *BusinessWeek International,* 31 July 1995, 36.
28. "Why Seoul Is Seething," *BusinessWeek,* European ed., 27 January 1997, 24.
29. Survol International (Bern), *La Vie Economique* (September 1997): 1.
30. A Case of Successful Adjustment, 3-6.
31. "Time to Focus," *Far Eastern Economic Review,* 8 April 1993, 60.
32. "Take That," *Far Eastern Economic Review,* 4 August 1994, 61.
33. "Sleek or Sluggish," *Far Eastern Economic Review,* 24 March 1994, 40.
34. "Control Freaks," *Far Eastern Economic Review,* 11 May 1995, 73.
35. *Trade Policy Review: Republic of Korea,* vol. I, 135.
36. Walter Kuemmerle, "Building Effective R&D Capabilities Abroad," *Harvard Business Review* (March–April 1997): 62.
37. "The Price of Success," *Far Eastern Economic Review,* 6 July 1995, 56.
38. *Trade Policy Review: Republic of Korea,* vol. I, 17.
39. "South Korea," *Economist,* 15 June 1996, 116.
40. *World Development Reports 1981, 1991, and 1997—Basic Indicators* (Washington, DC: World Bank).
41. *Trade Policy Review: Republic of Korea,* vol. I, 139-143, vol. II, 3.
42. Ibid., vol. I, 15, 16, 4.
43. Ibid., vol. I, 4, 29-34; vol. II, 78.
44. "South Korea: Trade and Investment," *Far Eastern Economic Review,* 22 June 1995, 4.
45. "Why Seoul Is Seething," 24.
46. "South Korea—Trade and Investment," *Far Eastern Economic Review,* 26 May 1994, 52, 53.
47. "Helping Handouts," *Economist,* 16 December 1995, 95.
48. "The Battle of the Belly Button," *Economist,* 24 September 1994, 69.
49. *Trade Policy Review: Republic of Korea,* vol. I, 11, 28.
50. "Free Markets," *Economist,* 11 February 1995, 124.
51. *Trade Policy Review: Singapore,* vol. 1 (Geveva: GATT, 1992), 11.
52. W.G. Huff, *The Economic Growth of Singapore* (Cambridge: Cambridge University Press, 1994), 33, 289.
53. Ibid., 290.
54. Ibid., 34.
55. *Trade Policy Review: Singapore,* vol. I, 1-5, 19.

56. "The Island Is Crazy for Chips," *BusinessWeek International,* 16 September 1996, 28.
57. *Economic Development of Singapore 1960-91* (Singapore: Economic Development Board, 1992).
58. "The Asian Miracle," *Economist,* 1 March 1997, 23.
59. *Singapore—Facts and Pictures 1987* (Singapore: Ministry of Communications and Information), 10-54.
60. *Trade Policy Review: Singapore,* vol. 1, 19-32, 57.
61. *Economic Development of Singapore,* 1-3.
62. *Singapore in Brief 1991* (Singapore: Ministry of Trade and Industry, 1992).
63. *Yearbook of Statistics Singapore 1991* (Singapore: Department of Statistics, 1992).
64. "The Island Is Crazy for Chips," 28.
65. *Rapport sur le développement dans le monde, 1997* (Washington, DC: Banque Mondiale), 237, 257.
66. "Capital Wars," *Newsweek,* 3 October 1994, 32-37.
67. "Economic Monitor: Singapore," *Far Eastern Economic Review,* 12 January 1995, 73.
68. *Economic Development of Singapore,* 2.
69. Ibid., 5.
70. *Trade Policy Review: Singapore,* vol. II, 38.
71. *Trade Policy Review: Singapore,* vol. I, 21.
72. "Ready for the Worst," *Newsweek,* 10 November 1997, 26.
73. *World Development Report, 1981, and 1997,* 135, 137-239.
74. Rajendra Sisodia, "Singapore Invests in the Nation Corporation," *Harvard Business Review* (May–June 1992): 40.
75. Huff, *The Economic Growth of Singapore,* 302, 318.
76. "Market in the Making," *Far Eastern Economic Review,* 22 July 1992, 38.
77. *Economic Development of Singapore 1960-91,* 2.
78. "One Happy, Culturally Superior Family," *Time,* 21 November 1994, 57.
79. "Democracy and Growth," *Economist,* 27 August 1994, 15.
80. "The Natural Resource Myth," *Economist,* 23 December 1995, 90.
81. "The Curse of Natural Resources," *Far Eastern Economic Review,* 26 October 1995, 40.
82. *World Development Reports, 1981, and 1997—Basic Indicators* (Washington, DC: World Bank).
83. Arnold Harberger, ed., *World Economic Growth* (Oakland: ICS Press, 1984), 322.
84. "Economic Reform and Social Regress," *Economist,* 20 September 1997, 70.

Chapter 5

1. "More than Meets the Eye," *Economist,* 26 December 1992, 87.
2. Kenneth Boulding, *The Structure of a Modern Economy* (London: Macmillan Press, 1993), 70.
3. Gabriel Kolko, *Main Currents in Modern American History* (New York: Harper and Row, 1976), 321, 327.
4. Jonathan Hughes, *American Economic History* (Glenview, IL: Scott, Foresman, 1987), 476.
5. D. Adams, *America in the 20th Century* (London: Cambridge University Press, 1967), 136.

6. Kolko, *Main Currents in Modern American History,* 318.
7. "Re-election by Numbers," *Economist,* 27 May 1995, 49.
8. Boulding, *The Structure of a Modern Economy,* 71.
9. J. Westwood, *Endurance and Endeavor* (Oxford: Oxford University Press, 1993), 411.
10. "Why We Went to the Moon?" *Time,* 25 July 1994, 52.
11. "A Space Odyssey or Just Pork Pie in the Sky," *BusinessWeek International,* 15 August 1994, 47.
12. *World Development Report, 1981* (Washington, DC: World Bank), 137.
13. "Cosmic Storms Coming," *Time,* 9 September 1996, 54.
14. Hans Mark, "Straight Up into the Blue," *Scientific American,* October 1997, 78.
15. "Biotechnology and Genetics," *Economist,* 25 February 1995, 7; Joel Swerdlow, "Exploration," *National Geographic,* 193, no. 2, February 1998, 24.
16. "The Technology Paradox," *BusinessWeek International,* 6 March 1995, 41.
17. Nathan Rosenberg, *Technology and American Economic Growth* (New York: Harper and Row, 1972), 182-184.
18. Ibid., 177.
19. Kolko, *Main Currents,* 319.
20. J. Fraser and G. Gerstle, *The Rise and Fall of the New Deal Order* (Princeton: Princeton University Press, 1989), 48.
21. Kolko, *Main Currents,* 324, 314.
22. *World Development Report, 1981,* 137.
23. Kolko, *Main Currents,* 384-386.
24. M. Bruno and J. Sachs, *Economics of Worldwide Stagflation* (Cambridge: Harvard University Press, 1985).
25. Boulding, *The Structure of a Modern Economy,* 177, 178.
26. Ibid., 44, 75.
27. *Rapport sur le développement dans le Monde, 1997* (Washington, DC: Banque Modiale), 239, 257.
28. "More of the Same," *Far Eastern Economic Review,* 5 October 1995, 88.
29. Kolko, *Main Currents,* 163.
30. "*The British Disease,*" *Economist,* 3 April 1993, 35.
31. Kolko, *Main Currents,* 379.
32. Boulding, *The Structure of a Modern Economy,* 54.
33. "Foreign Firms in Japan," *Economist,* 13 February 1993, 70.
34. "Long Term Growth," *Economist,* 28 August 1993, 95.
35. "The Fake Glitter of a Gold Standard," *BusinessWeek International,* 9 October 1995, 53.
36. "All the Tea in China," *BusinessWeek International,* 17 January 1994, 14.
37. Boulding, *The Structure of a Modern Economy,* 71.
38. *World Development Report 1981,* 137.
39. Swerdlow, "Exploration," 24.
40. "Tilling the Soil by Satellite," *BusinessWeek International,* 11 December 1994, 62.
41. "The Land of Tech and Money," *Newsweek,* 8 April 1996, 36.
42. Anders Aslund, *Gorbachev's Struggle for Economic Reform* (London: Pinter, 1991).
43. "No Free Lunch," *Economist,* 27 January 1996, 79.
44. *World Development Report, 1991—Basic Indicators* (Washington, DC: World Bank).
45. L. Davis, J. Hughes, and D. McDougall, *American Economic History* (New York: Richard Irwin, 1969), 414.

46. "Job One in America: Better Jobs," *BusinessWeek International,* 13 June 1994, 6.
47. Boulding, *The Structure of a Modern Economy,* 38.
48. Kolko, *Main Currents,* 327.
49. "The Burdensome National Debt," *Economist,* 10 February 1996, 70.
50. Boulding, *The Structure of a Modern Economy,* 38.
51. Paul Krugman, *The Age of Diminished Expectations* (Cambridge: MIT Press, 1991), 3.
52. "Rich Man—Poor Man," *Economist,* 24 July 1993, 65.
53. "Service Jobs," *BusinessWeek International,* 15 February 1993, 9.
54. Krugman, *The Age of Diminished Expectations,* 67.
55. Boulding, *The Structure of a Modern Economy,* 92.
56. Ibid., 80.
57. "American Banks," *Economist,* 7 November 1992, 92.
58. "Will China Be Next?" *Newsweek,* 1 December 1997, 33.
59. "*China's Economy,*" *Economist,* 14 February 1998, 63.
60. "Where Have All the Good Jobs Gone?" *BusinessWeek International,* 29 August 1994, 10.
61. "Robodocs and Mousecalls," *Newsweek,* 27 February 1995, 32.
62. "Fewer Bangs—More Bucks," *Economist,* 15 July 1995, 66.
63. "Can Defense Pain Be Turned to Gain?" *Fortune,* 8 March 1993, 40.
64. "History Lessons," *Newsweek,* 10 July 1995, 37.
65. Boulding, *The Structure of a Modern Economy,* 72.
66. "How They Stack Up," *Fortune,* 25 July 1994, 40.
67. "Goodbye to All That," *Far Eastern Economic Review,* 15 July 1993, 66.
68. "The Changing Face of Japan," *Newsweek,* 3 July 1995, 21.
69. *World Development Report, 1998—Basic Indicators* (Washington, DC: World Bank).
70. "The Peace Dividend," *Newsweek,* 26 January 1998, 2.
71. "The End of Corporate Welfare as We Know It?" *BusinessWeek,* European ed., 10 February 1997, 32.
72. "The Cancer Killer," *Newsweek,* 13 January 1997, 41; "Where Science and Religion Meet," *Scientific American,* February 1998, 18.
73. "The Disease Detective," *Time,* 30 December 1996.
74. "Trade's Great Irony," *Economist,* 2 November 1996, 17.
75. "What's Wrong," *BusinessWeek International,* 2 August 1993, 30.
76. *World Development Report, 1998—Basic Indicators,* Ibid.
77. "Why Job Growth Is Stalled," *Fortune,* 8 March 1993, 33.
78. "Manufacturing Technology," *Economist,* 5 March 994, 11.
79. "The Productivity Paradox," *BusinessWeek International,* 8 February 1993, 10.
80. "Why the Economy Isn't Heading Clinton's Call," *BusinessWeek International,* 26 July 1993, 8.
81. "What Has NAFTA Wrought?" *BusinessWeek International,* 21 November 1994, 16-18.
82. "Humble Pie," *Newsweek,* 16 January 1995, 38.
83. *World Development Report 1998—Basic Indicators.*
84. Lilia Dominguez and Flor Brown, "Mexico—Patrones de competencia y apertura economica," Comercio Exterior 47, no. 9 (Septiembre 1997): 695-704.
85. "USA Inc.," *Newsweek,* 6 March 1995, 15.
86. "How Clinton Is Shaking Up Trade," *Fortune,* 31 May 1993, 36.
87. "How Natural Is Unemployment?" *Economist,* 16 December 1995, 98.
88. Davis, Hughes, and McDougall, *American Economic History,* 412-414.

89. "Inflation," *Economist,* 22 March 1997, 142.
90. Krugman, The Age of Diminished Expectations, 12.
91. "The New Economy," *BusinessWeek,* European ed., 17 November 1997, 50.
92. "How to Raise U.S. Productivity," *BusinessWeek International,* 9 December 1996, 13.
93. "The Manufacturing Myth," *Economist,* 19 March 1994, 81.
94. "America's New Growth Economy," *BusinessWeek International,* 16 May 1994, 45.
95. "A Sprinkling of Herbs," *Economist,* 28 January 1995, 91.
96. "Time to Take Our Medicines Seriously," *Economist,* 24 May 1997, 26.
97. Anthony Perl, "Fast Trains: Why the U.S. Lags," *Scientific American,* October 1997, 74.
98. "Quick—Save the Ozone," *BusinessWeek International,* 17 May 1993, 81.
99. "Re-engineering the Engineers," *Economist,* 26 June 1993, 97.
100. "Why Boeing Abandoned Ship," *BusinessWeek International,* 17 June 1996, 35.
101. "Space Race," *Newsweek,* 15 July 1996, 35.
102. Kolko, *Main Currents,* 319; "The Knowledge Factory," *Economist,* 4 October 1997, 12.
103. "Jobs for Life," *Fortune,* 20 March 1995, 74.
104. "Biotechnology and Genetics," *Economist,* 4, 8.
105. "Computers that Think Are Almost Here," *BusinessWeek International,* 17 July 1995, 49.
106. "For Some a Necessity," *BusinessWeek,* European ed., 23 February 1998, 54.
107. "Detroit's Impossible Dream?" *BusinessWeek,* European ed., 2 March 1998, 36, 37.
108. "Money Lost in Space," *Economist,* 9 November 1996, 17.
109. Krugman, *The Age of Diminished Expectations,* 14.
110. Shigeto Tsuru, *Japan's Capitalism* (Cambridge: Cambridge University Press, 1993), 129.
111. "America's Edge in Wages," *BusinessWeek,* European ed., 30 June 1997, 11.
112. "Europe's Technology Policy," *Economist,* 9 January 1993, 21-23.
113. "Do Computers Slow Us Down?" *Fortune,* 30 March 1998, 18.
114. "The Tragedy of the Oceans," *Economist,* 19 March 1994, 23.
115. "Congratulations—You Struck Sand," *BusinessWeek International,* 18 December 1995, 56.
116. Krugman, *The Age of Diminished Expectations,* 67.
117. Boulding, *The Structure of a Modern Economy,* 208, 209; "Economic Mythmaking," *Newsweek,* 8 September 1997, 4.

Chapter 6

1. Roger Clarke and Dubravko Matko, *Soviet Economic Facts 1917-81* (London: Macmillan Press, 1983), 6, 7.
2. Robert Campbell, *Soviet Economic Power* (London: Macmillan, 1967), 26.
3. Maurice Dobb, *Soviet Economic Development Since 1917* (London: Routledge, 1960), 317.
4. William Blackwell, *The Industrialization of Russia* (New York: Thomas Crowell, 1970), 153.

5. Angus Madison, *Economic Growth in Japan and the USSR* (George Allen and Unwin, 1969), 108-110.
6. Y. Brenner, *A Short History of Economic Progress* (London: Frank Cass, 1969), 227.
7. J. Westwood, Endurance and Endeavour (Oxford: Oxford University Press, 1993), 395, 404.
8. Dobb, *Soviet Economic Development,* 322.
9. Clarke and Matko, *Soviet Economic Facts, 1917-81,* 7-12.
10. Basile Kerblay, *Modern Soviet Society* (London: Methuen, 1983), 190.
11. Blackwell, *The Industrialization of Russia,* 164.
12. Westwood, *Endurance and Endeavour,* 411.
13. John Keep, *Last of the Empires* (Oxford: Oxford University Press, 1995), 84, 88.
14. Madison, *Economic Growth in Japan and the USSR,* 125.
15. Blackwell, *The Industrialization of Russia,* 157, 158.
16. Westwood, *Edurance and Endeavour,* 411.
17. Blackwell, *The Industrialization of Russia,* 154.
18. Keep, *Last of the Empires,* 135, 235.
19. Blackwell, *The Industrialization of Russia,* 159.
20. *The Fontana Economic History of Europe: Contemporary Economics* (London: William Collins and Sons, 1976), 591.
21. *Economic Reforms in the European Centrally Planned Economies* (New York: Economic Commission for Europe, United Nations, 1989), 231.
22. Westwood, *Edurance and Endeavour,* 428.
23. Clarke and Matko, *Soviet Economic Facts, 1917-81,* 12.
24. Michael Kort, *The Soviet Colossus* (New York: Unwin Hyman, 1990), 257-259.
25. Westwood, *Endurance and Endeavour,* 432.
26. Stephen Cohen, *Rethinking the Soviet Experience* (New York: Oxford University Press, 1985).
27. Anders Aslund, *Gorbachev's Struggle for Economic Reform* (London: Pinter, 1991), 16-18.
28. Kort, *The Soviet Colossus,* 277.
29. Gerard Duchene, *L'économie de l'URSS* (Paris: Editions La Decouvert, 1987), 7, 8, 115.
30. Westwood, *Endurance and Endeavour,* 433, 472.
31. Kort, *The Soviet Colossus,* 291.
32. Duchene, *L'économie de l'URSS,* 14.
33. Kort, *The Soviet Colossus,* 297.
34. Westwood, *Endurance and Endeavour,* 480.
35. *Economic Reforms in the European Centrally Planned Economies,* 203.
36. Clarke and Matko, *Soviet Economic Facts, 1917-81,* 7-12.
37. Keep, *Last of the Empires,* 88.
38. Aslund, Gorbachev's *Struggle for Economic Reform,* 17.
39. Keep, *Last of the Empires,* 89, 233.
40. Kort, *The Soviet Colossus,* 260.
41. "La Situation Conjoncturelle dan la CEI," *Nouveaux Mondes* (Automne 1993): 89.
42. Aslund, *Gorbachev's Struggle for Economic Reform,* 19.
43. "The State of Russia," *Economist,* 28 March 1998, 22.
44. Keep, *Last of the Empires,* 393-400.
45. Ibid., 391.
46. *Statistiques de base de la Communaute* (Brussels: Eurostat, 1992).
47. "Russian Privatisation," *Economist,* 18 February 1995, 82.

48. "Call That a Reshuffle, Boris?" *BusinessWeek International,* 12 November 1994, 33.

49. "Sowing Reform in Russia's Fields," *BusinessWeek International,* 23 May 1994, 25.

50. "Toss Another Match into the Russian Tinder Box," *BusinessWeek International,* 6 June 1994, 25.

51. "Less Poor—Less Democratic," *Economist,* 22 April 1994, 34.

52. "Mass Privatisation in Eastern Europe," *Economist,* 25 November 1995, 82.

53. "Tired of Capitalism?" *Economist,* 21 January 1995, 69.

54. "Don't Just Sit There," *Newsweek,* 6 June 1994, 48.

55. Aslund, *Gorbachev's Struggle for Economic Reform,* 191.

56. Keep, *Last of the Empires,* 392.

57. "Latin American and Asian Inflation," *Economist,* 8 April 1995, 128.

58. "After Communism," *Economist,* 3 December 1994, 23.

59. "Vietnam Beats China at Its Own Game," *Economist,* 4 November 1994, 61.

60. Jeffrey Sachs and Wyne The Woo, "Structural Factors in the Economic Reforms of China, Eastern Europe, and the Former Soviet Union," *Economic Policy* (April 1994): 110, 111.

61. *China at a Threshold of a Market Economy* (Washington, DC: IMF, September 1993), 8.

62. "Spotlight in Laos," *BusinessWeek International,* 12 June 1995, 5.

63. *China at the Threshold of a Market Economy*, 8.

64. Sachs and Woo, *Structural Factors,* 111, 108.

65. Ibid., 120, 123.

66. Ibid., 102, 103.

67. *World Development Report, 1997—Basic Indicators* (Washington, DC: World Bank).

68. Sachs and Woo, *Structural Factors,* 104-109.

69. "Visegrad Growth," *Economist,* 15 April 1995, 100.

70. "Reconversion du complexe militaro-industriel Russe," *Nouveaux Mondes* (Automne 1993): 47.

71. "Fewer Bangs—More Bucks," *Economist,* 15 July 1995, 66.

72. "Russia's Defense Industry," *Economist,* 2 December 1995, 76.

73. "Russia's Emerging Market," *Economist,* 8 April 1995, 3.

74. "Economic Monitor—Vietnam," *Far Eastern Economic Review,* 29 December 1994, 96.

75. "After Communism," 23.

76. "Thumbling Dice," *Economist,* 28 August 1993, 63.

77. "Another Russian Shortage," *BusinessWeek International,* 27 February 1995, 19.

78. "Russian Banks Try Something New," *BusinessWeek International,* 5 February 1996, 40.

79. "Tourism," *Economist,* 15 March 1997, 120.

80. "The State of Russia," ibid. 22.

81. "Emerging Market Indicators," *Economist,* 28 March 1998, 114.

82. *Short Term Economic Indicators: Transition Economies,* Supplement 4/1997 (Paris: OECD), 16.

83. "The State of Russia," *Time,* 27 May 1996, 36.

84. "The Right Moves Aren't Working," *BusinessWeek International,* 23 June 1997, 4.

85. *Economie Européenne* 61 (1996): 70.

86. "China's Economy," *Economist,* 14 February 1998, 63.

87. "Russian Land Reform," *Economist,* 16 March 1996, 84.

88. "The New Economics of Food," *BusinessWeek International,* 20 May 1996, 33.
89. "The State of Russia," *Time,* ibid. 22.
90. "Russia's Emerging Market," 13.
91. "The Struggle for Vietnam's Soul," *Economist,* 24 June 1995, 63.
92. Walter Kirchner, A History of Russia (New York: Barnes and Noble, 1963), 358-360.
93. Madison, *Economic Growth in Japan and the USSR,* 108.
94. W. Easterly and S. Fischer, *The Soviet Economic Decline* (Washington, DC: World Bank, April 1994), 31.
95. Westwood, *Endurance and Endeavour,* 601.
96. "2000 or Bust?" *Newsweek,* 6 April 1998, 20-23.
97. "Up from the Wreckage of Russian Science," *BusinessWeek,* European ed., 27 October 1997, 78.
98. Ibid.
99. Keep, *Last of the Empires,* 313.
100. Madison, *Economic Growth in Japan and the USSR,* 119.
101. "Russia's Emerging Market," 15.
102. Madison, *Economic Growth in Japan and the USSR,* 57-59.
103. Keep, *Last of the Empires,* 260.
104. Blackwell, *The Industrialization of Russia,* 154.
105. Westwood, *Endurance and Endeavour,* 412.
106. Keep, *Last of the Empires,* 268-271.

Chapter 7

1. *German Unification* (Washington, DC: IMF, 1990), 28.
2. Roger Munting and B. Holderness, *Crisis, Recovery and War* (New York: St. Martin Press, 1991), 142.
3. Simon Kuznets, *Modern Economic Growth* (New Haven: Yale University Press, 1966), 237.
4. Ernest Bramsted, *Germany* (New York: Prentice-Hall, 1972), 12.
5. *GDR and Eastern Europe* (Gover, 1989), 12, 20.
6. Roger Clarke and Dubravko Matko, *Soviet Economic Facts 1917-81* (London: Macmillan, 1983), 6, 7.
7. Arnold Harberger, ed., *World Economic Growth* (Oakland: ICS Press, 1984), 95.
8. Guy Roustang, *Dévelopement Economique de l'Allemagne Orientale* (Paris: Société d'Edition d'Enseignement Superieur, 1963), 36-41.
9. Henry Krisch, *The German Democratic Republic* (Boulder, CO: Westview Press, 1985), 80-91.
10. GDR and Eastern Europe, 20, 12.
11. "The Man and the Plan," Newsweek, 26 May 1997, 12-16.
12. "A Lifeline to Europe," Newsweek, 26 May 1997, 18-20.
13. Bramsted, Germany, 9, 239.
14. Harberger, ed., World Economic Growth, 98.
15. Karl Hardach, Wirtschafts Geschichte Deutschlands im 20 Jahrhundert (Frankfurt: Vandenhoeck und Ruprecht, 1979), 192.
16. "Taking Revenge," Newsweek, 8 May 1995, 18, 19.
17. Harberger, World Economic Growth, 97, 98.
18. Wolfgang Heisenberg, ed., German Unification in European Perspective (Belgium: Brassey's, 1991), 200.
19. Munting and Holderness, Crisis, Recovery and War, 248.

20. Ibid., 242, 243.
21. L'economie Suisse 1946-1986 (Geneva: Union de Banques Suisses, 1987), 22-24.
22. Pierre Gaxotte, Histoire de l'Allemagne (Paris: Flammarion, 1975), 711.
23. Charles Bettelheim, L'economie Allemande sous le Nazisme, vol. II (Paris: Francois Maspero, 1971), 84, 127.
24. Martin Kitchen, The Political Economy of Germany 1815-1914 (Montreal: McGill-Queen's University Press, 1978), 132.
25. Gaxotte, Histoire de l'Allemagne, 714.
26. V. Berghahn, Modern Germany (Cambridge: Cambridge University Press, 1985), 184, 239, 208.
27. Bramsted, Germany, 239.
28. "The Euro," BusinessWeek, European ed., 27 April 1998, 24-29.
29. Heisenberg, ed., German Unification in European Perspective, 201.
30. "East of Eden," Time, 27 April 1998, 22-26.
31. Robert Wade, Governing the Market (Princeton: Princeton University Press, 1990), 43.
32. Bramsted, Germany, 11.
33. World Development Report, 1998—Basic Indicators (Washington, DC: World Bank).
34. Munting and Holderness, Crisis, Recovery, and War, 250.
35. "Economie Européenne," Commission Européenne 61 (1996): 70.
36. German Unification, 56.
37. Roustang, Développement economique de l'Allemagne Orientale, 41.
38. Ibid., 36.
39. Hardach, Wirtschafts Geschichte Deutschlands im 20 Jahrhundert, 255.
40. Roustang, Développement economique de l'Allemagne Orientale, 40, 94.
41. Ibid., 42.
42. Krisch, The German Democratic Republic, 91.
43. GDR and Eastern Europe, 6.
44. Hardach, Wirtschafts Geschichte Deutschlands im 20 Jahrhundert, 192.
45. Berghahn, Modern Germany, 239.
46. Heisenberg, ed., German Unification in European Perspective, 166.
47. GDR and Eastern Europe, 20.
48. German Unification, 70.
49. Ibid., 131.
50. Harberger, World Economic Growth, 108-110.
51. Ibid., 103.
52. Ibid., 100.
53. Germany: OECD Economic Surveys (Paris: OECD, 1992), 26.
54. "Suddenly German Labour Is as Meek as a Kitten," BusinessWeek International, 29 April 1996, 19.
55. World Development Report, 1981—Basic Indicators (Washington, DC: World Bank).
56. "Trade's Great Irony," Economist, 2 November 1996, 17.
57. M. Bruno and J. Sachs, Economics of World Stagflation (Cambridge: Harvard University Press, 1985).
58. Harberger, ed., World Economic Growth, 97.
59. A Case of Successful Adjustment: Korea's Experience During 1980-84 (Washington, DC: IMF, August 1985), 1, 2.
60. German Unification, 28.

61. "Fiscal Consolidation in Germany," Economist, 29 June 1996, 70.
62. Munting and Holderness, Crisis, Recovery, and War, 142.
63. Charles Bettelheim, L'Economie Allemande sous le Nazisme, vol. I (Paris: François Maspero, 1971), 14, 15.
64. The Federal Republic of Germany (Washington, DC: IMF, January 1991), 21.
65. "The Asian Miracle," Economist, 1 March 1997, 23.
66. "Spotlight on Singapore," BusinessWeek, European ed., 4 May 1998, 4.
67. "Central Bankers," Economist, 3 December 1994, 90.
68. World Development Report, 1998—Basic Indicators, ibid.
69. The Federal Republic of Germany, 21.
70. GDR and Eastern Europe, 12.
71. Krisch, The German Democratic Republic, 46.
72. Bramsted, Germany, 240.
73. German Unification, 131.
74. The Federal Republic of Germany, 76.
75. German Unification, 28.
76. The Federal Republic of Germany, 21.
77. German Unification, 3.
78. "More of the Same," Far Eastern Economic Review, 5 October 1995, 88.
79. Trade Policy Review: Republic of Korea, vol. I (Geneva: GATT, 1992), 23, 15.
80. Jonathan Hughes, American Economic History (Glenview, IL: Scott Foresman, 1987), 476-482.
81. The Federal Republic of Germany, 40, 47-51.
82. Germany: OECD Economic Surveys, 70.
83. Berghahn, Modern Germany, 276.
84. German Unification, 57, xv.
85. GDR and Eastern Europe, 12.
86. German Unification, 49, 6.
87. Heisenberg, ed., German Unification in European Perspective, 168-170.
88. Horst Siebert, "German Unification," Economic Policy 13 (October 1991): 310.
89. Heisenberg, ed., German Unification in European Perspective, 167.
90. "Farewell Sweet Treuhand," Economist, 24 December 1994, 85.
91. "Das Ist Mein Haus," BusinessWeek International, 22 March 1993, 20.
92. German Unification, 131.
93. Report of the Deutsch BundesBank 1991 (Frankfurt), 23.
94. "Finally Germany Is Paring the Fat," BusinessWeek International, 17 October 1994, 16.
95. "Eastern Germany's Subsidy Mountain," Economist, 29 July 1995, 59.
96. "Employment Turns Upward," Far Eastern Economic Review, 22 September 1994, 54.
97. East of Eden, 24.
98. "Is Eastern Germany Really Bouncing Back?" Economist, 6 August 1994, 55.
99. Trade Policy Review: Hong Kong, vol. I (Geneva: GATT, 1990), 17.
100. "China's Yuan," Economist, 20 March 1993, 86.
101. Heisenberg, ed., German Unification in European Perspective, 200, 201; East of Eden, ibid. 24.
102. Short Term Economic Indicators: Transition Economies, Supplement 4, 1997 (Paris: OECD); "Visegrade Growth," Economist, 15 April 1995, 100.
103. "A Survey of Germany," Economist, 9 November 1996, 9.
104. "Out of Africa: A Smoother Ride," Economist, 10 June 1995, 84.
105. Jeffrey Sachs and Wyne Thee Woo, "Structural Factors in the Economic Re-

forms of China, Eastern Europe, and the Former Soviet Union," Economic Policy (April 1994): 102-122.

106. Trade Policy Review: Hong Kong, vol. I (Geneva: GATT, 1994).

107. W.G. Huff, The Economic Growth of Singapore (Cambridge: Cambridge University Press, 1994).

108. "The Worst Is Finally Over in Eastern Germany," BusinessWeek International, 19 June 1995, 20, 21.

109. Deutsch BundesBank Annual Report 1993 (Frankfurt), 28-30.

110. Germany: OECD Economic Surveys, 20.

111. Siebert, "German Unification," 293.

112. "East German Industry," Economist, 27 March 1993, 84.

113. "Proved Right," Newsweek, 11 November 1996, 12.

114. "Money Lost in Space," Economist, 9 November 1996, 17.

115. A Survey of Germany, 9.

116. East of Eden, ibid. 24.

117. "Shock Treatment for the East," BusinessWeek, European ed., 17 November 1997, 33.

118. L. Davis, J. Hughes, and D. McDougall, American Economic History (New York: Richard Irwin, 1969), 136-138, 141-145, 414, 441.

119. "On the Sidelines but in the Game," BusinessWeek, European ed., 27 April 1998, 46.

120. German Unification, A Survey of Germany, ibid.

121. "Fiscal Consolidation in Germany," ibid.

122. World Development Report, 1998—Basic Indicators, ibid.

123. "What's Wrong," BusinessWeek International, 2 August 1993, 30; "Fewer Bangs—More Bucks," Economist, 15 July 1995, 66.

124. "The Addicts in Europe," Economist, 22 November 1997, 81.

125. "Trade's Great Irony," ibid. 17; The Federal Republic of Germany, 1, 86.

126. "Garbage Gap Alert," Newsweek, 28 October 1996, 17.

127. "When Virtue Pays a Premium," Economist, 18 April 1998, 65.

128. "Europe Is Catching Biotech Fever," BusinessWeek, European ed., 21 July 1997, 18.

129. "When Virtue Pays a Premium," ibid. 65.

130. "When Science and Religion Meet," Scientific American, February 1998, 18.

131. "The New Age of the Train," Economist, 21 February 1998, 25.

132. The Federal Republic of Germany, 36.

133. Bramsted, Germany, 11.

134. Harberger, ed., World Economic Growth, 102.

135. "The French Might Be on to Something," BusinessWeek, European ed., 23 June 1997, 14.

136. "The Manufacturing Myth," Economist, 19 March 1994, 81.

137. "Put Away Childish Things," Economist, 5 July 1995, 14.

138. "Science and Technology," Economist, 4 March 1995, 95.

139. "When Virtue Pays a Premium," ibid. 65.

140. Bramsted, Germany, 11.

141. The Federal Republic of Germany, 34, 3.

142. "Market in the Making," Far Eastern Economic Review, 22 July 1992, 38.

143. "Investment Ratios," Economist, 6 May 1995, 114.

144. Roustang, Développement economique de l'Allemagne Orientale, 69.

145. GDR and Eastern Europe, 12; World Development Reports, 1981, and 1991—Basic Indicators, ibid.

146. World Development Report, 1998—Basic Indicators, ibid.
147. Munting and Holderness, Crisis, Recovery, and War, 149.
148. "How Europe Can Create Jobs," Fortune, 9 August 1993, 18.
149. "The Right Track," Time, 8 December 1997, 34.
150. "The Changing Face of the Welfare State," Economist, 26 August 1995, 23.
151. "The European Community," Economist, 3 July 1993.
152. "European Monetary Union," Economist, 20 May 1995, 78.
153. "An Awfully Big Adventure," Economist, 11 April 1998.
154. "The Euro's Contribution to Growth," Time, 11 May 1998, 11.
155. "Sunshine and Showers," Economist, 29 April 1995, 96.
156. "Sick at Heart," Newsweek, 18 March 1996, 21.

Chapter 8

1. Andrew Gamble, *Britain in Decline* (London: Macmillan Education, 1985), 11.
2. Bernard Elbaum and William Lazonick, eds., *The Decline of the British Economy* (New York: Oxford University Press, 1986), 2.
3. G.C. Allen, *A Short Economic History of Modern Japan* (London: George Allen and Unwin, 1972).
4. *World Development Report, 1981, and 1998—Basic Indicators* (Washington, DC: World Bank).
5. Elbaum and Lazonick, eds., *The Decline of the British Economy,* 15.
6. Ibid., 11.
7. Roger Munting and B. Holderness, *Crisis, Recovery, and War* (New York: St. Martin Press, 1991), 246, 249.
8. *Britain 1984* (London: Central Office of Information, 1984), 181.
9. B. Alford, *British Economic Performance, 1945-1975* (Cambridge: Cambridge University Press, 1995), 26.
10. Ibid., 5.
11. G. Allen, *The British Disease* (London: Institute of Economic Affairs, 1979), 21.
12. Charles More, *The Industrial Age* (New York: Longman, 1989), 261.
13. Ibid., 266.
14. Munting and Holderness, *Crisis, Recovery, and War,* 249.
15. Roger Clarke and Dubravko Matko, *Soviet Economic Facts 1917-81* (London: Macmillan Press, 1983), 7-12.
16. Elbaum and Lazonick, eds., *The Decline of the British Economy,* 266, 14.
17. J. Allen and D. Massey, eds., *The Economy in Question* (London: SAGE, 1988), 14, 15.
18. Sidney Pollard, *The Development of the British Economy* (London: Routledge, 1972), 355.
19. John Keep, *Last of the Empires* (Oxford: Oxford University Press, 1995), 22, 84.
20. John Brunett, *Plenty and Want* (London: Routledge, 1989), 288-302.
21. Allen, *The British Disease,* 23.
22. More, *The Industrial Age,* 266.
23. Allen, *The British Disease,* 14-16.
24. Alford, *British Economic Performance, 1945-1975,* 26, 5.
25. Allen and Massey, eds., *The Economy in Question,* 14; *World Development Report, 1991—Basic Indicators,* ibid.
26. More, *The Industrial Age,* 267.
27. Allen, *The British Disease,* 22.

28. "Japan's Aerospace Giants Are Flying Low," *Newsweek,* 5 December 1994, 31.
29. More, *The Industrial Age,* 265.
30. "The Burdensome National Debt," *Economist,* 10 February 1996, 70.
31. Derek Aldcroft, *The British Economy,* vol. I (London: Wheatsheaf Books, 1986), 170.
32. Elbaum and Lazonick, eds., *The Decline of the British Economy,* 15.
33. More, *The Industrial Age,* 260.
34. "More of the Same," *Far Eastern Economic Review,* 5 October 1995, 88.
35. *Britain 1984,* 181.
36. Pollard, *The Development of the British Economy,* 367.
37. Elbaum and Lazonick, eds., *The Decline of the British Economy,* 190.
38. John Muellbauer and Anthony Murphy, "Is the UK Balance of Payments Sustainable?" *Economic Policy* 11 (October 1990): 386.
39. *The Cambridge History of Japan,* vol. 6 (Cambridge: Cambridge University Press, 1988), 508.
40. Allen, *The British Disease,* 18.
41. Paul Krugman, *The Age of Diminished Expectations* (Cambridge: MIT Press, 1991), 67.
42. More, *The Industrial Age,* 256.
43. Allen and Massey, eds., *The Economy in Question,* 16.
44. Alford, *British Economic Performance 1945-1975,* 5.
45. Derek Aldcroft and Harry Richardson, *The British Economy, 1870-1939* (London: Macmillan, 1969), 4, 65, 105.
46. More, *The Industrial Age,* 271.
47. Robert Wade, *Governing the Market* (Princeton: Princeton University Press, 1990), 96, 97, 43.
48. *Eurostat, 1975 and 1981* (Brussels: Statistiques de Base de la Communaut,), 104 and 107; *World Development Report, 1981—Basic Indicators,* ibid.
49. Pollard, *The Development of the British Economy,* 403.
50. More, *The Industrial Age,* 268.
51. "Inflation," *Economist,* 22 March 1997, 142.
52. "At Last—A Real Payoff from the Real Plan," *BusinessWeek,* European ed., 25 May 1998, 30.
53. *Trade Policy Review: Brazil,* vol. I (Geneva: GATT, 1992), 12, 23, 182.
54. More, *The Industrial Age,* 270.
55. "Why Seoul Is Seething," *BusinessWeek,* European ed., 27 January 1997, 24.
56. Pollard, *The Development of the British Economy,* 307, 422.
57. *Études economiques de l'OCDE 1996 Royaume-Uni* (Paris: OCDE), 22.
58. Pollard, *The Development of the British Economy,* 230-232.
59. Elbaum and Lazonick, eds., *The Decline of the British Economy,* 267.
60. More, *The Industrial Age,* 348.
61. Ibid., 272.
62. M.W. Kirby, *The Decline of British Economic Power Since, 1870* (London: George Allen and Unwin, 1981), 148-156.
63. Pollard, *The Development of the British Economy,* 377.
64. "Should We Praise Maggie or Bury Her?" *BusinessWeek International,* 28 August 1995, 6.
65. "Britain," *Economist,* 16 January 1993, 35.
66. "*The British Disease,*" *Economist,* 3 April 1993, 35.
67. "European Work Habits," *Economist,* 4 February 1995, 108.
68. "Money Supply Figures," *Economist,* 21 October 1995, 91.

69. Pollard, *The Development of the British Economy,* 417.
70. *Études economiques de l'OCDE 1996 Royaume-Uni,* 8
71. "Kiwi Fruits," *Far Eastern Economic Review,* 6 May 1993, 43.
72. *World Development Reports 1981 and 1991—Basic Indicators,* ibid.
73. "Garbage Gap Alert," *Newsweek,* 28 October 1996, 17.
74. *World Development Report 1998—Basic Indicators,* ibid.
75. "Pull Me Up—Weigh Me Down," *Economist,* 24 July 1993, 33.
76. "Divines Opine," *Economist,* 12 April 1997, 42.
77. "Workers of the World—Compete," *Economist,* 2 April 1994, 79, 80.
78. "Capital Wars," *Newsweek,* 3 October 1994, 32-37.
79. "The Global Economy," *Economist,* 1 October 1994, 7.
80. "Economic Viewpoint," *BusinessWeek International,* 5 April 1993, 10.
81. *World Development Report 1991 and 1998—Basic Indicators,* ibid.
82. "Eringo Boom," *Time,* 1 June 1998, 34.
83. *Trade Policy Review: Hong Kong* (Geneva: GATT, 1994), 3, 54-62.
84. "The Island Is Crazy for Chips," *BusinessWeek International,* 16 September 1996, 28.
85. "Attack the Frontiers," *Economist,* 8 April 1995, 18.
86. "Jekyll and Hyde," *Far Eastern Economic Review,* 13 July 1995, 71.
87. "American Banks," *Economist,* 7 November 1992, 92.
88. "Plugging the Holes in Nordic Banks," *Economist,* 9 January 1993, 67.
89. "Scandinavia," *Economist,* 10 September 1994, 31.
90. "Bank Alarms Are Blaring," *BusinessWeek International,* 26 June 1995, 16, 17.
91. "Asia: The Global Impact," *BusinessWeek,* European ed., 1 June 1998, 19-34.
92. "Capital Controversies," *Economist,* 23 May 1998, 76.
93. Aldcroft, The British Economy, vol. I, 190.
94. Pollard, *The Development of the British Economy,* 352.
95. Ibid., 351.
96. "The Knowledge Factory," *Economist,* 4 October 1997, 12.
97. "The Economics of European Integration," *Economist,* 22 May 1993, 28.
98. Gamble, *Britain in Decline,* 16.
99. Alford, *British Economic Performance, 1945-1975,* 5, 6, 26.
100. Ibid., 7.
101. Pollard, *The Development of the British Economy,* 297.
102. Muellbauer and Murphy, *Is the UK Balance of Payments Sustainable?* 386.
103. Pollard, *The Development of the British Economy,* 231; *World Development Report, 1991, and 1998—Basic Indicators,* ibid.
104. More, *The Industrial Age,* 333, 328.
105. Pollard, *The Development of the British Economy,* 236.
106. "The New Economy," *BusinessWeek,* European ed., 17 November 1997, 50.
107. "The Price of Success," *Far Eastern Economic Review,* 6 July 1995, 56.
108. "The French Might Actually Be on to Something," *BusinessWeek,* European ed., 23 June 1997, 14.

Bibliography

Books

Aberbach, Joel; Dollar, David; and Sokoloff, Kenneth, eds. *The Role of the State in Taiwan's Development.* Armonk, NY: M.E. Sharpe, 1994.
Adams, D. *America in the 20th Century.* London: Cambridge University Press, 1967.
Aldcroft, Derek. *The British Economy,* vol. I. London: Wheatsheaf Books, 1986.
Aldcroft, D., and Fearron, P., eds. *British Fluctuations, 1790–1939.* London: Macmillan, 1972.
Aldcroft, Derek, and Richardson, Harry. *The British Economy, 1870–1939.* London: Macmillan, 1969.
Alford, B. *British Economic Performance, 1945–1975.* Cambridge: Cambridge University Press, 1995.
Allen, G. *The British Disease.* London: Institute of Economic Affairs, 1979.
Allen, G.C. *A Short Economic History of Modern Japan.* London: George Allen and Unwin, 1972.
Allen, J., and Massey, D., eds. *The Economy in Question.* London: SAGE, 1988.
Almanach der Schweiz. Zurich: Verlag Peter Lang AG, 1978.
Ashworth, William. *An Economic History of England, 1870–1939.* London: Methuen, 1965.
Aslund, Anders. *Gorbachev's Struggle for Economic Reform.* London: Pinter, 1991.
Bagwell, Philip, and Mingay, G. *Britain and America, 1850–1939.* London: Routledge Kegan Paul, 1970.
Bairoch, Paul. *Economics and World History.* London: Harvester Wheatsheaf, 1993.
Bergere, M., Bianco, L., and Domes, J. *La Chine au Xxe Siècle.* Paris: Fayard, 1989.
Berghahn, V. *Modern Germany.* Cambridge: Cambridge University Press, 1985.
Bergier, Jean. *Naissance et croissance de la Suisse industrielle.* Lausanne: Francke Editions, 1974.
Bettelheim, Charles. *L'économie Allemande sous le Nazisme,* vols. I and II. Paris: François Maspero, 1971.
Blackwell, William. *The Beginnings of Russian Industrialization.* Princeton: Princeton

Blackwell, William. *The Beginnings of Russian Industrialization.* Princeton: Princeton University Press, 1968.

———. *The Industrialization of Russia.* New York: Thomas Crowell, 1970.

Bogart, Ernest, and Kemmerer, Donald. *Economic History of the American People.* New York: Longmans Green, 1947.

Bolton, J.L. *The Medieval English Economy, 1150–1500.* London: Garden City Press, 1980.

Boulding, Kenneth. *The Structure of a Modern Economy.* London: Macmillan Press, 1993.

Bramsted, Ernest. *Germany.* New York: Prentice-Hall, 1972.

Breach, R., and Hartwell, R., eds. *British Economy and Society, 1870–1970.* Oxford: Oxford University Press, 1972.

Bremond, J.; Chalaye, C.; and Loeb, M. *L'économie du Japon.* Paris: Hatier, 1987.

Brenner, Y. *A Short History of Economic Progress.* London: Frank Cass, 1969.

Brown, Richard. *Society and Economy in Modern Britain, 1700–1850.* London: Routledge, 1991.

Bruno, M., and Sachs, J. *Economics of Worldwide Stagflation.* Cambridge: Harvard University Press, 1985.

Burnett, John. *Plenty and Want.* London: Routledge, 1989.

Burns, Arthur. *Production Trends in the United States Since 1870.* National Bureau of Economic Research, 1934.

Cain, P.J. *Economic Foundations of British Overseas Expansion, 1815–1914.* London: Macmillan Press, 1980.

The Cambridge Economic History of Europe, vol. I (1970), vol. II (1987), vol. III (1978), vol. VII (1978). Cambridge: Cambridge University Press.

The Cambridge Economic History of India, vol. II. Cambridge: Cambridge University Press, 1983.

The Cambridge History of China, vol. I (1986), vol. III (1979), vols. VI, VII (1988). Cambridge University Press, Press, Cambridge, 1986.

The Cambridge History of Japan, vol. V. Cambridge: Cambridge University Press, 1989.

Campbell, Robert. *Soviet Economic Power.* London: Macmillan, 1967.

Central Office of Information. *Britain 1984.* London: 1984.

Chambers, J.D. *The Workshop of the World.* Oxford: Oxford University Press, 1961.

———. *Population, Economy, and Society in Pre-Industrial England.* Oxford: Oxford University Press, 1972.

Chandler, Lester. *America's Greatest Depression.* New York: Harper and Row, 1970.

Chaudhuri, K. *Asia Before Europe.* Cambridge: Cambridge University Press, 1990.

Chevalier, François. *L'Amérique Latine.* Paris: Presses Universitaires de France, 1993.

Chou Chin-sheng. *An Economic History of China.* Bellingham, WA: Western Washington State College, 1974.

Clapham, John. *A Concise Economic History of Britain.* Cambridge: Cambridge University Press, 1949.

Clark, George. *The Wealth of England from 1496 to 1760.* Oxford: Oxford University Press, 1946.

Clark, Peter, and Slack, Paul. *English Towns in Transition 1500–1700.* Oxford: Oxford University Press, 1976.

Clarke, Roger. *Soviet Economic Facts 1917–70.* London: Macmillan, 1972.

Clarke, Roger, and Matko, Dubravko. *Soviet Economic Facts 1917–81.* London: Macmillan, 1983.

Clarkson, Leslie. *Death, Disease, and Famine in Pre-Industrial England.* London: Gill and Macmillan, 1975.

Cleere, Henry, and Crossley, David. *The Iron Industry of the Weald.* London: Leicester University Press, 1985.
Clough, Shepard. *Histoire économique des Étas-Unis.* Paris: Presses Universitaires de France, 1953.
Clough, S.B. Grandeur et décadence des civilisations. Paris: Payot, 1954.
Cohen, Stephen. *Rethinking the Soviet Experience.* Oxford: Oxford University Press, 1985.
Coleman, D.C. *The Economy of England, 1415–1750.* Oxford: Oxford University Press, 1977.
Cowan, C.D., ed. *The Economic Development of China and Japan.* London: George Allen and Unwin, 1964.
Cullen, L., and Smout, T., eds. *Comparative Aspects of Scottish and Irish Economic and Social History, 1600–1900.* Edinburgh: John Donald, 1976.
Davis, Lance; Easterlin, Richard; and Parker, William, eds. *American Economic Growth.* New York: Harper and Row, 1972.
Davis, Lance, and Huttenback, Robert. *Mammon and the Pursuit of Empire.* Cambridge: Cambridge University Press, 1988.
Davis, L.; Hughes, J.; and McDougall, D. *American Economic History.* New York: Richard Irwin, 1969.
Deane, Phyllis. *The First Industrial Revolution.* Cambridge: Cambridge University Press, 1979.
Deiss, Joseph. *Economie politique et politique économique de la Suisse.* Geneva: Editions Fragniere, 1979.
Deutsche Bundesbank Annual Report 1992–95. Frankfurt.
Devin, J., and Dickson, D., eds. *Ireland and Scotland, 1600–1850.* Edinburgh: John Donald, 1983.
Dobb, Maurice. *Soviet Economic Development Since 1917.* London: Routledge, 1960.
Dore, R. *Structural Adjustment in Japan, 1970–82.* Geneva: International Labor Organization, 1986.
Duchéne, Gerard. *L'économie de l'URSS.* Paris: Editions La Decouvert, 1987.
Eberhard, Wolfram. *A History of China.* Berkeley: University of California Press, 1977.
Economic Development Board. *Economic Development of Singapore 1960–91.* Singapore: 1992.
L'économie Suisse, 1946–1986. Union de Banque Suisses, 1987.
Elbaum, Bernard, and Lazonick William, eds. *The Decline of the British Economy.* New York: Oxford University Press, 1986.
Etudes économiques de l'OCDE-1996-Royaume-Uni. Paris: OCDE, 1996.
Fairbank, John. *China: A New History.* Cambridge: Harvard University Press, 1992.
Faulkner, Harold. *Histoire économique des États-Unis d'Amérique,* vols. I, IV. Paris: Presses Universitaires de France, 1958.
Fite, Gilbert, and Reese, Jim. *An Economic History of the United States.* Boston: Houghton Mifflin, 1965.
Floud, Roderick, and McCloskey, Donald, eds. *The Economic History of Britain Since 1700,* Vol. I (1994) and II (1981). Cambridge: Cambridge University Press.
The Fontana Economic History of Europe: The Emergence of Industrial Societies. London: William Collins and Sons, 1976.
The Fontana Economic History of Europe: Contemporary Economies. London: William Collins and Sons, 1976.
Fraser, J., and Gerstle, G. *The Rise and Fall of the New Deal Order.* Princeton: Princeton University Press, 1989.
Gadgil, D.R. *The Industrial Evolution of India in Recent Times, 1860–1939.* London: Oxford University Press, 1972.

Gatrell, Peter. *The Tsarist Economy, 1850–1917.* London: B.T. Batsford, 1986.
Gaxotte, Pierre. *Histoire de l'Allemagne.* Paris: Flammarion, 1975.
George, Pierre. *L'économie de l'URSS.* Paris: Presses Universitaires de France, 1965.
Germany-OECD–Economic Surveys. Paris: OECD, 1992.
Gernet, Jacques. *La Chine ancienne.* Paris: Presses Universitaires de France, 1992.
Goodman, D., and Segal, G. *China in the Nineties.* New York: Oxford University Press, 1991.
Grassby, Richard. *The Business Community of Seventeenth Century England.* Cambridge: Cambridge University Press, 1995.
Gray, Jack. *Rebellions and Revolutions.* New York: Oxford University Press, 1990.
Gray, J., and White, G., eds. *China's New Development Strategy.* London: Academic Press, 1982.
Guroff, Gregory, and Cartensen, Fred, eds. *Entrepreneurship in Imperial Russia and the Soviet Union.* Princeton: Princeton University Press, 1983.
Hacker, Louis. *Major Documents in American Economic History,* vol. I. New York: D. Van Nostrand, 1961.
Harberger, Arnold, ed. *World Economic Growth.* Oakland: ICS Press, 1984.
Hardach, Karl. *Wirtschafts Geschichte Deutschlands im 20 Jahrhundert.* Goettingen: Vandenhoeck und Ruprecht, 1979.
Harding, Harry. *China's Second Revolution.* Washington, DC: Brookings Institution, 1987.
Harris, Seymour. *American Economic History.* New York: McGraw-Hill, 1961.
Heisenberg, Wolfgang, ed. *German Unification in European Perspective.* Brussels: Brassey's, 1991.
Hill, C.P. *British Economic and Social History.* London: Edward Arnold, 1977.
Hobday, Michael. *Innovations in East Asia.* Cheltenham, UK: Edward Elgar, 1997.
Hobsbawn, E.J. *Industry and Empire.* London: Weidenfeld and Nicolson, 1968.
Holborn, Hajo. *A History of Modern Germany,* vol. I. New York: Alfred A. Knopf, 1967.
Holderness, B.A. *Pre-Industrial England.* London: J.M. Dent and Sons, 1976.
Hopkins, Keith, ed. *Hong Kong: The Industrial Colony.* London: Oxford University Press, 1971.
Hou Chi-ming. *Foreign Investment and Economic Development in China, 1840–1937.* Cambridge: Harvard University Press, 1965.
Howe, Christopher. *China's Economy.* London: Granada, 1978.
Huff, W.G. *The Economic Growth of Singapore.* Cambridge: Cambridge University Press, 1994.
Hughes, Jonathan. *American Economic History.* Glenview, IL: Scott Foresman, 1987.
International Monetary Fund. *A Case of Successful Adjustment: Korea's Experience During 1980-84.* Washington, DC: August 1985.
———. *The Federal Republic of Germany.* Washington, DC: January 1991.
———. *Thailand: Adjusting for Success.* Washington, DC: August 1991.
———. *Mexico: The Strategy to Achieve Sustained Economic Growth.* Washington, DC: 1992.
———. *China at the Threshold of a Market Economy.* Washington, DC: 1993.
Jack, Sybil. *Trade and Industry in Tudor and Stuart England.* London: George Allen and Unwin, 1977.
Jao, Hungdah Chiu, and Wu, Yuan-li, eds. *The Future of Hong Kong.* New York: Quorum Books, 1987.
Japan Times. Japan's Economy and Japan–U.S. Trade. Tokyo: 1982.
Le Japon d'aujourd'hui. Tokyo: Ministère des Affaires Etrangères, 1971.
Jones, E.L., ed. *Agriculture and Economic Growth in England, 1650–1815.* London:

Jones, E.L., ed. *Agriculture and Economic Growth in England, 1650–1815.* London: Methuen, 1967.
Kahan, Arcadius. *The Plow, the Hammer, and the Knout.* Chicago: University of Chicago Press, 1985.
———. *Russian Economic History.* Chicago: University of Chicago Press, 1989.
Kaldor, Nicholas. *Causes of Growth and Stagnation in the World Economy.* Cambridge: Cambridge University Press, 1996.
Kasar, Michael. *Soviet Economics.* London: Weidenfeld and Nicolson, 1970.
Keep, John. *Last of the Empire.* Oxford: Oxford University Press, 1995.
Kellenbenz, Herman. *The Rise of the European Economy.* London: Weidenfeld and Nicolson, 1976.
Kenwood, A., and Lougheed, A. *The Growth of the International Economy, 1820–1990.* London: Routledge, 1992.
Kerblay, Basile. *Modern Soviet Society.* London: Methuen, 1983.
King, Frank. *Money and Monetary Policy in China.* Cambridge: Harvard University Press, 1965.
Kirby, M.W. *The Decline of British Economic Power Since 1870.* London: George Allen and Unwin, 1981.
Kirchner, Walter. *History of Russia.* New York: Barnes and Noble, 1965.
Kitchen, Martin. *The Political Economy of Germany, 1815–1914.* Montreal: McGill-Queen's University Press, 1978.
Kneschaurek, Francesco. *La Suisse face à une Nouvelle phase de son developpement.* Lausanne: Centre de Récherches Europées, 1975.
Kolko, Gabriel. *Main Currents in Modern American History.* New York: Harper and Row, 1976.
Kort, Michael. *The Soviet Colossus.* New York: Unwin Hyman, 1990.
Krisch, Henry. *The German Democratic Republic.* Boulder, CO: Westview Press, 1985.
Krout, John. *The United States Since 1865.* New York: Barnes and Noble, 1965.
———. *The United States to 1877.* New York: Barnes and Noble, 1966.
Krugman, Paul. *The Age of Diminished Expectations.* Cambridge: MIT Press, 1991.
Kuznets, Simon. *Modern Economic Growth.* New Haven: Yale University Press, 1966.
La Grande Bretagne à la fin du XXe siàcle: Notes et ètudes documentaires, Decembre 1994.
Landes, David. *The Wealth and Poverty of Nations.* New York: W.W. Norton, 1998.
Lee, C.H. *The British Economy Since 1700: A Macroeconomic Perspective.* Cambridge: Cambridge University Press, 1986.
Levine, Salomon, and Kawada, Hisashi. *Human Resources in Japanese Industrial Development.* Princeton: Princeton University Press, 1980.
Lombard, Denys. *La Chine Imperiale.* Paris: Presses Universitaires de France, 1967.
The Long Debate on Poverty. London: Institute of Economic Affairs, 1972.
Luck, Murray, ed. *Modern Switzerland.* Sposs, 1978.
Madison, Angus. *Economic Growth in Japan and the USSR.* London: George Allen and Unwin, 1969.
Major, J., and Kane, A. *China Briefing, 1987.* Boulder, CO: Westview Press, 1987.
Mann, Golo. *The History of Germany Since 1789.* London: Chatto and Windus, 1968.
Mathias, Peter. *The First Industrial Nation.* Methuen, 1983.
McCloskey, Donald, ed. *Essays on a Mature Economy: Britain After 1840.* London: Methuen, 1971.
Merrick, Thomas, and Graham, Douglas. *Population and Economic Development in Brazil.* Baltimore: Johns Hopkins University Press, 1979.
Millward, R., and Singleton, J., eds. *The Political Economy of Nationalization in Britain, 1920–1950.* Cambridge: Cambridge University Press, 1995.

Milward, A., and Saul, S.B. *The Development of the Economies of Continental Europe, 1850–1914.* London: George Allen and Unwin, 1977.

Ministry of Trade and Industry. *Singapore in Brief, 1991.* Singapore: 1992.

More, Charles. *The Industrial Age.* New York: Longman, 1989.

Morishima, Michio. *Why Has Japan Succeeded?* Cambridge: Cambridge University Press, 1982.

Munting, Roger, and Holderness, B. *Crisis, Recovery, and War.* New York: St. Martins Press, 1991.

Murphy, Brian. *A History of the British Economy, 1086–1740.* London: Longman Group, 1973.

Musson, A.E. *The Growth of British Industry.* London: B.T. Batsford, 1978.

Nafziger, Wayne. *Learning from the Japanese.* Armonk, NY: M.E. Sharpe, 1995.

Needham, Joseph. *Clerks and Craftsmen in China and the West.* Cambridge: Cambridge University Press, 1970.

New Encyclopaedia Britannica—Marcropaedia, vol. 29. Chicago: Encyclopaedia Britannica, 1988.

Niemi, Albert. *U.S. Economic History.* Skokie, IL: Rand McNally, 1975.

North, Douglas. *The Growth of the American Economy to 1860.* New York: Harper and Row, 1968.

Nove, Alec. *An Economic History of the USSR.* London: Penguin Press, 1969.

———. *Political Economy and Soviet Socialism.* London: George Allen and Unwin, 1979.

OECD Economic Outlook. Paris: OECD, 1990.

OECD Economic Surveys: Switzerland. Paris: OECD, 1992.

OECD Economic Surveys: Japan. Paris: OECD, 1992.

Parker, Geoffrey, ed. *The General Crisis of the Seventeenth Century.* London: Routledge and Kegan Paul, 1978.

Parker, W.H. *A Historical Geography of Russia.* London: University of London Press, 1968.

Perkins, Dwight, ed. *China's Modern Economy in Historical Perspective.* Stanford: Stanford University Press, 1975.

Pinson, Roppel. *Modern Germany.* New York: Macmillan, 1966.

Pollard, Sidney. *The Development of the British Economy.* London: Routledge, 1992.

Pope, Rex, ed. *Atlas of British Social and Economic History Since 1700.* London: Routledge, 1989.

Porter, Michel. *The Competitive Advantage of Nations.* London: Macmillan Press, 1990.

Ramsey, Peter. *The Price Revolution in Sixteenth Century England.* London:: Methuen, 1971.

Rawski, Thomas. *Economic Growth and Employment in China.* New York: Oxford University Press, 1979.

———. *Economic Growth in Pre-War China.* Berkeley: University of California Press, 1989.

Reynaud, Alain. *Une geohistoire: La Chine de printemps et des automnes.* France: Reclus, 1992.

Riado, Pierre. *L'Amérique Latine.* Paris: Presses Universitaires de France, 1993.

Robock, Stefan. *Brazil: A Study in Development Progress.* Lexington, MA: D.C. Heath, 1976.

Rose, Michael. *The Relief of Poverty, 1834–1914.* London: Macmillan Press, 1972.

Rosenberg, Nathan. *Technology and American Economic Growth.* New York: Harper and Row, 1972.

Rossabi, Morris, ed. *China Among Equals.* Berkeley: University of California Press, 1983.
Rostow, Walt. *The World Economy.* Austin: University of Texas Press, 1978.
Roustang, Guy. *Développement économique de l'Allemagne Orientale.* Paris: Société d'Edition d'Enseignement Superiéur, 1963.
Smith, R., and Christian, D. *Bread and Salt.* Cambridge: Cambridge University Press, 1984.
Statistics de base de la Communauté, Eurostat. Bruxelles, 1992.
Supple, Barry, ed. *The Experience of Economic Growth.* New York: Random House, 1963.
Swiss National Bank. *Lage und Probleme der Schweizerischen Wirtschaft.* Bern: 1977.
Tabb, William. *The Postwar Japanese System.* New York: Oxford University Press, 1995.
Tames, Richard. *Economy and Society in Nineteenth Century Britain.* London: George Allen and Unwin, 1972.
Tanaka, Heizo. *Contemporary Japanese Economy and Economic Policy.* Ann Arbor: University of Michigan Press, 1991.
Taylor, Arthur. *Laissez-Faire and State Intervention in Nineteenth Century Britain.* Hong Kong: Macmillan, 1972.
Taylor, Arthur, ed. *The Standard of Living in Britain in the Industrial Revolution.* London: Methuen, 1975.
Thane, Pat, and Sutcliffe, Anthony, eds. *Essays in Social History.* New York: Oxford University Press, 1986.
Thompson, F., ed. *Landowners, Capitalists, and Entrepreneurs.* Oxford: Oxford University Press, 1994.
Tomlinson, Jim. *Problems of British Economic Policy, 1870–1945.* London: Methuen, 1981.
Trade Policy Review: Brazil, vols. I and II. Geneva: GATT, 1992.
Trade Policy Review: Hong Kong. Geneva: GATT, 1990 and 1994.
Trade Policy Review: Indonesia, vols. I and II. Geneva: GATT, 1991.
Trade Policy Review: Japan, vols. I and II. Geneva: GATT, 1992.
Trade Policy Review: Republic of Korea, vols. I and II. Geneva: GATT, 1992.
Trade Policy Review: Singapore, vols. I and II. Geneva: GATT, 1992.
Trade Policy Review: Switzerland, vols. I and II. Geneva: GATT, 1991.
Trevelyan, G.M. *A Shortened History of England.* New York: Longmans Green, 1959.
Tsuru, Shigeto. *Japan's Capitalism.* Cambridge: Cambridge University Press, 1993.
Vie, Michael. *Le Japon contemporain.* Paris: Presses Universitaires de France, 1971.
Vogel, Erza. *Japan as No. 1.* Cambridge: Harvard University Press, 1979.
Wade, Robert. *Governing the Market.* Princeton: Princeton University Press, 1990.
Westwood, J. *Endurance and Endeavor.* Oxford: Oxford University Press, 1993.
Will, Pierre. *Bureaucratie et famine en Chine au 18e Siècle.* Paris: Mouton Editeur, 1980.
World Bank. *East Asian Miracle, Economic Growth, and Public Policy.* Washington, DC: IBRD, 1993.
Woronoff, Jon. *Hong Kong: Capitalist Paradise.* Hong Kong: Heinemann Educational Books, 1980.
Wright, Chester. *Economic History of the United States.* New York: McGraw-Hill, 1941.
Wrigley, E. *Continuity, Chance, and Change.* Cambridge: Cambridge University Press, 1988.
Yearbook of Statistics Singapore 1991. Singapore: Department of Statistics, 1992.
Zysman, John, and Cohen, Stephen. *Manufacturing Matters.* New York: Basic Books, 1987.

Articles

Ackerman, Jennifer. "Islands at the Edge." *National Geographic,* 2 August 1997, 2, 31.

Ainsworth, Martha. "Fertility in Sub-Saharan Africa." *World Bank Economic Review* 10, no. 1 (January 1996): 81–84.

Ainsworth, Martha, and Over, Mead. "Aids and African Development." *Research Observer* 9, no. 2 (July 1994): 203–239.

Alesina, Alberto, and Perutti, Roberto. "Fiscal Adjustment." *Economic Policy* 21 (October 1995): 205–248.

Alley, Richard, and Bender, Michael. "Greenland Ice Cores—Frozen in Time." *Scientific American,* February 1998, 66–71.

Anderson, Dennis. "La Energia y el Medio Ambiente." *Finanzas and Desarrollo* 33, no. 2 (Junio 1997): 10–13.

Anderson, Kym. "Multilateral Trade Negotiations, European Integration and Farm Policy Reform." *Economic Policy* 18 (April 1994): 13–52.

Annez, Patricia, and Friendly, Alfred. "Ciudades del Mundo en Desarrollo." *Finanzas and Desarrollo* 33, no. 4 (Diciembre 1996): 12–14.

Antremont, Antoine; Corpet, Denis; and Courvalin, Patrice. "La résistance des bactéries aux antibiotiques." *Pour La Science,* Février 1997, 66–73.

Ayres, Wendy, and Mccalla, Alex. "Desarrollo Rural, Agricultura y Seguridad Alimentaria." *Finanzas and Desarrollo* 33, no. 4 (Diciembre 1996): 8–11.

Balcerowicz, Leszek. "Common Fallacies in the Debate on the Transition to a Market Economy." *Economic Policy* 19 (Supplement December 1994): 16–50.

Ballard, Robert. "High-Tech Search for Roman Shipwrecks." *National Geographic,* April 1998, 32–41.

Barr, Nicholas. "Dimensiones Humanas de la Transicion: Reforma de la Educacion y la Salud." *Finanzas and Desarrollo* 33, no. 3 (Septiembre 1996): 24–27.

Bartolini, Leonardo; Razin, Assaf; and Symansky, Steve. "G-7 Fiscal Restructuring in the 1990s." *Economic Policy* 20 (April 1995): 111–146.

Beck, Gregory, and Habicht, Gail. "Le Système Immunitaire des Invertébrés." *Pour la Science,* Janvier 1997, 46–51.

Benchley, Peter. "French Polynesia." *National Geographic,* June 1997, 2–29.

Bentolila, Samuel, and Dolado, Juan. "Labor Flexibility and Wages." *Economic Policy* 18 (April 1994): 53–100.

Berns, Michael. "Laser Scissors and Tweezers." *Scientific American,* April 1998, 52–57.

Birdsall, Nancy; Ross, David; and Sabot, Richard. "Inequality and Growth Reconsidered." *World Bank Economic Review* 9, no. 3 (September 1995): 477–508.

Blackhurst, Richard. "The WTO and the Global Economy." 20, no. 5 (August 1997): 527–544.

Blanc, Paul; Gauthier, François; and Ledoux, Emmanuel. "Déchets nucléaires naturels." *Pour La Science,* Juillet 1997, 82–87.

Blanchard, Olivier, and Muet, Pierre. "Competitiveness Through Disinflation." *Economic Policy* 16 (April 1993): 11–56.

Borish, Michael, and Noël, Michel. "Fomento del Sector Privado en los Paises de Visegrad." *Finanzas and Desarrollo* 33, no. 4 (Diciembre 1996): 43–46.

Brazaitis, Peter; Watanabe, Myrna; and Amato, George. "The Caiman Trade." *Scientific American,* March 1998, 52–58.

Buckley, Robert, and Gurenko, Eugene. "Housing and Income Distribution in Russia." *Research Observer* 12, no. 1 (February 1997): 19–33.

Buiter, Willem; Corsetti, Giancarlo; and Roubini, Nouriel. "Excessive Deficits." *Economic Policy* 16 (April 1993): 57–100.

Burda, Michael. "Unemployment, Labor Markets, and Structural Change in Eastern Europe." *Economic Policy* 16 (April 1993): 101–138.

Campbell, Colin, and Laherrère, Jean. "The End of Cheap Oil." *Scientific American,* March 1998, 60–65.

Caprio, Gerard, and Levine, Ross. "Reforming Finance in Transitional Socialist Economies." *Research Observer* 9, no. 1 (January 1994): 1–25.

Chadwick, Douglas. "Blue Refugees." *National Geographic,* March 1998, 2–31.

———. "Planet of the Beetles." *National Geographic,* March 1998, 101–118.

Claessens, Stijn. "The Emergence of Equity Investment in Developing Countries." *World Bank Economic Review* 9, no. 1 (January 1995): 1–17.

Clements, Benedict; Gupta, Sanjeev; and Schiff, Jerald. "Que Paso con los Dividendos de la Paz?" *Finanzas and Desarrollo* 34, no. 1 (Marzo 1997): 17–19.

Cohen, Daniel; Lefranc, Arnaud; and Saint–Paul, Gilles. "French Unemployment." *Economic Policy* 25 (October 1997): 265–292.

Coiteux, Martin. "El Tipo de Cambrio Real en Argentina."

Comercio Exterior. "Perfiles de la Industria Maquiladora." Vol. 47, no. 5 (Mayo 1997): 367–374

———. Vol. 47, no. 7 (Julio 1997): 522–531.

Curtisinger, Bill, and Kristof, Emory. "Testing the Waters of Rongelap." *National Geographic,* April 1998, 63–75.

De Grauwe, Paul. "Towards European Monetary Union Without the EMS." Economic Policy 18 (April 1994): 147–185.

Dehaene, Stanislas. "Comment Notre Cerveau Calcule-t-il?." *Pour La Science,* Juin 1997, 50–57.

De la Fuente, Angel, and Vives, Xavier. "Infrastructure and Education as Instruments of Regional Policy." *Economic Policy* 20 (April 1995): 13–51.

De Masi, Paula. "El Dificil Arte de Formular Proyecciones Economicas." *Finanzas and Desarrollo,* 33, no. 4 (Diciembre 1996): 29–31.

De Melo, Martha; Denizer, Cevdet; and Gelb, Alan. "Patterns of Transition from Plan to Market." *World Bank Economic Review* 10, no. 3 (September 1996): 397–423.

Deininger, Klaus, and Squire, Lyn. "Crecimiento Economico y Desigualdad en el Ingreso." *Finanzas and Desarrollo* 34, no. 1 (Marzo 1997): 36–39.

Demery, Lionel, and Squire, Lyn. "Macroeconomic Adjustment and Poverty in Africa." *Research Observer* 11, no. 1 (February 1996): 39–59.

Demirgüç-Kunt, Asli, and Levine, Rose. "Stock Markets, Corporate Finance, and Economic Growth." *World Bank Economic Review* 10, no. 2 (May 1996): 223–239.

Dini, Marco, and Katz, Jorge. "Nuevas Formas de Encarar las Politicas Technologicas." *Comercio Exterior* 47, no. 8 (Agosto 1997): 607–624.

Dixon, John, and Hamilton, Kirk. "Ampliacion de los Criterios de Evaluacion de la Riqueza." *Finanzas and Desarrollo* 33, no. 4 (Diciembre 1996): 15–18.

Dodsworth, John. "El Despegue Econmico en Indochina." *Finanzas and Desarrollo* 34, no. 1 (Marzo 1997): 20–23.

Dolowitz, David. "British Employment Policy in the 1980s." *Governance* 10, no. 1 (January 1997): 23–42.

Dominguez, Lilia, and Brow, Flor. "México-Patrones de Competencia y Apertura Economica." *Comercio Exterior* 47, no. 9 (Septiembre 1997): 695–704.

Dooley, Michael; Fernàndez, Eduardo; and Kietzer, Kenneth. "Is the Debt Crisis History?" *World Bank Economic Review* 10, no. 1 (January 1996): 27–49.

Dunning, John. "The European Internal Market Programme and Inbound Foreign Direct Investment." *Journal of Common Market Studies* 35, no. 1 (March 1997): 1–30.

Easterly, William, and Fischer, Stanley. "The Soviet Economic Decline." *World Bank Economic Review* 9, no. 3 (September 1995): 341–371.

Easterly, William, and Schmidt-Hebbel, Klaus. "Fiscal Deficits and Macroeconomic

211–237.

Eberlein, Burkard. "French Center-Periphery Relations and Science Park Development." *Governance* 9, no. 4 (October 1996): 351–374.

Economic Commission for Europe-UNO. "Economic Reforms in the European Centrally Planned Economies." New York: 1989.

Edwars, Mike. "China's Gold Coast." *National Geographic,* March 1997, 2–31.

Eichengreen Barry. "A Payment Mechanism for the Former Soviet Union." *Economic Policy* 17 (October 1993): 309–354.

Eichengreen, Barry. "Financing Infrastructure in Developing Countries." *Research Observer* 10, no. 1 (February 1995): 75–91.

Eichengreen Barry; Rose, Andrew; and Wyplosz, Charles. "Exchange Rate Mayhem." *Economic Policy* 17 (October 1995): 249–312.

Eliot, John. "Polar Bears," *National Geographic* 193, no. 1 (January 1998): 52–70.

Enoch, Charles, and Quintyn, Marc. "Union Monetaria Europea." *Finanzas and Desarrollo* 33, no. 3 (Septiembre 1996), 28–31.

Ernst, Andreas. "Psychologie des Unweltverhaltens." *Spektrum der Wissenschaft,* April 1998, 70–75.

Eskeland, Gunnar, and Jimeney, Emmanuel. "Policy Instruments for Pollution Control in Developing Countries." *Research Observer* 7, no. 2 (July 1992): 145–169.

Evans, John. "New Satellites for Personal Communications." *Scientific American,* April 1998, 60–67.

Ewen, John. "La Synthèse Ordonnée des Matières Plastiques." *Pour La Science,* Juillet 1997, 76–81.

Fabian, Katalin. "Privatization in Central Europe." *Governance* 8, no. 2 (April 1995): 218, 242.

Felgner, Phili. "La Thérapie Génique sans Virus." *Pour La Science,* Novembre 1997, 66–70.

Feltestein, Andrew, and Ha, Jiming. "The Role of Infrastructure in Mexican Economic Reform." *World Bank Economic Review* 9, no. 2 (May 1995): 287–304.

Ferdows, Kasra. "Making the Most of Foreign Factories." *Harvard Business Review* (March–April 1997): 73–88.

Ferrer, Aldo. "El Mercosur: Entre el Consenso de Washington y la Integracion Sustenable." *Comercio Exterior* 47, no. 5 (Mayo 1997): 347–354.

Fischer, Stanley. "Mantenimiento de la Estabilidad de Precios." *Finanzas and Desarrollo* 33, no. 4 (Diciembre 1996): 32–35.

———. "La Solidez del Systema Financiero." *Finanzas and Desarrollo* 34, no. 1 (Marzo 1997): 14–16.

Folkers, Gerd, and Kubinyi, Hugo. "Vers une conception rationnelle de médicaments." Pour La Science, Novembre 1997, 40–47.

Freeman, Richard. "Towards an Apartheid Economy?" *Harvard Business Review* (September–October 1996): 114–126.

Friedmann, Theodore. "La Thérapie Génetique: Promesses et Limites." *Pour La Science,* Novembre 1997, 60–65.

Gang, Fan. "Incremental Changes and Dual–Track Transition." *Economic Policy* 19 (Supplement December 1994): 99–122.

Genschel, Philip. "The Dynamics of Inertia." *Governance* 10, no. 1 (January 1997): 43–66.

George, Richard. "Mining for Oil." *Scientific American,* March 1998, 66–77.

Giaimo, Susan. "Health Care Reform in Britain and Germany." *Governance* 8, no. 3 (July 1995): 354–379.

Goldstein, Joshua; Huang, Xia-oming; and Akan, Burcu. "Energy in the World Economy

Goldstein, Joshua; Huang, Xia-oming; and Akan, Burcu. "Energy in the World Economy 1950–1992." *International Studies Quarterly* 41, no. 2 (June 1997): 241–266.
Gore, Rick. "Expanding Worlds." *National Geographic,* May 1997, 86–109.
———. "The Dawn of Humans." *National Geographic,* July 1997, 96–113.
Gourley, Paul. "Nanolasers." *Scientific American,* March 1998, 40–45.
Guerra, Alfredo. "Globalizaci6n de la Regionalizacion en Am,rica Latina." *Comercio Exterior* 46, no. 6 (Junio 1996): 436–442.
Halstead, Lauro. "Post-Polio Syndrome," *Scientific American,* April 1998, 36–41.
Harvard Business Review. The Challenge of Going Green." (July–August 1994): 37–50.
———. "Finding a Lasting Cure for U.S. Health Care." (September–October 1994): 45–63.
Haseltine, William. "Gensuche fur medizinische Entwicklunger." *Spektrum der Wissenschaft,* Mai 1997, 64–70.
Havrylyshyn, Oleh. "Reviving Trade Amongst the Newly Independent States." *Economic Policy* 19 (Supplement December 1994): 171–190.
Hedin, Lars, and Zikens, Gene. "Atmosph,,rischer Staub und saurer Regen." *Spektrum der Wissenschaft,* April 1997, 52–55.
Helpman, Elhanan; Leiderman, Leonardo; and Bufman, Gil. "A New Breed of Exchange Rates." *Economic Policy* 19 (October 1994): 259–306.
Henderson, Rebecca. "Managing Innovation in the Information Age." *Harvard Business Review* (January–February 1994): 100–105.
Henderson, Vernon, and Kuncoro, Ari. "Industrial Centralization in Indonesia." *World Bank Economic Review* 10, no. 3 (September 1996): 513–539.
Herzlinger, Regina. "Can Public Trust in Nonprofits and Governments Be Restored?" *Harvard Business Review* (March–April 1996): 97–107.
Hollister, Charles, and Nadis, Steven. "Burial of Radioactive Waste Under the Seabed." *Scientific American,* January 1998, 40–45.
Howard, Alice, and Magretta, Joan. "Surviving Success." *Harvard Business Review* (September–October 1995): 109–118.
Hubbard, Michael. "Bureaucrats and Markets in China." *Governance* 8, no. 3 (July 1995): 354–379.
Hubschmid, Claudia, and Moser, Peter. "The Co-operation Procedure in the EU." *Journal of Common Market Studies* 35, no. 2 (June 1997): 173–188.
Hyde, William; Amacher, Gregory; and Magrath, William. "Deforestation and Forest Land Use." *Research Observer* 11, no. 2 (August 1996): 223–247.
Isham, Jonathan; Kaufmann, Daniel; and Pritchett, Lant. "Civil Liberties, Democracy and the Performance of Government Projects." *World Bank Economic Review* 11, no. 2 (May 1997): 219–241.
Johnson, Simon, and Loveman, Gary. "Starting Over: Poland After Communism." *Harvard Business Review* (March–April 1995): 44–57.
Karl, Thomas; Nicholls, Neville; and Gregory, Jonathan. "Le Climat de Demain." *Pour La Science,* Juillet 1997, 38–43.
Kikeri, Sunita; Nellis, John; and Shirley, Mary. "Privatization: Lessons from Market Economies." *Research Observer* 9, no. 2 (July 1994): 241–271.
Klum, Mathias. "Malaysia's Secret Realm." *National Geographic,* August 1997, 122–131.
Kovacs, Kit. "Bearded Seals." *National Geographic,* March 1997, 124–136.
Krugman, Paul."A Country Is Not a Company." *Harvard Business Review* (January–February 1996): 40–51.
———. "Does Third World Growth Hurt First World Prosperity?" *Harvard Business Review* (July–August 1994): 113–121.

Kuemmerle, Walter. "Building Effective R&D Capabilities Abroad." *Harvard Business Review* (March–April 1997): 61–70.
Kvint, Vladimir. "Don't Give Up on Russia." *Harvard Business Review* (March–April 1994): 62, 74.
Landry, Donald. "Katalytische Antikürper Gegen Kokainsucht." *Spektrum der Wissenschaft,* April 1997, 56–59.
Lang, Kenneth. "Das Sonnenobservatorium SOHO." *Spektrum der Wissenschaft* (Mai 1997): 44–52.
Leonard, Jonathan, and Audenrode, Marc Van. "Corporatism Run Amok." *Economic Policy* 17 (October 1993): 355–400.
Levinson, Marc. "Capitalism with a Safety Net?" *Harvard Business Review* (September–October 1996): 173–181.
Levy, Stuart. "The Challenge of Antibiotic Resistance." *Scientific American,* March 1998, 32–39.
Lindauer, David, and Valenchik, Ann. "Government Spending in Developing Countries." *Research Observer* 7, no. 1 (January 1992): 59–77.
Long, Michael. "The Grand Managed Canyon." *National Geographic,* July 1997, 114–135.
———. "The Vanishing Prairie Dog." *National Geographic,* April 1998, 118–131.
Losick, Richard, and Kaiser, Dale. "Wie und Warum Bakterien Kommunizieren." *Spektrum der Wissenschaft,* April 1997, 78–84.
Lucid, Shannon. "Six Months on Mir." *Scientific American,* May 1998, 26–35.
Lusting, Nora. "NAFTA: Setting the Record Straight." *World Economy* 20, no. 5 (August 1997): 605–614.
Madigan, Michael, and Marrs, Barry. "Extremophiles." *Scientific American,* April 1997, 66–71.
Marks, Gary; Hooghe, Liesbet; and Blank, Kermit. "European Integration from the 1980s." *Journal of Common Market Studies* 34, no. 3 (September 1996): 341–378.
McCarry, John. "The Promise of Pakistan." *National Geographic,* October 1997, 48–73.
McDonald, Kevin. "Russian Raw Materials." *Harvard Business Review* (May–June 1994): 54–64.
Mitchell, John. "Our National Forests." *National Geographic,* March 1997, 58–87.
———. "Oil on Oil." *National Geographic,* April 1997, 104–131.
Mody, Ashoka, and Wang Fang-Yi. "Explaining Industrial Growth in Coastal China." *World Bank Economic Review* 11, no. 2 (May 1997): 293–325.
Moffett, Mark. "Tree Giants of North America." *National Geographic,* January 1997, 44–61.
Moin, Parviz, and Kim, John. "Les Superorrdinateurs Analysent la Turbulence." *Pour La Science,* Mars 1977, 46–52.
Monastersky, Richard. "Life Grows Up." *National Geographic,* April 1998, 100–114.
Montaigne, Fen. "Nenets: Surviving on the Siberian Tundra." *National Geographic,* March 1998, 121–137.
Musgrove, Phili. "Feeding Latin America's Children." *Research Observer* 8, no. 1 (January 1993): 23–45.
National Geographic. "In Focus—Central Africa." June 1997, 124–133.
Nava, Gabriela. "Anàlisis Comparativo de las Capacidades Technológicas de México y Korea." *Comercio Exterior* 47, no. 2 (Febrero 1997): 132–144.
Neven, Damien, and Seabright, Paul. "European Industrial Policy." *Economic Policy* 21 (October 1995): 313–358.
Nucci, Mary, and Abuchowski, Abraham. "The Search for Blood Substitutes." *Scientific American,* February 1998, 60–65.

O'Brien, Stephen. "The Human–Cat Connection." *National Geographic,* June 1997, 77–85.

Ohmae, Kenichi. "Putting Global Logic First." *Harvard Business Review* (January–February 1995): 119–125.

———. "Letters from Japan." *Harvard Business Review* (May–June 1995): 154–163.

O'neal, John, and Russett, Bruce. "The Classical Liberals Were Right." *International Studies Quarterly* 41, no. 2 (June 1997): 267–294.

Ott, Wayne, and Roberts, John. "Everyday Exposure to Toxic Pollutants." *Scientific American,* February 1998, 72–77.

Pelton, Joseph. "Telecommunications for the 21st Century." *Scientific American,* April 1998, 68–73.

Perl, Anthony, and Dunn, James. "Fast Trains: Why the U.S. Lags." *Scientific American,* October 1997, 74–76.

Pisano, Gary, and Wheelwright, Steven. "The New Logic of High–Tech R&D." *Harvard Business Review* (September–October 1995): 93–105.

Pisarides, Christopher. "Learning by Trading and the Returns to Human Capital in Developing Countries." *World Bank Economic Review* 11, no. 1 (January 1997): 17–31.

Plomin, Robert, and Defries, John. "The Genetics of Cognitive Abilities." *Scientific American,* May 1998, 40–47.

Plunkett, Mathew, and Ellman, Jonathan. "Chimie combinatoire et nouveaux médicaments." *Pour La Science,* Juin 1997, 72–77.

Porter, Michael, and Van der Linde, "Green and Competitive." *Harvard Business Review* (September–October 1995): 120–134.

Posner, Bruce, and Rothstein, Lawrence. "Reinventing the Business of Government." *Harvard Business Review* (May–June 1994): 133–143.

Potier, Pierre. "La corne d'abondance est encore pleine." *Pour La Science,* Novembre 1997, 30–32.

Prud'homme, R,my. "The Dangers of Decentralization." *Research Observer* 10, no. 2 (August 1995): 201–219.

Rabe, Barry, and Zimmerman, Janet. "Beyond Environmental Regulatory Fragmentation." *Governance* 8, no. 1 (January 1995): 58–77.

Ramesh, M. "Economic Globalization and Policy Choices." *Governance* 8, no. 2 (April 1995): 243–260.

Ramirez, Miguel. "Crecimiento y Desarrollo en Am,rica Latina." *Comercio Exterior* 47, no. 6 (Junio 1997): 473–482.

Ranis, Gustav. "Another Look at the East Asian Miracle." *World Bank Economic Review* 9, no. 3 (September 1995): 509–534.

Raoul, Jean–Claude. "How High–Speed Trains Make Tracks." *Scientific American,* October 1997, 68–73.

Ravallion, Martin, and Datt, Gaurav. "How Important to India's Poor Is the Sectoral Composition of Economic Growth?" *World Bank Economic Review* 10, no. 1 (January 1996): 1–25.

Remnick, David. "Moscow: The New Revolution." *National Geographic,* April 1997, 78–102.

Rice, Richard; Gullison, Raymond; and Reid, John. "Peut-on sauver les forêts tropicales?" *Pour La Science,* Juin 1997, 58–62.

Richelson, Jeffrey. "Scientists in the Black." *Scientific American,* February 1998, 38–45.

Roberts, Mark, and Tybout, James. "Producer Turnover and Productivity Growth in Developing Countries." *Research Observer* 12, no. 1 (February 1997): 1–18.

Rodrik, Dani. "Getting Interventions Right." *Economic Policy* 20 (April 1995): 55–107.

Rosen, Harold, and Castleman, Deborah. "Flywheels in Hybrid Vehicles." *Scientific American,* October 1997, 49–57.

Ross, Fiona. "Cutting Public Expenditures in Advanced Industrial Democracies." *Governance* 10, no. 2 (April 1997): 175–200.

Sachs, Jeffrey, and Woo, Wing Thye. "Structural Factors in the Economic Reforms of China, Eastern Europe and the Former Soviet Union." *Economic Policy* 18 (April 1994): 101–146.

Schadler, Susan. "Que Exito Tienen los Programas de Ajuste Respaldados por el FMI?" *Finanzas and Desarrollo* 33, no. 2 (Junio 1996): 14–17.

Schafer, Andreas, and Victor, David. "The Past and Future of Global Mobility." *Scientific American,* October 1997, 36–43.

Scheierling, Susanne. "La Contaminacion del Agua de Origen Agricola en la Union Europea." *Finanzas and Desarrollo* 33, no. 3 (Septiembre 1996): 32–35.

Schiavo–Campo, Salvatore. "La Reforma de la Administracion P—blica." *Finanzas and Desarrollo,* 33, no. 3 (Septiembre 1996): 10–13.

Schlehofer, Jôrg. "Parvoviren-Krebshemmende Symbionten?" Spektrum der Wissenschaft, April 1997, 44–50.

Schwab, Klaus, and Smadja, Claude. "Power and Policy: The New World Economic World Order." *Harvard Business Review* (November–December 1994): 40–50.

Serageldin, Ismael. "El Desarrollo Sostenible: De la Idea a la Accion." *Finanzas and Desarrollo* 33, no. 4 (Diciembre 1996): 3–7.

Shirley, Mary. "Son los Contratos con las Empresas un Instrumento Eficaz?" *Finanzas and Desarrollo* 33, no. 3 (Septiembre 1996): 6–9.

Siebert, Horst. "German Unification." *Economic Policy* 13 (October 1991): 287–339.

Siegel, Richard. "Les Nanomatériaux." *Pour La Science,* Février 1997, 5863.

Sisodia, Rajendra. "Singapore Invests in the Nation Corporation." *Harvard Business Review* (May–June 1992): 40–50.

Smith, Edward, and Marsden, Richard. "The Ulysses Mission." *Scientific American,* January 1998, 52–57.

Smyrl, Marc. "Does European Community Policy Empower the Regions?" *Governance* 10, no. 3 (July 1997): 287–310.

Sobel, Alan. "Television's Bright New Technology." *Scientific American,* May 1998, 48–55.

Sorensen, Peter. "Public Finance Solutions to the European Unemployment Problem?" Economic Policy 25 (October 1997): 221–264.

Spektrum der Wissenschaft. "Biopatente." April 1998, 28–36.

———. "Grundwasser." April 1998, 86–102.

———. "Transistoren." März 1998, 80–92.

Sperling, Daniel. "Les V,hicules Electriques." Pour La Science, Janvier 1997, 58–63.

Squire, Lyn, and Suthiwart Narueput, Sethaput. "The Impact of Labor Market Regulations." *World Bank Economic Review* 11, no. 1 (January 1997): 119–143.

Steinmo, Sven. "Why Is Government So Small in America?" *Governance* 8, no. 3 (July 1995): 303–334.

Stiglitz, Joseph. "Some Lessons from the East Asian Miracle." *Research Observer* 11, no. 2 (August 1996): 151–177.

———, and Uy, Marilou. "Financial Markets, Public Policy and the East Asian Miracle." *Research Observer* 11, no. 2 (August 1996): 249–275.

Stotsky, Janet. "Desigualdades en el Tratamiento de Hombres y Mujeres en los Systemas Fiscales." *Finanzas and Desarrollo* 34, no. 1 (Marzo 1997): 28–31.

Summers, Lawrence, and Thomas, Vinod. "Recent Lessons of Development." *Research Observer* 8, no. 2 (July 1993): 241.

Suplee, Curt. "Robot Revolution." *National Geographic,* 192, no. 1 July 1997, 76–95.

Swanson, Timothy. "Regulating Endangered Species." *Economic Policy* 16 (April 1993): 183–205.

Swerdlow, Joel. "Exploration." *National Geographic,* February 1998, 2–39.

Swerdlow, Joel. "Making Sense of the Millennium." *National Geographic,* January 1998, 2–35.

Tanzi, Vito, and Schuknecht, Ludger. "Reforma del Estado en Pa¡ses Industriales." *Finanzas and Desarrollo* 33, no. 3 (Septimebre 1996): 2–5.

Tattersal, Ian. "Out of Africa Again and Again?" *Scientific American,* April 1997, 46.

Teranishi, Juro. "Economic Recovery, Growth and Policies." *Economic Policy* 19 (Supplement December 1994): 137–154.

Theroux, Peter. "The Imperiled Nile Delta." *National Geographic,* January 1997, 2–35.

———. "Beirut Rising." *National Geographic,* September 1997, 100–123.

———. "Down the Zambezi." *National Geographic,* October 1997, 2–31.

Thimann, Christian. "El Programa de Asistencia Social de Alemania." *Finanzas and Desarrollo* 33, no. 3 (Septiembre 1996): 40–43.

Thurow, Lester. "Needed: A New System of Intellectual Property Rights." *Harvard Business Review* (September–October 1997): 94–101.

Ul Haque, Nadeem; Mathieson, Donald; and Sharma, Sunil. "Causas de la Afluencia de Capital y Medidas de Politica Pertinentes." *Finanzas and Desarrollo* 34, no. 1 (Marzo 1997): 3–6.

Vanhonacker, Wilfried. "Entering China: An Unconventional Approach." *Harvard Business Review* (March–April 1997), 130–140.

Vasiliev, Sergei. "Market Forces and Structural Change in the Russian Economy." *Economic Policy* 19 (Supplement December 1994), 123–136.

Velander, William; Lubon, Henryk; and Drohan, William. "Des animaux trans géniques produisent des médicaments." *Pour La Science,* Mars 1997, 70–75.

Visilind, Priit. "Sri Lanka." *National Geographic,* January 1997, 110–133.

———. "Why Explore?" *National Geographic,* March 1998, 2–31.

Vittas, Dimitri, and Cho Yoon Je. "Credit Policies: Lessons from Japan and Korea." *Research Observer* 11, no. 2 (August 1996): 277–298.

Walley, Noah, and Whitehead, Bradley. "It's Not Easy Being Green." *Harvard Business Review* (May–June 1994): 46–52.

Ward, Geoffrey. "India." *National Geographic,* May 1997, 3–57.

Webster, Donovan. "The Orinoco." *National Geographic,* April 1998, 8–31.

Williamson, Jeffrey. "Globalization and Inequality." *Research Observer* 12, no. 2 (August 1997): 117–135.

Williamson, Peter. "Asia's New Competitive Game." *Harvard Business Review* (September–October 1997): 55–67.

Wood, Adrian. "Openness and Wage Inequality in Developing Countries." *World Bank Economic Review* 11, no. 1 (January 1997): 33–57.

Yan, Rick. "To Reach China's Consumers—Adapt to Guo Qing." *Harvard Business Review* (September–October 1994): 66–74.

Yeats, Alexander; Amjadi, Azita; Reincke, Ulrich; and Ng, Francis. "Que Ha Provocado la Marginacion de Africa en el Comerico Mundial?" *Finanzas and Desarrollo* 33, no. 4 (Diciembre 1996): 36–39.

Youdim, Moussa, and Riederer, Peter. "La maladie de Parkinson." *Pour La Science,* Mars 1997, 60–67.

Zhen, Kun Wang. "Integracion de las Economias en Transicion en la Economia Mundial." *Finanzas and Desarrollo* 33, no. 3 (Septiembre 1996): 21–23.
Zich, Arthur. "China's Three Gorges." *National Geographic,* September 1997, 2–33.

And also numerous articles in:

Bilan, Switzerland
BusinessWeek, New York
Cambio, Madrid
Der Spiegel, Frankfurt
EU Magazin, Germany
Fortune, New York
Le Nouvel Economist, Paris
L'Hebdo, Switzerland
Newsweek, New York
The Economist, London
The Far Eastern Economic Review, Hong Kong
Time, New York

Index

About the Author

Carlos Sabillon is currently a freelance researcher at the University of Geneva working on the development of South Asia, Africa, and Latin America. Dr. Sabillon earned his PhD in international relations in 1990 from the Graduate Institute of International Studies (GIIS) in Geneva. He is a former consultant on foreign direct investment and economic growth for the Ministry of Economics in Tegucigalpa, Honduras. Dr. Sabillon has also completed extensive postdoctoral research on the subject of economic growth at Ludwig Maximilians University (Munich, Germany), the University of Geneva, and the GIIS; his new book is based on this work.